FIB Conference: Sustainable Concrete Materials and Structures 2018

IOP Conference Series: Materials Science
and Engineering Volume 442

Valleta, Malta
10 April 2018

ISBN: 978-1-5108-7657-6
ISSN: 1757-8981

Printed from e-media with permission by:

Curran Associates, Inc.
57 Morehouse Lane
Red Hook, NY 12571

Some format issues inherent in the e-media version may also appear in this print version.

Printed by Curran Associates, Inc. (2019)

Institute of Physics
Dirac House, Temple Back
Bristol BS1 6BE UK

Phone: 44 1 17 929 7481
Fax: 44 1 17 920 0979

techtracking@iop.org

Additional copies of this publication are available from:

Curran Associates, Inc.
57 Morehouse Lane
Red Hook, NY 12571 USA
Phone: 845-758-0400
Fax: 845-758-2633
Email: curran@proceedings.com
Web: www.proceedings.com

TABLE OF CONTENTS

Sustainable Concrete: Materials and Structures

Ruben Paul Borg[1], **Petr Hajek**[2], **David Fernandez-Ordonez**[3]

[1] Faculty for the Built Environment, University of Malta, Malta
Member, *fib* The International Federation for Structural Concrete, Commission 7 Sustainable Concrete

[2] Czech Technical University in Prague, Czech Republic,
Chairman, *fib* The International Federation for Structural Concrete, Commission 7 Sustainable Concrete

[3] Secretary General, *fib* The International Federation for Structural Concrete

E-mail: ruben.p.borg@um.edu.mt

Sustainability of Concrete Materials and Structures refers to materials and structures which not only have sufficient strength and durability but which meet these requirements throughout their intended life time in intended applications. The term sustainability, indeed *durable* in French, reflects the need to address environmental, societal and economic priorities. Cement and Concrete have a significant large impact on the environment with 1 ton of CO_2 resulting for every ton of cement produced and with the cement production industry contributing to a significant quantity of the global carbon emissions.

Economic development puts significant larger demands on the exploitation of natural resources. Indeed the construction industry relies particularly on concrete for new buildings and infrastructure to address these growing demands, with large quantities of natural resources required to address these needs. Sustainable construction promotes effective use of resource and waste, therefore exploiting the potential for waste material and industrial by-products not only to reduce demands on natural resources but also exploiting the recycling of materials therefore reducing the volume of waste disposal. The exploitation of waste and industrial by-products present an opportunity in the concrete industry. Industrial by-products have the potential for cement replacement, to improve the properties of the materials even contributing to durability. These measures directly promote sustainable use of the materials through environmental friendly practices, which become even more effective when considering the large quantities of concrete produced.

The innovative use of materials together with the development of new materials and innovation in cement based materials and composites, for more challenging applications in Civil and Structural Engineering, also lead to more durable concrete and reinforced concrete structures with better performance throughout their intended life-time. Advanced materials help achieve more economic structures when considering the whole life cycle; with increased safety of structures during their service life with lower demands for repair and maintenance leading to significant economic advantages.

IOP Conf. Series: Materials Science and Engineering **442** (2018) 011001 doi:10.1088/1757-899X/442/1/011001

Quality concrete is key for the sustainable use of materials. Concrete is a very complex composite which can be engineered to address specific performance requirements and which relies on a variety of constituent materials including locally available materials in a specific region. Quality assurance systems are necessary to ensure the adequate use of constituents, mix proportions and production processes and application.

Sustainable use of concrete is not only achieved through innovation in materials and structural systems but also through an appreciation of the optimum use of materials for an intended application. This can be achieved equally through innovation in research and developments in industry, therefore working through effective research and development frameworks as indeed promoted by *fib*, The International Federation for Structural Concretes. Academic and industry collaboration is key towards effective solutions to the demands for greater efficiency and economy, safety of structures and environmental protection.

In this direction, the conference Sustainable Concrete: Materials and Structures, which was the first *fib* conference in Malta, was organised by *fib* The International Federation for Structural Concrete and the University of Malta as the higher academic institution in Malta, with the support of Industrial partners BASF and Philip A. Tabone Ltd, in association with the leading Construction Industry organisations and entities in the Maltese Islands; the Planning and the Property Market Secretariat, Ministry for Transport, Infrastructure and Capital Projects, the Building Industry Consultative Council of the Government of Malta, the Kamra tal-Periti (Malta Chamber of Architects and Civil Engineers), The Malta Developers Association, The Malta Group of Professional Engineering Institutions (Institution of Civil Engineers, UK), Sustainable Built Environment Malta, the International Initiative for a Sustainable Built Environment (iiSBE). The Conference was also organised in association with Concrete Plant International CPI-Worldwide.

The papers presented in the first Malta *fib* Conference, held on the 10th April 2018 at the Valletta Campus of the University of Malta, cover a range of advances in cement and concrete. Yet there is a unifying element and a focus which is evident; the promotion of the sustainable use of concrete materials with regards to environmental protection and resource efficiency, addressing societal needs and economic demands, while achieving safe reinforced concrete structures which perform effectively throughout their lifetime.

IOP Conf. Series: Materials Science and Engineering **442** (2018) 011002 doi:10.1088/1757-899X/442/1/011002

Peer review statement

All papers published in this volume of *IOP Conference Series: Materials Science and Engineering* have been peer reviewed through processes administered by the proceedings Editors. Reviews were conducted by expert referees to the professional and scientific standards expected of a proceedings journal published by IOP Publishing.

Sustainable construction materials for concrete: A question of responsible use

Ravindra K Dhir OBE

Chao Qun Lye, Ciarán J Lynn and Abdurrahaman A Elgalhud

School of Civil Engineering, University of Birmingham, UK

E-mail: r.k.dhir@bham.ac.uk

Abstract. This paper has been inspired by the principal author's life-time research at the cutting edge of the concrete industry and is propelled by recent research undertaken with his co-authors, which is referenced fully in this paper. In the main, this work has been created to show that the use of sustainable construction materials, in making concrete, should be carefully evaluated for their rheological, engineering and long-term durability effects, in the relevant exposure classes, as well as environment impacts, in the form of leaching, as a material and whilst in use. Failure to do so, can, in the long-term prove to be disappointingly counterproductive in achieving the estimated sustainability benefits from the use of such materials in concrete, and may even prove disastrous.

Keywords: Sustainable construction materials, environmental impact

1. Introduction

The concrete industry has been growing rapidly, ever since Portland cement was invented, presumably in 1824. The consumption of concrete as a construction material is now second only to water by volume and the statistics for concrete use world-wide are awesome, accounting for about 2 ½ tonnes per person of the global population per annum, with cement consumption reaching close to 5 billion tonnes and with associated aggregate consumption in the region of 30 billion tonnes per annum.

These are huge numbers for natural resources usage, given the associated consumption of energy and pollution of the environment. Thus, it is not surprising that the concrete industry world over has been a major target for lowering CO_2 emissions and promoting sustainable construction, to comply with global sustainability agendas, such as the recently agreed Paris Climate Accord 2015, signed by 195 member states of the United Nations Framework Convention on Climate Change. This agreement aims to keep the global temperature rise this century below 2.0°C, and possibly 1.5°C.

The emphasis on sustainability is bringing about major changes in the construction industry, particularly regarding the promotion of sustainable materials and how concrete is specified and produced, with emphasis rapidly shifting towards the use of waste materials arising from industrial (e.g. fly ash and ground granulated blastfurnace slag (GGBS)) and domestic (e.g. incinerated ash) sectors. In this respect national, regional and international Standards, such as EN 197-1 [1] and EN 12620 [2], which were adopted in Europe at the turn of the century, are classic examples of the determination of the construction industry to bring about changes in the way cement and aggregates are selected for manufacturing concrete.

Beyond the standards, in efforts to accommodate the use of sustainable construction materials (waste materials), the concrete industry is continually refining the production processes, including an ever-increasing use of chemical admixtures. A shift from prescriptive to performance-based specifications is now part of the evolution that is currently sweeping through the construction industry, albeit slowly at present.

2. Sustainable Construction Materials

The list of sustainable construction materials that can be aimed at the concrete industry has increased noticeably since the turn of this century and this trend is expected to continue as innovation takes hold and pressure grows on the construction industry to further reduce its carbon foot print. It is estimated that at present there is 10 to 20 billion tonnes of waste that can be potentially processed for use.

On the other hand, although the promotion of sustainability is becoming more popular, it must still be recognised that this objective should not grant free reign for any and all use of waste materials in construction. From this perspective, a breakdown of the key aspects of sustainability is presented in Figure 1. Future practices in the construction industry should strive to fulfil each of the social, environmental and economic requirements, to ensure that waste materials are utilised responsibly.

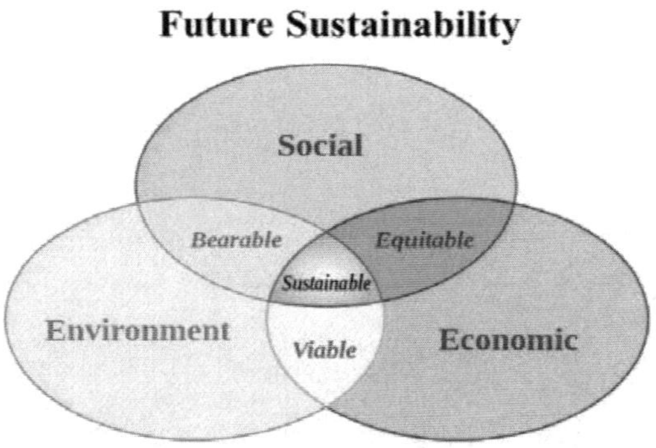

Figure 1: Breakdown of the key aspects of sustainability

The cement industry is already well into the use of first and second generations of Portland-pozzolanic combinations as composite cements, such as those in EN 197-1 [1], namely fly ash, GGBS, pozzolana and burnt shale as bulk replacements of Portland clinker and silica fume in small quantities to refine the pore structure to improve the performance of the cement paste. Pressure to do more in this respect, is encouraging researchers to explore the potential use of a wide-range of the next generation of pozzolanic materials, including incinerated ashes (e.g. sewage sludge ash and municipal incinerated bottom ash), metallic slags (e.g. copper slag) and ceramics (e.g. glass cullet). A comprehensive assessment of these first three as sustainable construction materials is presented in recently published books for each [3. 4 and 5], whilst fourth and fifth books in this series on glass cullet and recycled aggregate are upcoming in 2018. Furthermore, research is also presently ongoing with other researchers to develop CEM-free cements, such as geopolymers in different forms.

In the use of aggregate, the research shows that coarse recycled concrete aggregate (coarse RCA) can replace natural aggregates. Greater emphasis is being placed on refining the production of coarse RCA to further improve its performance in concrete to gain the confidence of engineers. In parallel, other wastes materials such glass, metallic slags, incinerated ashes, quarry fines and marble cuttings

IOP Conf. Series: Materials Science and Engineering **442** (2018) 012001 doi:10.1088/1757-899X/442/1/012001

etc. are being developed to further reduce the extraction of natural resources. However, at present, the progress is painfully slow and more concentrated and organised efforts are needed to bring about meaningful fundamental changes.

3. Designing Sustainable Concrete

This is the area where engineering skills, cost-effective and focused R&D and performance-based construction specifications, can all help with the development of mixes that can routinely offer real sustainable concrete, ensuring a long-term durable performance of structures with whole-life costs sustained. This is, in effect, the real challenge facing us all!

3.1. Fresh Concrete Properties

Although often ignored in many ways, the fresh concrete can greatly affect the performance of concrete in the hardened state, including strength development and durability and therefore, the potential sustainability of the resultant product.

Apart from fly ash, GGBS and ground limestone, which have a long history of producing a workable stable mix, free from bleeding and segregation in the fresh state, the suitability of new materials remains to be determined for many. Thus, the performance in the fresh state, with emerging materials replacing the currently used constituents, in the form of cement and/or aggregate as coarse, fine or filler, can often be a critical factor in deciding their acceptance for use. The specifications for the performance of fresh concrete must not be ignored in developing the use of new materials, as in some cases, these may limit their use. For example, with sewage sludge ash (SSA) and municipal incinerated bottom ash (MIBA), extra water demands arising from their morphology and high porosity [6] would have to be accounted for when using the ashes as aggregate and in ground form, as a cement component in concrete [7-9].

3.2. Strength

Whilst, in principle, applying Abrams law relating strength to water/cement ratio, a concrete mix can be designed with any set of materials by adjusting the water/cement ratio, and thereby the cement content, to comply with a given specified strength; however, an increased cement content may negate the potential environmental benefits of using materials sourced as recycled, industrial by-products, or incinerated domestic waste, Figure 2 [10]. This would often require coming up with a solution, which may include the use of a water-reducing admixture, to make it feasible to use the new materials, whilst retaining potential environment benefits. In a commercial sense, the material variability, together with increased rate of testing for quality control would also have to be considered in assessing the viability of the new materials in practice. However, this may not rule out the use of new materials with low grade concrete mixes, which could still offer some environmental benefits.

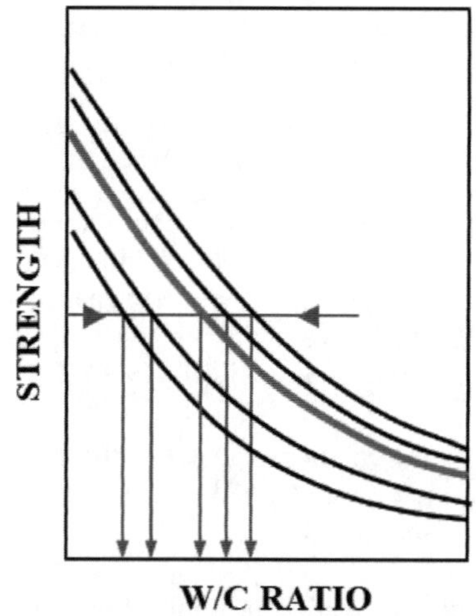

Figure 2: Relationship between strength and water/cement ratio

In assessing the potential use of constituent materials, a pragmatic approach to their sustainable application can greatly widen the scope for the new materials to be adopted in concrete construction, as illustrated in Figure 3 [11].

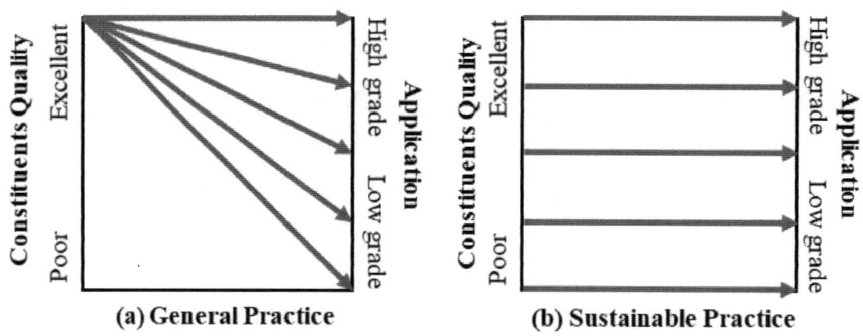

Figure 3: (a) General practice and (b) sustainable practice in dealing with constituent materials

3.3. Deformation Properties

Structural concrete also has to be designed for its elastic, creep and shrinkage deformation properties (Eurocode 2) [12] (though in practice these parameters are largely estimated from its compressive strength). Figure 4 has been prepared to show the outcome of an analysis of a large volume globally sourced data matrix (50,000 data points) based on the research undertaken since 1977 by 860 researchers at 537 institutions in 46 countries, to assess the effect of using coarse RCA, and glass cullet and copper slag as fine aggregates (fine GCA and CSA) on the modulus of elasticity, creep and shrinkage [13-17].

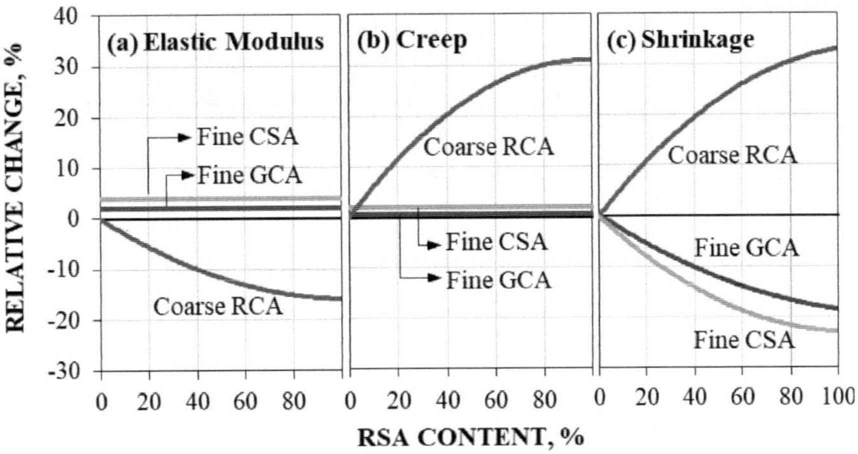

Figure 4: (a) Elastic modulus, (b) creep and (c) shrinkage of recycled and secondary aggregate (RSA) concrete

These studies concluded that, whilst fine GCA and fine CSA can replace fine natural aggregate (NA), the use of coarse RCA in place of coarse NA in structural concrete, in its current form, will adversely affect the deformation properties. These studies also suggested that whilst the use of fly ash and a water-reducing admixture alongside the coarse RCA, to a certain extent closed the gap towards obtaining the required deformation properties, additional material costs would have to be sustained.

The use of a blended aggregate, in this case, coarse RCA mixed with fine CSA, may allow sustainability benefits to be retained, as the fine CS would improve the consistence (workability) and thereby allow the mix water content to be reduced, without having to use a water-reducing admixture [13]. However, further study would be required to formulate how best certain materials, such as CS, with low water demand, can as fine aggregate, be blended with coarse RCA to minimise its negative impact on the deformation properties of concrete. This would be a challenging research project.

3.4. Durability

For sustainable materials to be considered suitable for use as an aggregate and/or cement in manufacturing concrete, the durability performance, in all aspects relating to the application, should not be adversely affected. Taking the case of a pozzolanic material used as a component of Portland clinker-based blended cement, for the pozzolanic to perform, some of the alkalinity released from the Portland cement hydration will be taken up to activate the pozzolanic reaction to form calcium silicate hydrates (the cementitious products).

In a CO_2-environment, the use of a Portland-pozzolanic blended cement would lower the pH of the concrete, unless compensated with moist curing forming a densified cement paste pore structure, which helps to reduce the permeation properties, and in turn, control the alkalinity of the concrete.

To assess the carbonation effect of fly ash, GGBS and limestone, a global data-matrix with 79000 entries, built from 583 studies, published since 1968, and undertaken by 983 researchers from 494 institutions in 41 countries were analysed [18-20]. The trend lines obtained are shown in Figure 5.

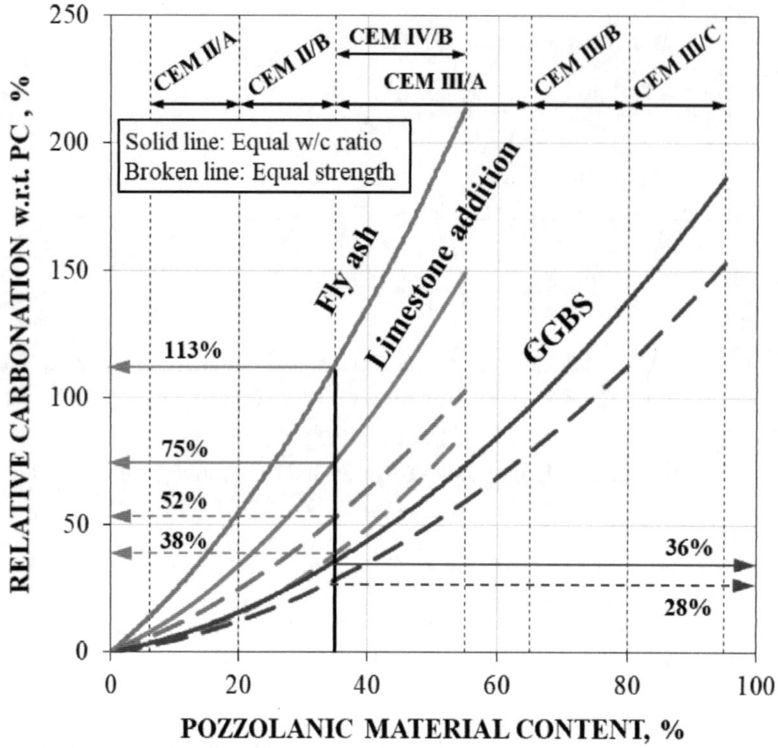

Figure 5: Effect of pozzolanic materials on the carbonation of concrete

This shows that concrete designed using EN 197-1 [1] Portland-based, fly ash, GGBS and limestone cements can be expected to show higher rates of carbonation than the corresponding Portland cement (PC) concrete. The increase in carbonation depth varies with the addition type, with fly ash giving the highest increase. For example, reading from Figure 5, with an equal water cement ratio and 35% fly ash addition, the concrete exhibited a 113% increase in carbonation compared to PC mix. Corresponding increases of 75% for limestone and 36% for GGBS were also determined. However, when the concrete is specified for a given strength, for the same 35% addition, the corresponding carbonation increases drop to 52% for fly ash, 38% for limestone and 28% for GGBS.

This increase in the depth of carbonation presents a risk of early carbonation-induced corrosion and reduces the concrete durability, which in turn adversely affects the sustainability impact gained from using these materials. As all pozzolanic cementitious materials consume $Ca(OH)_2$, released upon hydration of Portland cement, their use as Portland-Pozzolanic cement combinations could potentially present a serious durability risk. It has also been established that combined effects of carbonation and chloride ingress further increase the susceptibility of the concrete to corrosion [21].

Focusing on the use of limestone, a comparison of its overall effects, relative to PC concrete, on the pore structure, durability and strength is presented in Figure 6. Simple modifications have been carried out to the trend lines in this figure, changing from previously developed polynomial regression curves [20, 22, 23] to linear regression, as in this case, the goal was to produce a useful tool that could be more easily adopted in practice. For practical purposes, it is feasible to accept that for limestone contents up to 15%, the effects may be considered constant and close to neutral. Increasing limestone beyond 15% gives rise to a progressive reduction of the concrete properties, in this case, the pore structure, durability and strength. As such, it is concluded that limestone can be incorporated at contents up to 15% without significantly comprising the concrete performance, which is below the maximum 20% limit for CEM II/A for Portland limestone cement in EN 197-1 [1].

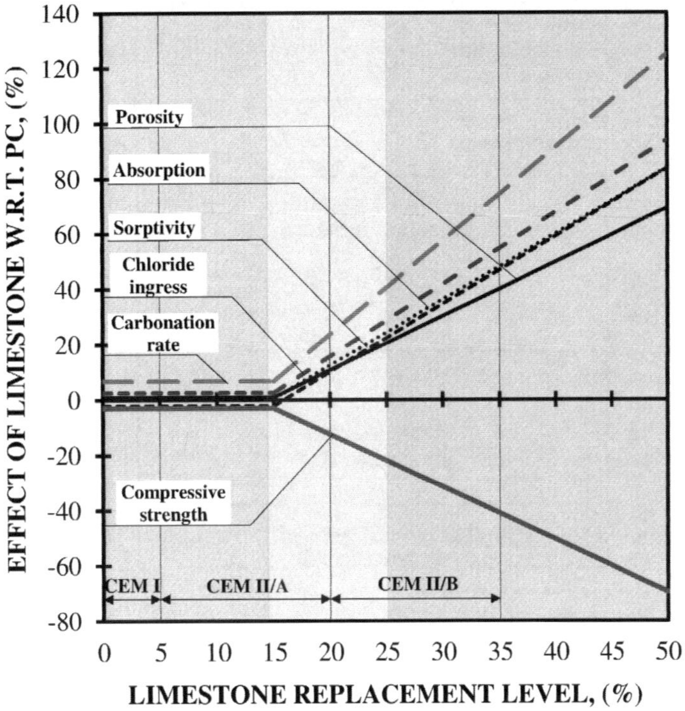

Figure 6: Limestone effects on pore structure and related properties, compressive strength, carbonation rate and chloride ingress of concrete [23].

Regarding the use of municipal incinerated bottom ash as aggregate in concrete, a ranking of the effects on various aspects of durability is presented in Figure 7 [8]. On the permeation properties, measured as absorption and initial surface absorption, decreasing durability was evident with MIBA. For aggregate replacement levels greater than 25%, initial surface absorption of the concrete drifted above the 0.5 ml/m^2/s upper limit for normal-strength concrete. This suggests that the ash contents may have to be controlled to lower levels. Increases in the chloride diffusion coefficient occurred with MIBA, owing to the higher overall mix porosity. The effect on sulphate attack susceptibility was classed as neutral, as concrete containing 25%–100% MIBA as fine or coarse aggregate exhibited no extra expansion.

Carbonation resistance improved slightly with MIBA (reduced carbonation depths). As carbonation is sensitive to moisture conditions, the extra water absorbed by MIBA appeared to be beneficial in slowing down the carbonation ingress, despite higher permeation. MIBA mixes displayed comparable or superior freeze/thaw resistance compared to the control, as the higher porosity of the aggregates can act as an air-entrainer. Alkali-silica expansion potential of MIBA concrete exceeded that of the natural aggregate mix; however, the validity of this outcome appears questionable, as the control limestone blend also greatly exceeded the limit that signifies potential alkali-silica reactivity. Indeed, a second test series, involving both laboratory and field work, revealed no evidence of silicate gel expansion.

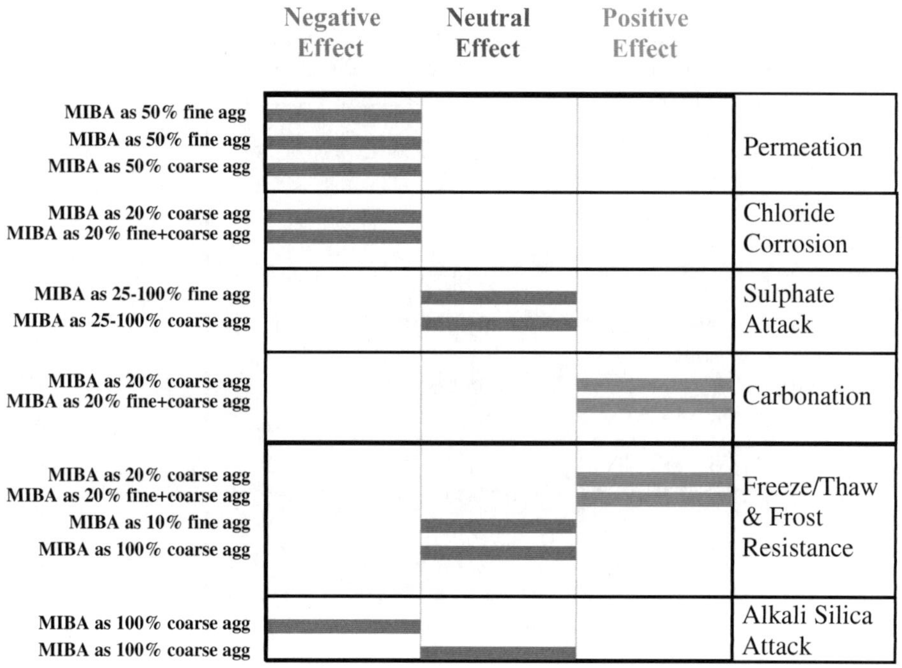

Figure 7: Effects of MIBA as aggregate on concrete durability

3.5. Environmental Impacts

Use of sustainable construction materials must be predicated on ensuring that there are no detrimental impacts on the surrounding environment, arising from leaching. With a notable presence of heavy metals in MIBA and SSA, tests have been undertaken to assess their leaching in use.

However, as traditional construction materials generally do not contain harmful amounts of metals, most standards do not account for these issues. Countries such as Germany, Denmark and The Netherlands have developed guidelines for leaching limits for waste materials reuse, though with MIBA and SSA use in construction products, there was a large variety of leaching test methods adopted and heavy metal limits applied. The emerging consensus was that the restriction of the ash-products' leaching to suitable levels requires certain conditions to be adopted regarding the pre-processing of the ash (e.g. ageing and washing treatments) and limitations to avoid unfavourable leaching conditions such as strongly acidic conditions or high rainfall regions [24 and 25].

Covering the various applications in which incinerated ash has been explored, an estimated ranking of the leaching potential in each is presented in Figure 8. The potential is highest in the ground as fill or subbase, as the ash is unbound and there is a direct pathway for the metals to travel into sensitive receptors such as water sources. In lightweight aggregate production, the heat treatment and hard outer layer formed is somewhat beneficial in restricting the mobility of the heavy metals, though with the extra inclusion of a binder fraction (cement or clay) in the aggregate blend, the leaching reduces greatly. In mortar and concrete mixes, as well as hydraulically bound pavements, the cement provides solidification (physical encapsulation) and stabilisation (chemical binding) effects, though there are some questions regarding leaching after the product's deterioration or demolition and increased solubility of some elements in alkaline conditions. The hydrophobic nature and binding ability of bitumen translated into low leaching potential in these pavement mixtures. Low heavy metal leaching was measured for tiles and glass ceramics, owing to the restricting effect of clay and the very high temperatures involves in the production process. Combined positive effects of the heat treatment, cement binding and low ash contents adopted in cement clinker production also led to low leaching.

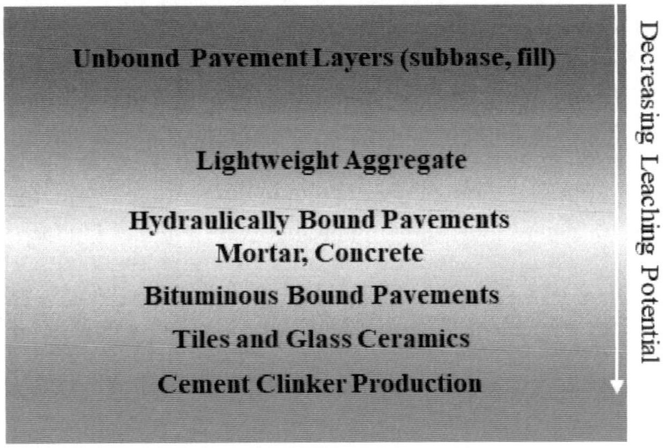

Figure 8: In-use leaching of incinerated ash

4. Cost Considerations

As economic viability is a key pillar of sustainability, a realistic audit of any sustainable material use (in the form of reuse, recycle or reconstitution) in concrete construction should to begin with, be made on an equal strength basis, allowing for factors relating to variability in product performance. There are already some suggestions for moving the specification of concrete characteristic strength from 28 days to 56 days, to promote the use of certain pozzolanic materials, such as coal fly ash and GGBS. It is not certain whether this change would help the suitability of other pozzolanic materials for use in bulk, such as rice husk ash.

The role of other engineering properties in designing structures, such as the load related and load independent deformation, is critical and cannot be ignored. Indeed, it can be argued that given the results presented in Figure 4, the proportion of some sustainable materials that can be used, in their present form, is likely to be limited e.g. in the case of recycled concrete aggregate it may be about 20%, beyond which additional costs will be incurred.

Additionally, all aspects of increased costs arising from changes in durability should be taken into consideration, for example, a potential reduction in the resistance of the reinforcement to corrosion, as illustrated in Figures 5-7. In this case it would be fair to say that the sustainable materials have not been subjected to the rigour of a systematic evaluation, though a large number of studies have been undertaken and an even larger volume of publications have been produced.

Finally, there are issues of environment impact that need to be addressed, where there is a lack of harmonisation in the evaluation processes and the information available is rather sparse, not commonly understood and often does not inspire confidence. Indeed, it is a difficult field to explore and it could be argued that perhaps the construction industry at present is not well prepared to handle such environment issues.

5. Balanced Approach

A balanced approach, based on a realistic assessment, as opposed to short-term commercial gain is needed in developing the use of sustainable construction materials, both as cement and aggregate components in concrete. This is important, as failure do so can prove to be extremely expensive and do little to promote the use of sustainable materials in construction.

IOP Conf. Series: Materials Science and Engineering **442** (2018) 012001 doi:10.1088/1757-899X/442/1/012001

6. Conclusions

- The concrete sector has a significant role to play in the development of sustainable construction.
- Whilst the use of sustainable construction materials is potentially an attractive proposition, to minimise any long-term risk to durability with concrete construction, their use should be carefully evaluated and must not be based on strength only. In this respect, deformation properties should also be evaluated.
- The use of all pozzolanic materials and recycled materials with high porosity in structural concrete, in certain exposure classes, carry risks in the pursuit of sustainable construction.
- When using secondary materials that contain heavy metals and pose a potential leaching risk, the environmental aspect of sustainability must be carefully assessed.
- Clearly whilst further research is badly needed, given that the stakes can be high, it must be ethical, robust and thorough using standards which are well explained and understood.

- This research should also be:

(i) Reproducible, dependable and inspire confidence through international collaboration for universal endorsement.
(ii) Leaning on industrial involvement, though it is considered to be an essential requirement.
(iii) Innovative, imaginative, progressive, bold and deliverable.
(iv) Conclusive, explaining clearly the methodologies and test procedures adopted, and yield recommendations that are doable, clear and of practical relevance.

References

[1] EN 197-1. Cement. Composition, specifications and conformity criteria for common cement, European Committee for Standardization, Brussels, Belgium, 2011.
[2] EN 12620. Aggregate for concrete, Amended in 2008, European Committee for Standardisation, Brussels, Belgium, 2002.
[3] Dhir R K, Ghataora G S and Lynn C J. Sustainable construction materials: sewage sludge ash, Woodhead Publishing, 288p, 2016.
[4] Dhir R K, de Brito J, Lynn C J and Silva R. Sustainable construction materials: municipal incinerated bottom ash, Woodhead Publishing, 458p, 2017.
[5] Dhir R K, de Brito J, Mangabhai R and Lye C Q. Sustainable construction materials: copper slag, Woodhead Publishing, 322p, 2016.
[6] Lynn C J, Ghataora G S and Dhir R K. Municipal incinerated bottom ash (MIBA) characteristics and potential for use in road pavements and geotechnical applications, International Journal of Pavement Research and Technology, 10(2), 185-201, 2017.
[7] Lynn C J, Dhir R K, Ghataora G S, and West R P. Sewage sludge ash characteristics and potential for use in concrete, Construction and Building Materials, 98, 767-779, 2015.
[8] Lynn C J, Dhir R K and Ghataora G S. Municipal incinerated bottom ash characteristics and potential for use as aggregate in concrete, Construction and Building Materials, 127, 504-517, 2016.
[9] Lynn C J, Dhir R K and Ghataora G S. Municipal incinerated bottom ash use as a cement component in concrete, Magazine of Concrete Research, 69(10), 512-525, 2017.
[10] Jackson N and Dhir R K. Civil engineering materials, 5th ed., Palgrave MacMillan, London, 1996.
[11] Dhir R K, Paine K A, Dyer T D and Tang M C. Value-added recycling of domestic, industrial and construction waste arising as concrete aggregate, Concrete Engineering International, 8 (1), 43–48, 2004.
[12] BS EN 1992-1-1. Eurocode 2 - Design of concrete structures. General rules and rules for buildings, British Standards Institute, London, 2004.
[13] Lye C Q, Dhir R K, Mangabhai R and Koh S K. Use of copper slag and washed copper slag as sand in concrete: a state-of-the-art review, Magazine of Concrete Research, 67 (12), 665-679, 2015.
[14] Lye C Q, Dhir R K and Ghataora G S. Elastic modulus of concrete made with recycled aggregates, Structures and Buildings 169 (5), 314-339, 2016.

[15] Lye C Q, Dhir R K, Ghataora G S and Li H. Creep strain of recycled aggregate concrete, Construction and Building Materials, 102, 244-259, 2016.
[16] Lye C Q, Dhir R K and Ghataora G S. Shrinkage of recycled aggregate concrete. Structures and Buildings, 169 (12), 867-891, 2016.
[17] Lye C Q, Dhir R K and Ghataora G S. Deformation properties of concrete made with glass cullet, Structures and Buildings, 170 (5), 321-335, 2017.
[18] Lye C Q, Dhir R K and Ghataora G S. Carbonation resistance of fly ash concrete, Magazine of Concrete Research, 67 (21), 1150-1178, 2015.
[19] Lye C Q, Dhir R K and Ghataora G S. Carbonation resistance of GGBS concrete, Magazine of Concrete Research, 68 (18), 936-969, 2016.
[20] Elgalhud A A, Dhir R K and Ghataora GS. Carbonation resistance of concrete: limestone addition effect, Magazine of Concrete Research, 69 (2), 84-106, 2017.
[21] Zhu X, Zi G, Cao Z and Cheng X. Combined effect of carbonation and chloride ingress in concrete, Construction and Building Materials, 110, 369-390, 2016.
[22] Elgalhud A A, Dhir R K and Ghataora G S. Limestone addition effects on concrete porosity, Cement and Concrete Composites, 72, 222-234, 2016.
[23] Elgalhud A A, Dhir R K and Ghataora G S. Chloride ingress in concrete: limestone addition effects, Magazine of Concrete Research, 70(6), 292-313, 2018.
[24] Lynn C J, Ghataora G S and Dhir R K. Environmental impacts of MIBA in geotechnics and road applications, Environmental Geotechnics. DOI: http://dx.doi.org/10.1680/jenge.15.00029, 2016.
[25] Lynn C J, Dhir R K and Ghataora G S. Sewage sludge ash characteristics and potential for use in bricks, tiles and glass ceramics, Water Science & Technology, 74(1), 17-29, 2016.

Sorptivity of self-compacting concrete with high volume fly ash and its eco-mechanical-durability performance

S A Kristiawan[1], Sunarmasto[1] and M M Ridlo[2]

[1] SMARTCrete Research Group, Civil Engineering Department, Universitas Sebelas Maret, Indonesia
[2] Civil Engineering Department, Universitas Sebelas Maret, Indonesia

E-mail: s.a.kristiawan@ft.uns.ac.id

Abstract. The durability of concrete is governed by the ability of the concrete to resist the penetration of aggressive agents from the environment. The capillary pores of concrete play an important role in the absorption of water, through which aggressive agents may enter into the body of concrete. The rate and amount of absorbed water can be measured by the sorptivity test. It is also recognised that the composition of concrete itself influences the pore structure of the concrete. Hence, in a non-conventional concrete with a very different mixture such as self-compacting concrete (SCC) incorporating a high volume of fly ash, the sorptivity is expected to be different from that of the concrete. This paper aims to determine the influence of high volume fly ash contents on the sorptivity of SCCs, as measured by the method of ASTM C1858. The balanced performance of the concrete – in terms of ecological, mechanical, and durability performance – is assessed using the eco-mechanical-durability index (*EMDI*). The index is developed as a combined value of eco-mechanical index (*EMI*) and eco-durability index (*EDI*). The sorptivity is adopted as the durability parameter in developing the *EDI*. The results confirm that an optimum value of sorptivity at an early age is obtained when about 55% of the OPC is replaced with fly ash. At a later age, however, the optimum value is changed to the 65% replacement level. A higher volume of fly ash tends to decrease the *EMI* but increase the *EDI*. Consequently, the effect of fly ash content on *EMDI* depends on the relative importance between *EMI* and *EDI*. Where *EMI* and *EDI* are considered to be equally important, the effect of fly ash content on *EMDI* is controlled by the greater sensitivity of the *EDI* to the fly ash. Thus, it is shown in this particular case that at later age i.e. when the effect of fly ash is maximal, the highest *EMDI* is found in the SCC with 70% of the cement replaced by fly ash.

Keywords: eco-mechanical-durability index, high volume fly ash, SCC, sorptivity

1. Introduction

In general, durability may be defined as the ability of a product to serve its intended purpose for a period of time (service life) while preserving the minimum required engineering properties without excessive maintenance. The service life of a product is terminated when its engineering properties decrease below the required limits [1].

In terms of concrete, a decrease in the engineering properties over time can be related to deterioration as a result of physical processes, chemical attack, or a combination of both [2, 3]. Chemical attack can occur when aggressive agents penetrate into concrete. There are many mechanisms by which penetration of aggressive agents into concrete may occur [4, 5]. One of these is the absorption and transmission of liquids, which may contain aggressive agents, into the porous concrete structure by capillary action. The extent to which a material absorbs water in this way can be quantified by its sorptivity.

In concrete, the characteristics of the pores play an important role in determining the rate and amount of absorbed water. The relevant characteristics include the open porosity (inter-connection/network and tortuosity of the pores) and pore size distribution. It is a common perception that concrete with a higher total volume of pores (porosity) will absorb more water. However, one may recognise that concretes with similar porosity can have different sorptivity since pore size, network, and tortuosity are also influential factors. The size of pores in concrete may be classified into three categories: macro pores, capillary pores, and gel pores [6]. Capillary pores induce strong capillary forces that absorb water into the concrete.

Capillary pores are formed when the spaces between cement particles are not occupied by cement hydration products. These pores are usually about 1 μm in diameter. Capillary pores are influenced primarily by the water/cement (w/c) ratio, the degree of cement hydration, and the particle size distribution of the cement. The larger the w/c ratio, the greater the space between the cement particles. The higher the degree of hydration, the more the hydration products will fill these spaces. Consequently, as hydration proceeds, the capillary pores will decrease with time, both in terms of size (pore diameter) and total volume [7]. Meanwhile, the spaces between large particles tend to be wider than the spaces between fine particles. These spaces can be reduced by using particles that have a continuous size distribution. This particle size distribution allows smaller particles to occupy the spaces between larger particles, thus acting as fillers and reducing pore size. Once hydration is complete and the solid phase has formed, what matters is the characteristics of the pores. Discontinuous pores, saturated pores, and narrow-opening pores do not affect the transport properties [8].

Lockington et al [9] developed an analytical method to estimate the water penetration profile into an unsaturated concrete from the sorptivity. The method was developed from the basis of equations describing unsaturated flow through porous media, and is applied to one-dimensional cases of water uptake. Meanwhile, Abyaneh et al [10] proposed a numerical computation to simulate capillary absorption in concrete. The numerical computation was validated by comparing the results with available experimental data in the literature [11] and the analytical method of Lockington et al [9]. In their simulation, aggregate particles are considered as non-sorptive constituents in the hardened cement paste. The main findings of their simulation are as follows: 1) the inclusion of aggregate particles dilutes the volume of porous media (hardened paste) through which absorption can occur; 2) the inclusion of aggregate particles also redirects the water to flow around them, and so increases the tortuosity of the transport path; 3) the amount of water absorbed into the concrete is therefore reduced when there is a higher volume fraction of aggregates; and 4) aggregate size does not significantly affect the rate or amount of absorbed water – while the use of larger aggregates provides additional tortuosity, the effect is negated by the lower number of aggregates, as the volume fraction remains the same. The findings highlight an important conclusion i.e. sorptivity is influenced by the relative volume fraction of the concrete constituents.

The relative volume fraction of self-compacting concrete (SCC) constituents is different to that of conventional concrete. Flowability, filling-ability, and passing-ability are fundamental characteristics of SCC, and can be achieved by maintaining moderate viscosity of the mixture. For this reason, the addition of fine constituents (maximum diameter of 0.125 mm) and the use of superplasticiser are essential. Consequently, in SCC, the volume fraction of fine constituents is higher and that of coarse aggregates is lesser than in conventional concrete [12]. The requirement for a high volume fraction of fine constituents leads to a demand for a higher cement content. Therefore, it is common to obtain a flowable concrete using a relatively high cement content [13]. However, a substantial reduction of cement content can be realised by using cementitious materials such as fly ash to replace the cement. It is possible to replace a high volume of cement in this way (more than 50% of the total binder). Considering the distinctive composition of SCC as given above, the sorptivity of this concrete is expected to be diverse than that of conventional concrete. Kanellopoulos et al [8] showed that the sorptivity of SCC is lower than that of conventional concrete, and other investigators have corroborated this conclusion [14–16]. However, these findings cannot be explained by the effect of the relative volume aggregate fraction as discussed in the preceding paragraph. Instead, it is suggested that this behaviour could be related to the pore refinement in the hardened paste of SCC. Kanellopoulos et al [8] confirmed that the sorptivity of SCC correlates with the open porosity. The lower the open porosity of SCC, the lower the sorptivity is. The lower open porosity of SCC may be related to a reduction of the open pore volume in the interfacial transition zone (ITZ). The ITZ is influenced by the packing density of the solid ingredients, the w/c ratio, the binder composition, the type of filler, etc. All these parameters are represented in the composition and proportion of the raw materials used for making concrete [17].

The effect of fly ash on the mechanical and durability performance of concrete varies depending on the physical and chemical properties of the fly ash. The variation of fly ash properties obtained from several Indonesian power plants has been identified by Ekaputri et al [18, 19] and Nurwidayati et al [20]. The source of the variation could be traced to the source of coal, mineralogy, coal burning conditions, the method of collecting fly ash, deposit duration, and the sampling period. It should be noted that the coal supplied to the power plants may also vary from time to time, which brings about inconsistent fly ash properties even when taken from the same power plant. For this reason, characterisation of fly ash is necessary before it is used as a cement substitute. With regard to the sorptivity, the use of fly ash to substitute cement can be expected to reduce the sorptivity of SCC, with significant reductions found when the percentage of fly ash is increased up to 50% [21]. It is interested to know how much further the fly ash percentage can be increased while still reducing the sorptivity. The reduction of sorptivity with the inclusion of fly ash is thought to be due to the filling of micro air voids by finer fly ash particles [21], and also due to the pozzolanic effect of fly ash at a later age [22, 23].

From a sustainability point of view, utilising a high volume of fly ash will bring beneficial effects in terms of reduced carbon dioxide emissions (e-CO_2) and a reduction of unused waste material. Compared to other constituents, cement is the main contributor of e-CO_2 in the production of concrete. The embodied CO_2 (on a mass basis i.e. kg CO_2 per kg of concrete) of cement is about 0.83 [24]. Hence, a partial substitution of cement with fly ash (and other cementitious minerals) becomes a practical way to reduce the e-CO_2. Fly ash itself is a by-product of coal-fired power plants and is considered a waste material. Indonesia has more than 85 coal-fired power plants, which contribute to million tonnes of fly ash every year [25]. If this material is left unused, it eventually becomes an environmental burden.

This research aims to develop an eco-mechanical-durability index that will be used to assess the balance of environment, mechanical and durability performance of SCC containing high volume fly ash. The environment, mechanical and durability performance will be quantified based on the e-CO_2, compressive strength and sorptivity coefficient of the concrete, respectively.

2. Eco-mechanical-durability performance index

The beneficial effects of using fly ash with respect to environmental performance must not compromise the mechanical and durability requirements limit. So, an optimum use of fly ash has to be sought by considering the balance of the environmental, mechanical, and durability performance. The first step to assess this balance is by developing performance indicators capable of measuring the combined three parameters. Fantilli and Chiaia [26] developed eco-mechanical indicators based on eco-mechanical ratios as shown in the following equation:

$$I = \frac{CO_2\ released}{mechanical\ properties} \tag{1}$$

where a lower value of *I* indicates a better eco-mechanical performance. In a further study by Chiaia et al [27], the inverse of equation (1) is preferred:

$$EMI = \frac{MI}{EI} \tag{2}$$

where *EMI*, *MI*, and *EI* are the eco-mechanical, mechanical, and ecological indices, respectively. In their paper, the numerator and denominator of equation (2) are expressed as *f(MI)* and *g(EI)* i.e. correlation function of mechanical index and ecological index, respectively. With equation (2), a higher value of *EMI* is desirable.

Long et al [28] used an indicator similar to equation (1), but the numerator representing the environmental impact was extended to include the embodied CO_2 (C_I), embodied energy (E_I), and embodied primary resources (R_I). Meanwhile, strength (σ) was adopted as the parameter representing the mechanical properties in the denominator of the equation. Hence, the following set of equations were suggested:

$$C_I = \frac{embodied\ CO_2\ (\text{kg}{\cdot}\text{m}^{-3})}{\sigma\ (\text{MPa})} \tag{3.a}$$

$$E_I = \frac{embodied\ energy\ (\text{MJ}{\cdot}\text{m}^{-3})}{\sigma\ (\text{MPa})} \tag{3.b}$$

$$R_I = \frac{embodied\ primary\ resources\ (\text{kg}{\cdot}\text{m}^{-3})}{\sigma\ (\text{MPa})} \tag{3.c}$$

In the previous equations, durability has not been considered for assessing the concrete performance. The eco-durability index (*EDI*) may be calculated using the following equation to account for the contribution of durability:

$$EDI = \frac{DI}{EI} \tag{4}$$

Finally, a combination of *EMI* and *EDI* may be expressed into the following equation:

$$EMDI = a\left(\frac{MI}{EI}\right)_n + b\left(\frac{DI}{EI}\right)_n \tag{5}$$

where *EMDI, MI, DI,* and *EI* are the eco-mechanical-durability index, mechanical index, durability index, and ecological index respectively; *a* and *b* are weighting coefficients of *MI/EI* and *DI/EI* respectively; and the subscript *n* indicates that the ratios of *MI/EI* and *DI/EI* are normalised to the values of the reference concrete. Hence, equation (5) presents non-dimensional parameters. The weighting coefficient is introduced to account for the relative importance between *MI/EI* and *DI/EI.* A higher value indicates that the index is considered relatively more important than the other index. The value of each weighting coefficient is between 0 to 1, and the summation of the two coefficients (*a* + *b*) must equal 1. The value of these coefficients could vary depending on certain circumstances; for example, if the concrete structure is exposed to a severely aggressive environment, the durability of the concrete holds greater importance, and thus the coefficient *b* would be greater than *a*. The assigned values of the coefficients may be determined either by an expert's judgment, the Delphi method, analytical hierarchy process (AHP), or other relevant methods.

3. Experimental investigation

3.1. Materials

The mix proportions of SCCs investigated in this paper are presented in table 1. The maximum aggregate size used in the mixture was 10 mm. The bulk specific gravity (on a saturated surface-dry basis) and the fineness modulus of the coarse aggregates were 2.51 and 6.03, respectively. The corresponding values for the fine aggregates were 2.50 and 2.92. The gradations of both aggregates conformed to the specification of ASTM C-33. The fly ash was supplied from the Cilacap Power Plant and the oxide composition of the fly ash is given in table 2. A modified polycarboxylate based superplasticiser was used in the mixture to aid the formation of flowable concrete. The strengths of the SCCs are presented in figure 1.

Table 1. The mix proportion of SCC with various percentage of fly ash per m^3.

ID	Cement (kg)	Fly ash (kg)	CA[a] (kg)	FA[b] (kg)	Water (kg)	Sp[c] (kg)
SCC-50	368.55	368.55	703.2	578.64	211	7.37
SCC-55	331.70	405.41	703.2	578.64	211	7.37
SCC-60	294.84	442.26	703.2	578.64	211	7.37
SCC-65	257.99	479.16	703.2	578.64	211	7.37
SCC-70	221.13	515.97	703.2	578.64	211	7.37

[a] Course aggregate
[b] Fine aggregate
[c] Superplasticiser

Table 2. Oxide composition of fly ash.

SiO_2	Al_2O_3	Fe_2O_3	TiO_2	CaO	MgO	K_2O	Na_2O	P_2O_5	SO_3	MnO_2
45.27	20.07	10.59	0.82	13.32	2.83	1.59	0.98	0.41	1	0.07

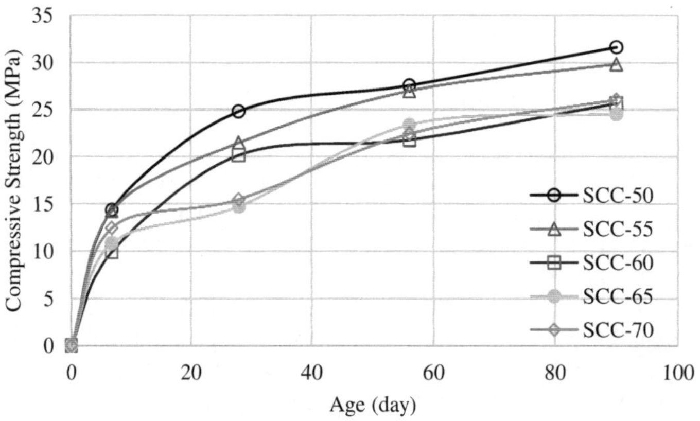

Figure 1. Compressive strength development of SCCs with varying amounts of fly ash.

3.2. Methods

Sorptivity was determined following the method given in ASTM C1858. The method measures the rate of water absorbed through one surface of dry concrete (one-dimensional case of water uptake). The increase of specimen weight over time is measured. The data are then analysed and plotted in graphical form. The vertical axis of the graph indicates the absorption (I, in mm) where the value is calculated using the following equation:

$$I = \frac{m_t}{A \cdot d} \tag{6}$$

where m_t is the change in the mass of the specimen at time t; A is the exposed surface area of the specimen through which the absorption takes place; and d is the density of water. Meanwhile, the horizontal axis represents the square root of time ($Sec^{1/2}$). The slope of the graph is determined, to indicate the sorptivity coefficient. In the method given by ASTM C1858, the sorptivity coefficient is separated into two phases: initial and secondary absorption. The initial absorption is determined from the start of the uptake of water until 6 hours, whereas the secondary absorption is determined from 6 hours until 8 days. In this current investigation, the sorptivity of concretes aged for 7, 28, 56, and 90 days were measured. The obtained sorptivity coefficients were calculated as average values from 5 measurements. Hence, a total of 100 specimens were prepared for this investigation.

4. Results and discussion

4.1. Sorptivity

The cumulative amount of absorbed water over time is presented in figure 2. The profiles of cumulative absorbed water by all SCCs reveal two distinctive rates of water absorption. At the start of the absorption process, a fast rate of water absorption can be identified as shown by the initially steep slope of the lines in figure 1. After 1 day (about 294 $Sec^{1/2}$), a clear change of the slopes can be observed. The slope of the lines decrease, corresponding to the slower rate of water absorptions at longer times, as the pores within the concrete are already partially saturated. In the method given by ASTM C1858, the initial and secondary slopes represent the values of the initial and secondary sorptivity coefficients, respectively.

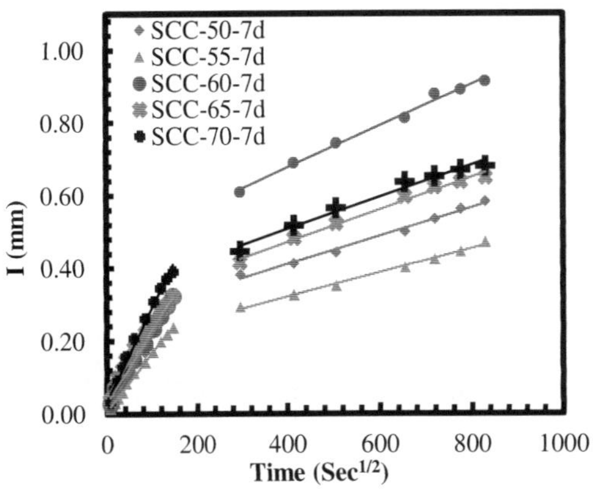

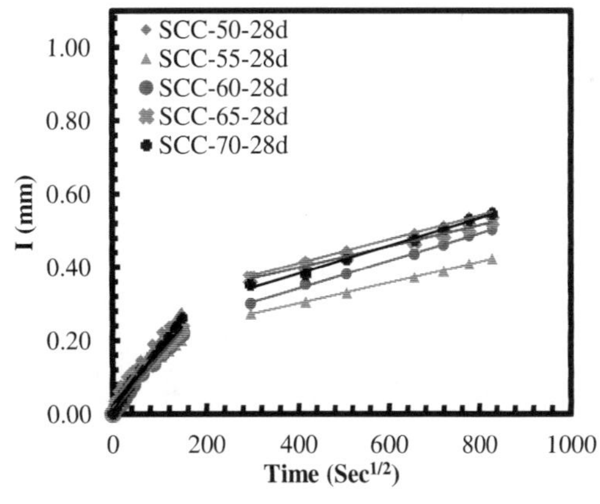

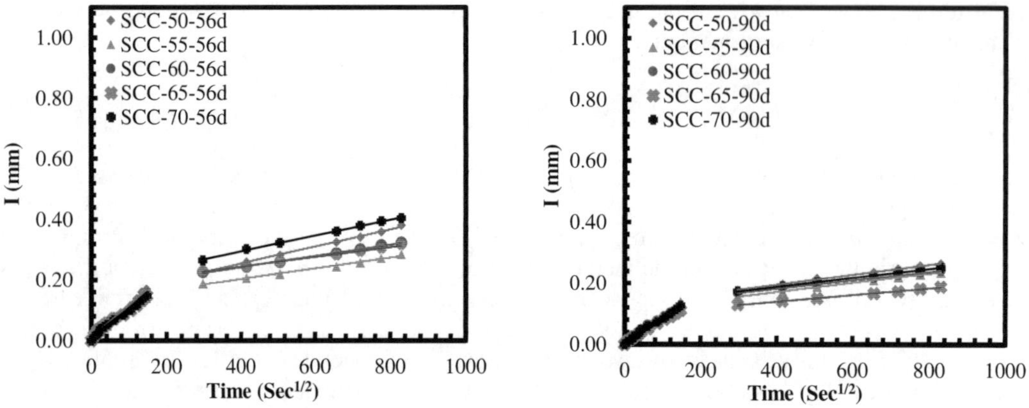

Figure 2. Absorption of water into SCCs at various ages (7, 28, 56 and 90 d).

The data of initial absorption presented in figure 2 are too crowded, making it difficult to compare between mixtures. However, a least square linear regression analysis has been performed for each set of absorption data mixture to study the effect of concrete age and fly ash content on the sorptivity coefficients. The results of the analysis are presented in figure 3. It is noticeable that both the initial and secondary sorptivity coefficients decrease with concrete age. The initial sorptivity coefficients of SCCs at 28 days are reduced by 18–34% compared to those at 7 days. The corresponding reduction of secondary sorptivity coefficients is 13–35%. Further decreases of both the initial and secondary sorptivity coefficients can be expected at later ages. At 90 days of age, the initial sorptivity coefficients are reduced by 59–68% compared to those at 7 days. For the secondary sorptivity, the reduction is counted at 54–76%. The reduction may be related to a change in the pores' structure as a result of the progressing hydration. When the concrete first hardens, a network of capillary pores is formed, through which water can easily be absorbed. As hydration progresses, these capillary pores are gradually occupied by cement hydration products, which disrupt the connectivity between these capillary pores. In addition, at later ages, a sufficient pozzolanic reaction between the fly ash and hydrated lime takes place, bringing a similar effect. This finding matches those observed in other literatures [22, 23].

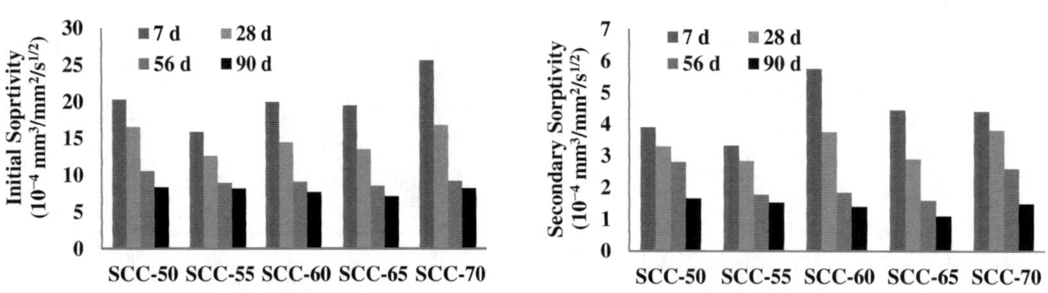

Figure 3. Initial and secondary sorptivity coefficients of SCCs at various ages.

The sorptivity coefficients are also affected by the amount of cement replaced with fly ash. At an early age (7 d), the lowest values of both the initial and secondary sorptivity coefficients occur at a 55% replacement level. At a later age (90 d), the lowest values are shifted to the 65% replacement level. The highest initial sorptivity was observed for the 7-day SCC with the 70% replacement level. The processes, mechanisms, and reactions that contribute to this change in sorptivity with both the amount of fly ash and time are explained as follows.

The capillary pores are formed when empty spaces between the solid particles are not occupied by the hydration products. Fly ash, with its finer particle size, can refine these spaces before hydration takes place; hence, replacing cement with fly ash is expected to reduce the pores. However, the replacement level reaches an optimum value when the lowest packing density of the mixture is achieved. For the concrete investigated in this study, this optimum value is found at a 55% replacement level. Meanwhile, at 7 d of age, the cement hydration is still progressing as indicated by the strength development in figure 1. The strength of SCC at 7 d only represents about 39–48% of the strength at 90 days. Thus, the hydration products have not fully occupied the inter particle spaces and so, the original packing density of the mixture becomes a crucial factor in determining the pore characteristics at early age. When the replacement level is too high (70%), fly ash tends to dilute the cement particles. Consequently, less hydration products are available to occupy the inter particles spaces. On the other hand, at a later age (90 d), the hydration is nearly completed and sufficient pozzolanic reaction has also taken place. The capillary pores that were formed at an early age are now occupied by the hydration and pozzolanic reaction products. The replenishment of empty capillary pores with hydration and pozzolanic reaction products at a later age is a significant factor in determining the pore characteristics. All of the processes (formation of capillary pores that occupy inter particles spaces, refinement of capillary pores in the progress of hydration, and replenishment of empty capillary pores with hydration and pozzolanic products) requires balanced quantities of cement and fly ash to achieve an optimal value of sorptivity. This study indicates that a 65% replacement level of cement with fly ash gives the lowest value of both initial and secondary sorptivity coefficients at 90 d.

4.2. Eco-mechanical-durability performance

Implementation of equation (5) for assessing the eco-mechanical-durability performance of concrete requires consideration in assigning relevant parameters to be used as indicators of eco, mechanical, and durability performance:

With respect to ecological performance, the following parameters may be considered to evaluate environmental impacts related to the production, use, maintenance, and demolition of concrete; environmental emissions, energy consumption, and resource depletion. Emission of carbon dioxide (e-CO_2) from concrete production and transportation is estimated to be responsible for approximately 10% of the total man-made CO_2 in the atmosphere [28]. Therefore, e-CO_2 becomes the most significant parameter to assess the environmental impact of concrete. Purnell and Black [24] and Long et al [28] reviewed embodied CO_2 values for major concrete constituents; cement itself accounts for about 0.83 kg CO_2 per kg of concrete. Other embodied CO_2 values for major concrete constituents have been summarised in [28] and these values are applied for this study.

In previous studies [26–28], the mechanical performance of concrete is represented by either strength or a combination of strength and ductility. Compressive strength (σ) has been chosen for this evaluation since this is the fundamental property required in the utilisation of concrete as structural elements.

There are several indicators that can be considered suitable for the purpose of assessing the durability performance of concrete, such as permeability, sorptivity, bulk resistivity, chloride diffusion, and abrasion resistance [8, 22, 23, 29, 30]. All or part of these parameters may be employed to comprehensively evaluate the durability performance of concrete. For the current study, it is simply sorptivity that is adopted as the indicator. Hence, for the SSCs evaluated in this investigation, the durability performance is only relevant to concrete that is in contact with water. It is noted that a more durable concrete would have a lower sorptivity coefficient i.e. durability index (*DI*) is inverse of the sorptivity coefficient.

Given the above considerations, equation (5) is now transformed as follows:

$$EMDI = a\left(\frac{\sigma}{embodied\ CO_2}\right)_n + b\left(\frac{s^{-1}}{embodied\ CO_2}\right)_n \tag{7}$$

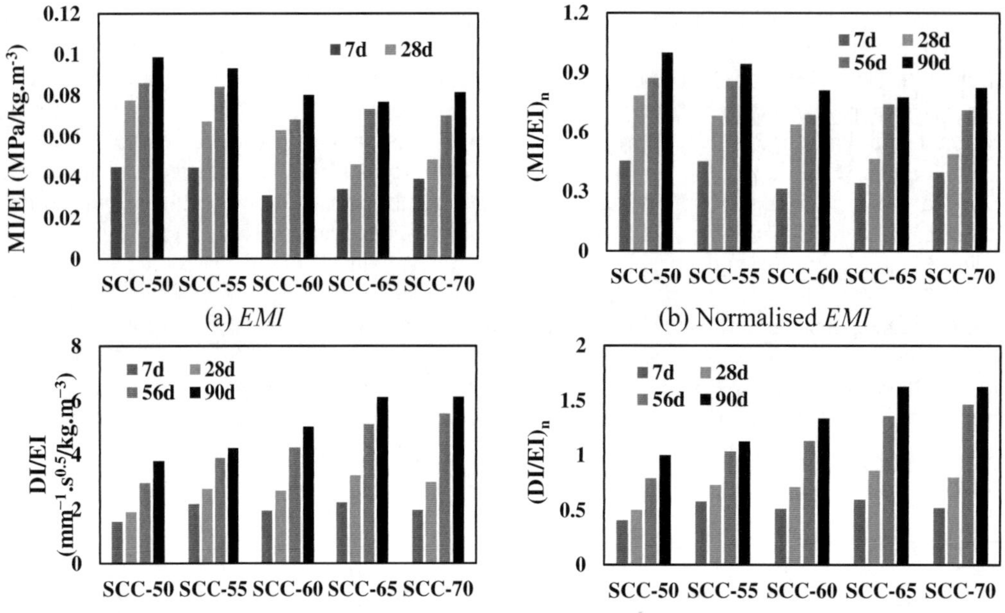

(a) *EMI* (b) Normalised *EMI* (c) *EDI* (d) Normalised *EDI*

Figure 4. Eco-mechanical index (EMI) and eco-durability index (EDI) with their respective normalised values

The eco-mechanical index (*EMI=DI/EI*) and the eco-durability index (*EDI=DI/EI*), along with their normalised values, are given in figure 4. The normalised values are defined as the ratio of the *EMI* and *EDI* of each SCC to the corresponding values of SCC-50 at 90 d. Thus, the values for the reference concrete (SCC-50 at 90 d) must be 1, and index values less than 1 indicate that the performance of the concrete is lower that this reference concrete, and vice versa. The later age (90 d) of concrete was chosen as a reference since at this age the effect of fly ash is maximal. The *EMI* is expressed in MPa/kg·m^{-3} while *EDI* is expressed in mm^{-1}·s$^{0.5}$/kg·m^{-3}. The denominators represent the kg of embodied CO_2 per m^3 of SCC; the compressive strength used to calculate the *EMI* is specified in MPa; and an inverse unit of sorptivity is used for determining *EDI* i.e. mm^{-1}·s$^{0.5}$. It is clear from figure 4 that the beneficial effects of fly ash increase as the concrete ages; both the *EMI* and *EDI* rise with time. Figure 4 also shows that SCC-50 gives the highest value of the *EMI*. Increasing the replacement level of cement with fly ash above 50% tends to decrease the *EMI*. On the other hand, the trend is reversed with respect to the *EDI*. An increase of fly ash content will result in an increase of the *EDI*.

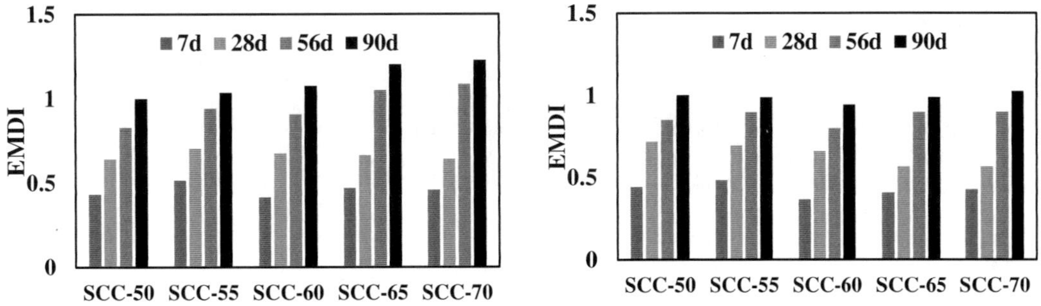

(a) $a = 0.5$; $b = 0.5$ (b) $a = 0.75$; $b = 0.25$

Figure 5. The eco-mechanical-durability index (EDMI).

The combined value of *EMI* and *EDI* i.e. *EMDI*, calculated using equation (7) with the normalised values, is presented in figure 5 for two cases. In the first case, both the *EMI* and *EDI* are considered to be equal factors. Hence, the assigned values of a and b in equation (7) are both 0.5. Figure 5.a shows that at a later age (90 d) i.e. when the beneficial effect of fly ash is maximised, the highest *EMDI* is found for SCC-70. A reduction of *EMI* with an increase of fly ash is counter balanced by the increase of *EDI*. Since the *EDI* is more sensitive to changes to the fly ash content than the *EMI*, the result is that the *EMDI* increases with increasing fly ash content. For the second case, where the *EMI* is considered to be relatively more important that the *EDI* by assigning values of $a = 0.75$ and $b = 0.25$, an increase of fly ash content does not significantly affect the *EMDI* of SCCs at 90 days (see figure 5.b).

The preceding paragraph illustrates the use of *EMDI* as a means to assess the eco-mechanical-durability performance of SCCs incorporating various amounts of fly contents. The *EMDI* is especially useful when the effects of fly ash on the mechanical and durability indicators are not aligned. The *EMDI* can serve in seeking the best composition of a particular concrete mixture in a specific case.

5. Conclusions

The main conclusions of this study are summarised as follows:

- The sorptivity of SCCs incorporating various high-volume fly ash contents decreases with time. At 90 days of age, the initial and secondary sorptivity coefficients are reduced by 59–68% and 54–76% respectively, compared to those at 7 days.
- At an early age (7 days), the lowest value of both the initial and secondary sorptivity coefficients occur when 55% of cement is replaced with fly ash. However at a later age (90 days), the lowest values are shifted to the 65% replacement level.
- An increase in fly ash content tends to decrease the eco-mechanical index (*EMI*) but increase the eco-durability index (*EDI*) of SSC.
- Given the same weighting factor of *EMI* and *EDI*, the highest eco-mechanical-durability index (*EMDI*) is found for the SCC where 70% of cement is replaced with fly ash.

References

[1] Cooper T 1994 *Beyond Recycling: The Longer Life Option* (London: The New Economic Foundation) p 5
[2] Portland Cement Association 2002 *Types and Causes of Concrete Deterioration* (Skokie, Illinois: Portland Cement Association) pp 1–15
[3] Dyer T 2014 *Concrete Durability* (Boka Raton, Florida: CRC Press) pp 7–182
[4] Kropp J and Hilsdorf H K 1995 *Performance Criteria for Concrete Durability* (London: E & FN Spon) pp 3–9

[5] Claisse P A 2014 *Transport Properties of Concrete: Measurement and Applications* (Cambridge: Woodhead Publishing) pp 1–16
[6] Comite Euro-International Du Beton 1997 *Durable Concrete Structures: Design Guides* (London: Thomas Telford) pp 3–6
[7] Neville A M 2004 *Properties of Concrete* (Essex: Pearson Education Limited) pp 31–2
[8] Kanellopoulos A, Petrou M F and Ioannou I 2012 *Constr. Build. Mater.* 37 320–5
[9] Lockington D, Parlange J Y and Dux P 1999 *Mater. Struct.* 32 342–7
[10] Abyaneh S D, Wong H S and Buenfeld N R 2014 *Comp. Mater. Sci.* 87 54–64
[11] Hall C 1989 *Mag. Conc. Res.* 41 51–61
[12] Okamura H and Ozawa K 1995 *Conc. Lib. JSCE.* 25 107–20
[13] Dinakar P 2012 *Mag. Conc. Res.* 64 401–9
[14] Assie S, Escadeillas G and Walter V 2007 *Constr. Build. Mater.* 21 1909–17
[15] Sonebi M and Nanukuttan S 2009 *ACI Mater. J* 106 161–6
[16] Zhu W and Bartos P J M 2003 *Mag. Conc. Res.* 33 921–6
[17] Kristiawan S A, Sunarmasto and Murti G Y 2017 *IOP Conf. Ser. Mater. Sci. Eng.* 176 012043
[18] Ekaputri J J, Triwulan, Priadana K A, Susanto T E and Junaedi S 2013 *Proc. The 2013 World Congress on Advances in Structural Engineering and Mechanic* (September 8–12, Jeju) pp 2988–96
[19] Ekaputri J J, Ulum M B, Triwulan, Ridho B, Susanto T E and Mohd Mustafa A B A 2015 *App. Mech. Mater.* 754–755 320–5
[20] Nurwidayati R, Ulum M B, Ekaputri J J, Triwulan and Suprobo P 2016 *Mater. Sci. For.* 841 118–25
[21] Leung H Y, Kim J, Naseem A, Jaganathan J and Anwar M P 2016 *Constr. Build. Mater.* 113 369–75
[22] Elahi A, Basheer P A M, Nanukuttan S V and Khan Q U Z 2010 *Constr. Build. Mater.* 24 292–9
[23] Uysal M and Akyuncu V 2012 *Constr. Build. Mater.* 34 170–8
[24] Purnell P and Black L 2012 *Cem. Conc. Res.* 42 874–7
[25] Wijaya A L, Ekaputri J J and Triwulan 2017 *MATEC Web Conf.* 138 01010
[26] Fantilli A P and Chiaia B 2013 *Constr. Build. Mater.* 40 189–96
[27] Chiaia B, Fantilli A P, Guerini A, Volpatti G and Zampini D 2014 *Constr. Build. Mater.* 40 189–96
[28] Long G, Gao Y and Xie Y 2015 *Constr. Build. Mater.* 84 301–6
[29] Nganga G, Alexander M and Beushausen 2013 *Constr. Build. Mater.* 45 251–61
[30] Siddique R 2013 *Constr. Build. Mater.* 47 1444–50

Quarry limestone dust as fine aggregate for concrete

Liborio Cavaleri[1], Ruben Paul Borg[2], F. P. La Mantia[1], Vincenzo Liguori[1]

[1] Dipartimento di Ingegneria Civile, Ambientale, Aerospaziale, dei Materiali, University of Palermo, Italia
[2] Faculty for the Built Environment, University of Malta, Malta

E-mail: liborio.cavaleri@unipa.it, ruben.p.borg@um.edu.mt

Abstract. In quarrying activities, rock is extracted and transformed into aggregate of various sizes for civil engineering applications. In this process waste fine aggregates (dust waste) are generated. The disposal of this type of waste is a further cost in the extraction process, but also a possible cause of environmental pollution (e.g. leaching into water reserves, atmospheric pollution as a result of small particles causing respiratory diseases or deposited on plants disrupting photosynthesis, affecting aquatic habitats, etc.). A strategy for the effective recycling of quarry dust does not only reduce waste generation and disposal, but also addresses protection of the environment. The Italian quarrying industry covers a relevant portion of global mineral extraction resulting in a significant production of fine waste. In some cases quarries are located close to ecological sensitive and protected areas and to the coast with higher risks for biodiversity (an example is the limestone extraction industry in Trapani, Sicily). In this context, the paper reviews the strategies proposed in the use of limestone fine waste, especially for concrete, and discusses an experimental program intended to assess the mechanical properties of concrete made with the fine limestone waste produced in the area of Trapani, as a partial substitute of fine aggregate (sand).

Keywords: Quarry limestone dust, limestone waste, fine aggregate, mechanical properties

1. Introduction

There has been an increase in awareness on the need for sustainable development to guarantee a rational use of the earth resources. Quarrying activities, as a result of large demands for mineral resources including aggregate for civil engineering works, have a significant impact on the environment and wellbeing in society.

Quarrying of stone intended for construction related and civil engineering applications include also the extract of stone for ornamental products, stone to be processed as dimensioned stone or as paving or cladding slabs resulting in significant amount of waste rock including coarse and fine material. The latter is generally the result of the cutting machines in quarries.

In the case of stone extracted for ornamental stone products the marble quarries are known to generate a particularly large quantity of waste. 18% of global marble extraction is concentrated in Italy. The main sites for the extraction of marble are located in Tuscany (in the province of Massa Carrara) and in Sicily (in the province of Trapani, in Custonaci). The marble in Custonaci is effectively a limestone and not a metamorphic stone as marble is by the geological point of view. Nevertheless, it is worked as it was a marble and for this reason it is improperly called marble being in all a marble by the physic point of view.

IOP Conf. Series: Materials Science and Engineering **442** (2018) 012003 doi:10.1088/1757-899X/442/1/012003

Approximately 200 quarries are located around Massa Carrara while about 70 quarries are located in Custonaci. Only 20% of the extracted material is used commercially, while the remaining 80% is waste generated through the activities. It is estimated that about 50% is constituted by rocks of various size while 30% is dust. 6 million tons/year are extracted in Italy of which 4.8 million tons consist of quarry waste including 1.8 million tons of waste dust.

As a consequence of the intense quarrying activity over the years, the marble basins of Custonaci (figures 1-3) and Massa Carrara are characterized by enormous volumes of accumulated discarded deposits resulting from the quarrying activity (in some cases these can reach dozens of hectares). Generally, waste is pushed towards the slopes close to the quarry areas, threatening the integrity of landscape and of the environment in particular with regards to ground water and surface water contamination, or even because of atmospheric contamination due to dust dispersion [1]. Furthermore, quarry waste represents one of the principal sources of hydro-geological instability, leading to rock-falls and debris flow.

a) b)

Figure 1. a) View of Sicily indicating the area of Custonaci (in the red circle); b) the area of Custonaci indicating (in white) the distribution of quarries.

Figure 2. Waste near a quarry in Custonaci

Figure 3. Quarry dust in Custonaci

In the past, different quarry waste deposits have provoked landslides. During the initial deposition of the quarry waste the material accumulates in heaps based on the angle of friction. The continuous storage and increment of the volume of waste results in alterations to the slope boundaries. Therefore, sudden landslides can be activated, even more frequently during periods of intense rainfall.

Over the years the accumulation of waste has become a significant problem due to the large volumes of waste generated. As a result different strategies for disposal of the material are investigated. In most cases the main focus is, first, how to possibly reduce the volume of waste generated, and then how to transform the waste into a resource. With reference to the marble industry, for a long time the economic interests have been a priority. The profits deriving from the 20% marketable extracted materials have distracted the attention of the industry from the waste generation problem. However there is a growing interest on the waste generated and how the large volumes of waste generated can be transformed into an opportunity through recycling.

Effectively, such a huge quantity of quarry waste may represent a resource of good quality. Different studies have addressed the quarry marble waste (at least the part that can be processed as coarse aggregate) suitable as inert materials for civil engineering applications including road and railway construction works and in the production of concrete and concrete products. Recycling of the material on one hand can solve the problem of disposal and, on the other hand, is considered to be favourable from an environmental point of view because it leads to a reduction in the demand of new quarries for the production of coarse aggregate for concrete or for civil engineering applications. Nevertheless, referring to marble quarries, a non-negligible part of the waste (30% of the material extracted) consists of dust waste or a dust-water mix (the water is used during the cutting and trimming of the blocks using the diamond wire cutting machines, to allow for cooling). The recycling of the quarry dust waste is more challenging.

Regulations impose the transportation to landfill of the dust or the water-dust mix, but the high cost involved, suggest recycling of the material for uses in different fields.

Many researchers have tried to exploit quarry dust for the production of concrete. The quarry dust has been used as partial or full substitute of the sand. Rai et al [2] carried out an experimental investigation based on the substitution of the sand with quarry dust for the production of self-compacting concrete. Different quantities of sand are substituted up to 100% and the strength of the concrete was assessed. Prakash et al [3] assessed the use of dust for normal concrete, reporting that a 40% replacement of fine aggregate by quarry dust resulted in the highest compressive strength. In the study reported by Bahoria et al [4] sand was replaced with quarry dust in combination with waste plastic. A similar investigation was carried out by Aishwaryalakshmi et al [5] where up to a 30% substitution of fine aggregate was assessed. Other studies have addressed the replacement of sand with quarry dust for the production of concrete (e. g. Jagadeesh et al [6] and Liebermann et al [7]). In some cases, the mechanical properties of non-conventional concrete was investigated, as reported by Lim at al [8] where the use of quarry waste for lightweight foam concrete was considered. A literature review of the strategies used in the replacement of aggregate for concrete with waste is presented by Bahoria et al [9] focusing mainly on the strength characteristics. However, some studies on the stress strain characteristics of concrete made using quarry dust waste are reported in literature (Kankam et al [10]). An interesting strategy refers to the use of quarry dust for the production of coarse aggregate through a cold bonding procedure as reported by Thomas et al [11-12].

With regards marble quarries, different strategies have been employed for the use of coarse aggregate in concrete as reported by André et al [13].

The potential recycling of waste marble dust in Sicily has not been explored yet. The objective of the research is to explore the potential use of waste marble dust derived from the Sicilian marble industry, through an experimental investigation.

The marble in Custonaci (that, as above mentioned, is improperly said marble because it is a limestone worked as a marble) has an age varying from the Cretaceous to the Eocene. It is a sedimentary rock constituted mainly by calcite (98%) and (2%) dolomite, apatite, illite, goethite and quartz. It originated in a marine environment, after some processes of alteration, erosion, transport and deposition of debris of pre-existing rocks and/or animal or plant remains. The marbles of Custonaci can be fossiliferous or non fossiliferous, stratified, sometimes coloured with different pigmentation. The physical-mechanical characteristics are as follows: unit volume weight 23-26 kN/m^3, thermal expansion coefficient 0.0047-0.0043 mm/m°C, compressive strength 160-200 MPa, compressive strength after freezing 155-195 MPa, tensile strength 13-17 MPa.

2. Experimental program

In this study the fine aggregate of a reference control mix has been progressively replaced with a single type of quarry dust sourced from Custonaci, at two different levels of substitution. The experimental study reported forms part of a more extensive research. The paper presents the first results, which are discussed and compared with data available in literature.

2.1 Materials

The reference concrete was produced using portland limestone cement Type 32.5 as the hydraulic binding agent, together with quarry rock of 2 mm nominal maximum size used as fine aggregate, and 20 mm crushed rock coarse aggregate. The particle size distribution of the fine aggregate is presented in figure 4.

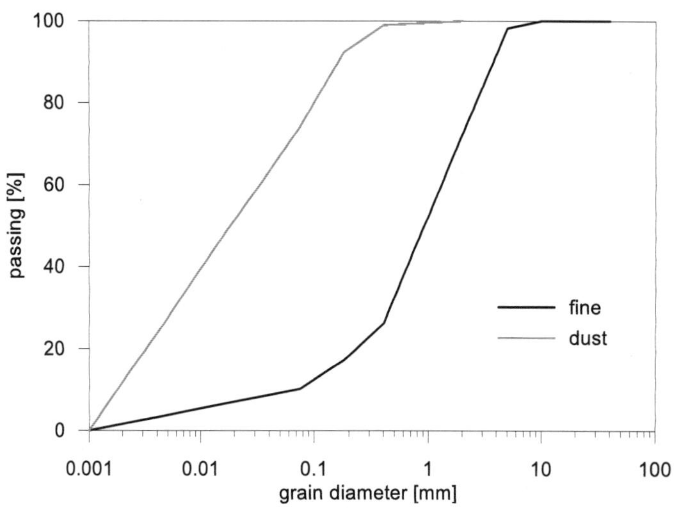

Figure 4. Fine aggregate and quarry dust: granulometric curves

IOP Conf. Series: Materials Science and Engineering **442** (2018) 012003 doi:10.1088/1757-899X/442/1/012003

The sand (having a specific weight of 2860 kg/m^3) was replaced by a marble quarry dust with granulometric characteristics as presented in figure 4.

2.2 Mix proportion and preparation of test specimens

The reference concrete was prepared using a mix typical for obtaining a 28 days characteristic strength of 30 MPa. The mix consisted of 315 kg of cement, 798 kg of coarse aggregate, 735 kg of sand per m^3 of concrete, with a water/cement ratio of 0.5.

The concrete mixes were prepared in the materials laboratory of the University of Palermo (each mix volume was approximately 1/21 of a m^3 in agreement to the characteristics of the mixer. Therefore, the reference mix was composed of 15 kg of cement, 38 kg of course aggregate, 35 kg of sand and a water/cement ratio of 0.5. In total, 11 cubic test samples (15 cm cubic specimen) were produced for each mix. Six mixes were prepared: two reference mixes and further four mixes with different levels of replacement of fine aggregate. The four mixes consisted of the same quantities of cement, water and coarse aggregate as for the reference mixes. However the fine aggregate portion was composed of 31 kg of sand and 4 kg of quarry dust for two of the four cases, and 27 kg of sand and 8 kg of quarry dust for the remaining two mixes. The concrete mixes had a similar slump class in the range 30-60 mm, but which was reduced with the increase in the quarry dust content in the concrete. The concrete mix proportions are presented in table 1 with reference to a cubic meter of fresh concrete. The laboratory mix proportions are presented in table 2.

Table 1. Mix proportions per cubic meter

Id. mix	Portland cement 32.5 kg/m^3	Coarse aggregate kg/m^3 (m^3)	Sand kg/m^3 (m^3)	Dust kg/m^3	Water l/m^3
1-2	315	798 (0.7)	735 (0.5)	-	157
3-4	315	798 (0.7)	651 (0.44)	84	157
5-6	315	798 (0.7)	567 (0.39)	168	157

Table 2. Mix proportions referred to 1/21 of a cubic meter

Id. Mix	Portland cement 32.5 kg per 1/21 m^3	Coarse aggregate kg per 1/21 m^3	Sand kg per 1/21m^3	Dust kg per 1/21 m^3	Water l per 1/21 m^3
1-2	15	38	35	-	7.5
3-4	15	38	31	4	7.5
5-6	15	38	27	8	7.5

The percentage of sand replacement in the concrete mixes was 13 % (mixes 3 and 4) and 26% (mixes 5 and 6) respectively. The cement and aggregates were first mixed together in a mixer. Water was then added and mixing was continued until a uniform and homogenous matrix was obtained. The fresh concrete was cast in 150 × 150 × 150 mm moulds and compacted with a concrete vibrator. The specimens were de-moulded after 24 hours and cured by immersion in water at 21 °C. The cubes from the mixes 1-3-5 were used to obtain the compressive strength of the concrete after three, seven and fourteen days. The cubes from the mixes 2-4-6 were used for tests carried out at 28 and 60 days.

2.3 Discussion

The results of the compressive strength tests on each cube at 3, 7, 14, 28, 60 days are presented in figure 5. It is immediately clear that the replacement of 13% of the sand produces an increase of the strength while the replacement of the 26% of the sand produce a decrease of strength variable with the days of curing. The strength of the mix characterized by a sand replacement of 26% is recovered at 60 days. Further, a greater dispersion of the compressive strength results was observed for the mixes with no replacement and replacement of 13 % of the sand.

Figure 6 presents the average results of the compressive strength tests showing the effect of sand replacement with Custonaci quarry dust. In spite of the replacement of 26% of sand resulting in a decrease of the strength with reference to the control mix, the reduction in strength is less at 14 days than at 7 days and negligible at 60 days. This is clearer in figure 7 where the increment factors of each mix at 3, 7 and 14, 28 and 60 days are normalized with respect to the strength at 3 days. This factor is obtained as a ratio between the strength obtained at each reference time and the strength obtained at 3 days. For each mix the starting point at 3 days is therefore 1 and greater values are obtained depending on the strengths reached. From figure 7 it can be observed that the reference mix and the mix characterized by a sand replacement of 13 % have a similar strength increment, indicating a 14-day strength approximately 1.7 times greater than the 3 day strength. On the other hand the mix characterized by a sand replacement of 26% has a higher gain in strength between 7 and 14 days, with a 14-day strength approximately 2 times greater than the 3 day strength. A similar consideration can be done at 28 and 60 days.

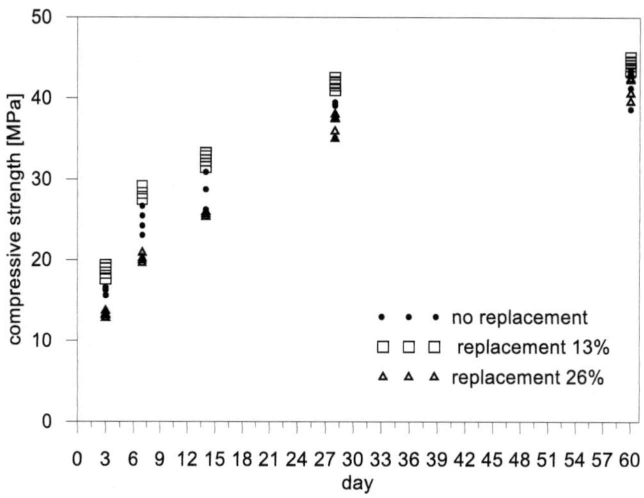

Figure 5. Strengths of each cube at 3,7, 14, 28, 60 days

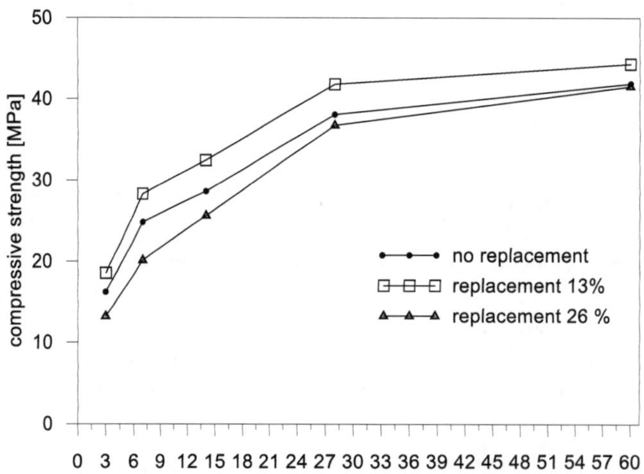

Figure 6. Average strengths at 3,7, 14, 28, 60 days

It appears that the dust slows the gain in strength of the concrete, at least in the first days after casting but there is an indication of a recovery in the gain in strength with time.. In figure 8 this concept is illustrated through the percentage of variation of strength at 3, 7, 14, 28 and 60 days with respect to the reference mix. It is shown that the mix with 13% sand replacement maintains a constant strength greater than 15% with respect to the reference mix during the first 28 days of curing. The mix characterized by a 26 % replacement of sand indicates a reduction in the gap from -18% to about -10 % (at the 14th day) with respect to the reference mix. This gap is further reduced at 60 days (2%).

In each case the above-mentioned figures indicate that replacement of sand with dust obtained from the quarries of Custonaci has potential for use in concrete. From 14 to 60 days curing, the increment in strength observed for the mix with 26% sand replacement results sufficient to obtain a strength very close to that of the control mix.

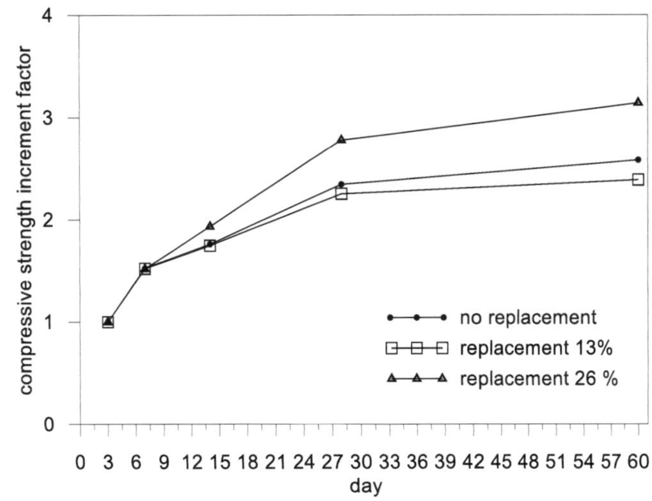

Figure 7. Strength increment factors

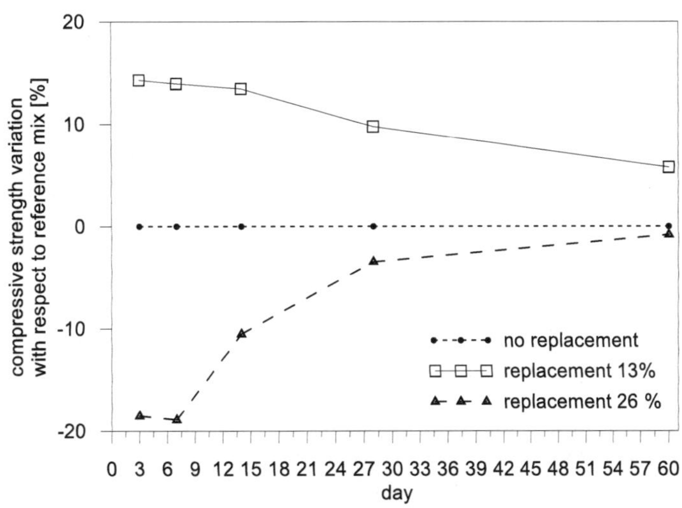

Figure 8. Percentage of strength increment with respect to the reference mix

IOP Conf. Series: Materials Science and Engineering **442** (2018) 012003 doi:10.1088/1757-899X/442/1/012003

3. Comparison with some experiences reported in literature

With regards to the replacement of fine aggregate with quarry dust in the production of concrete, the results reported in literature for mechanical properties of concrete, tend to vary from one study to another. The reason is that the final characteristics depend on the constituent material properties including the properties of the waste materials used and also on the mix design. Nevertheless, various studies reported in literature confirm the potential use of waste quarry dust in concrete as a partial replacement of the fine aggregate (sand) resulting in a concrete with mechanical characteristics that are acceptable for structural concrete.

Prakash et al [3] report on the use of the quarry dust of the Paritala region in India as a substitute of the fine aggregate for concrete. Different concrete grades are studied. The results reported for M30 and M20 concrete with a 53-grade cement are of particular interest. For the first case the results are included in figure 9 while for the second case the results are included in figures 10-11.

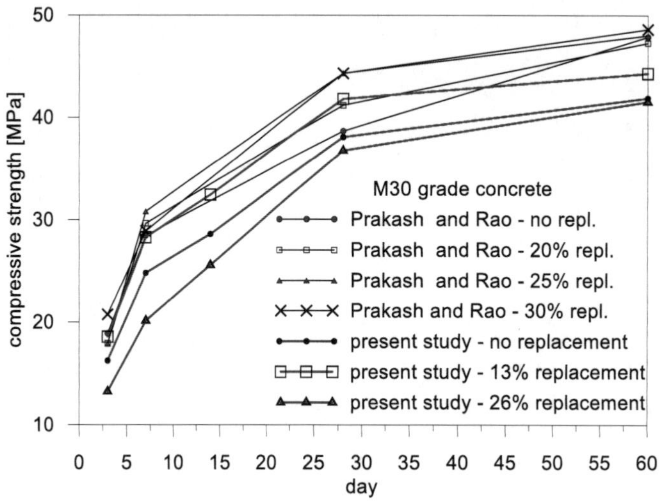

Figure 9. M30 concrete – Compressive Strength of the concrete mixes proposed by Prakash et al.

Figure 9 shows the results for concrete produced with sand replacement up to 30%. The results proposed are in contrast with the results obtained in the present study because the results reported by Prakash et al indicate a continuous increase in strength for increasing levels of replacement with respect to the reference mix. For sake of clarity the mean strength of the mixes analysed in the present study are inserted in figure 9, showing also that the strengths obtained in the present study during the curing are lower than those of the mixes reported by Prakash et al. This difference is possible also due to the different mix design and different cement used for the mixes. Prakash et al use a 53-grade cement while a 32.5 grade cement is used in the present study.

Prakash et al reported that larger amounts of dust (more than 50% replacement) cause a reduction of strength of the concrete as show in figures 10-11, which refer to a M20 grade concrete.

This fact is clearer in figure 11 where the strengths are directly correlated to the percentage of sand replacement and where a first reduction in strength is reported at 10% substitution of sand followed by an increase in strength.

Kankam et al in [10] present the results of a research carried out in Ghana where two levels of replacement of fine aggregate with quarry dust were investigated. In particular 25% and 100% replacement levels were considered showing the advantages of the substitution of fine aggregates with quarry dust in both cases.

Three target concrete strengths were investigated in the study. The compressive strengths for a M30 grade concrete at 28 days are presented in table 3.

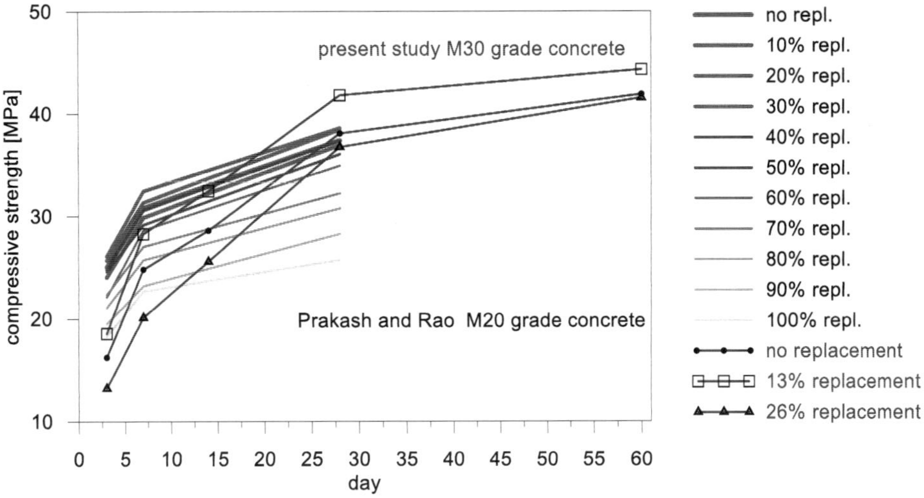

Figure 10. M20 concrete - Strengths of the mixes proposed by Prakash et al

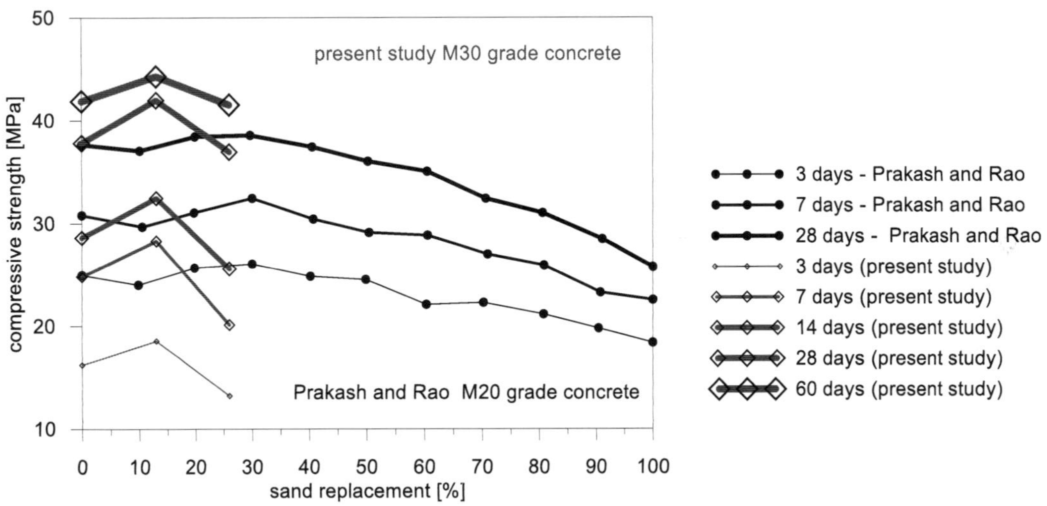

Figure 11. M20 concrete - Strengths of the mixes in Prakash et al, with varying percentage replacement

Table 3. Characteristics of some concrete mixes Kankam et al [10]

Concrete grade [MPa]	Fine aggregate replacement [%]	Cement in the mix [kg/m^3]	w/c	Fineness of sand	Fineness of quarry dust	Strength at 28 days [MPa]
30	0	397	0.56	2.66	3.54	30.50
30	25	416	0.56	2.66	3.54	31.20
30	100	429	0.56	2.66	3.54	30.17

In the study conducted by Kankam et al [10] it appears that the quarry dust used has a fineness modulus greater than that for the sand. It means that the particles constituting the waste used are coarser than those used in the present study. Therefore, the low reduction of strength exhibited with increase in quarry waste is not due to the introduction of a fine aggregate having the characteristics of a filler but for the introduction of a fine aggregate with grains greater than that of the sand. This demonstrates the importance and relevance of the characterisation of the waste quarry dust, which is transformed into a product, as substitute for industrial grade sand. Furthermore the quantity of cement used per cubic meter of concrete, (the cement grade is not given), is much higher than that employed in the present study and increases with increasing quarry waste in the mix. This strategy for compensating for the negative effect of inclusion of waste through the additional of cement gives better results, but also implies additional costs in production.

4. Conclusions

The paper presents preliminary results of an investigation intended to explore the potential of waste quarry marble dust from Custonaci (Sicily) for the production of concrete. The study has the objective of exploring practical options for the recycling of the large quantities of waste therefore addressing environmental concerns. The study refers to mechanical properties of the material and includes also a comparison with some experiences reported in literature.

The results indicate that the marble dust investigated in this research, having the characteristic of a filler, can be effectively used for the production of concrete, with improvements in mechanical characteristics if used in low quantities. Lower mechanical properties were reported when the waste is used in larger quantities. In the latter case the loss in performance is however such that it can be compensated with an increase in the cement content of the mix.

Similar experiences are reported in literature. In those cases negligible reduction of strength has been encountered even if a 100% replacement of fine aggregates has been done. However different results are also due to the different variables considered including varying materials and mix designs, the characteristics of the waste dust used and the cement type and cement content.

References

[1] Dey D, Kumar S, Chakraborty S 2017 Assessment of pulmonary health status among stone quarry workers at Kashipur, Silchar, Assam Indian Journal of Public Health Research and Development 8 382-86
[2] Rai B, Kumar S, Satish K 2016 Effect of Quarry Waste on Self-Compacting Concrete Containing Binary Cementitious Blends of Fly Ash and Cement Advances in Materials Science and Engineering 1326960
[3] Prakash K S and Rao C H 2016 Study on compressive strength of quarry dust as fine aggregate in concrete Advances in Civil Engineering 1742769
[4] Bahoria B V Parbat D K Nagarnaik P B Waghe U P 2017 Effect of characterization properties on compressive strength of concrete containing quarry dust and waste plastic as fine aggregate International Journal of Civil Engineering and Technology (IJCIET) 8 699–707
[5] Aishwaryalakshmi V, Kumar N, Prathap M.G. Haribabu S 2017 Experimental study on strength of concrete by partial replacement of fine aggregates with quarry dust International Journal of Civil Engineering and Technology (IJCIET) 8 12–17
[6] Jagadeesh P, Kumar P S, Prakash S S V 2016 Influence of quarry dust on compressive strength of concrete Indian Journal of Science and Technology 9 93663
[7] Lieberman R N, Knop Y, Querol X, Moreno N, Munoz-Quirós C, Mastai Y, Anker Y, Cohen H 2018 Environmental impact and potential use of coal fly ash and sub-economical quarry fine aggregates in concrete Journal of Hazardous Materials 344 1043–56
[8] Lim S K, Tan C S, Li B, Ling T, Hossain M U, Poon C S 2017 Utilizing high volumes quarry wastes in the production of lightweight foamed concrete Construction and Building Materials 151 441–48
[9] Bahoria B V Parbat D K Nagarnaik P B Waghe U P 2013 Comprehensive literature review on use of waste product in concrete International journal of Application or Innovation in Engineering & Management 2 387-94.
[10] Kankam C K, Meisuh B K, Sossou G, Buabin T K 2017 Stress-strain characteristics of concrete containing quarry rock dust as partial replacement of sand Case Studies in Construction Materials 7 66-72

[11] Thomas J, Harilal B 2016 Mechanical properties of cold bonded quarry dust aggregate concrete subjected to elevated temperature *Construction and Building Materials* **125** 724–30
[12] Thomas J, Harilal B 2016 Properties of cold bonded quarry dust coarse aggregates and its use in concrete *Cement & Concrete Composites* **62** 67–75
[13] André A, De Brito J, Rosa A, Pedro D 2014 Durability performance of concrete incorporating coarse aggregates from marble industry waste *Journal of Cleaner Production* **65** 389-96

Waste of granite dust utilization in ultra-light weight concrete

Evaldas Serelis[1], Vitoldas Vaitkevicius[1], Zymantas Rudzionis[1], Vidas Kersevicius[1]

[1]Faculty of Civil Engineering and Architecture, Kaunas University of Technology, Lithuania

E-mail: zymantas.rudzionis@ktu.lt

Abstract. In the article possibility to utilize granite dust in production of ultra-light weight concrete (≤800 kg/m3) is researched. Granite dust is a by-product of crushed granite, has very fine particles and high surface area. Due to very fine particles and high surface area, applicability of granite dust in concrete is limited. Mainly because granite dust increases W/C ratios and thus mechanical and durability properties tends to decrease. Ultra-light weight concrete, usually has very high W/C ratios, therefore in this type of concrete utilization of granite dust has favourable opportunity. In the research ultra-light weight concrete based on aluminium powder was created. Main mechanical and microstructure properties were researched and thermal conductivity coefficient was calculated. During experiment ultra-light weight concrete with 700 kg/m3 and up to 3.33 MPa was created. According to the experiment results, waste of granite dust can be properly utilised in production of ultra-light weight concrete blocks suited for one or two storages individual houses.

Keywords: ultra-light weight concrete, ULWC, granite dust

1. Introduction

Ultra-light weight concrete (ULWC) is a type of concrete which has density bellow 800 kg/m^3 [1]. Due to light density it relatively high strength it can serve as insulator and also will make lower load on the foundation. There are many types of ultra-light weight concrete: 1) lightweight aggregate concrete – this type of concrete is created by using porous lightweight aggregate of low apparent specific gravity (lower than 2000 kg/m^3); 2) Aerated, cellular, foamed or gas concrete – this type of concrete is created by introducing large voids within the concrete or mortar mass; these voids should be clearly distinguished from the extremely fine voids produced by air entrainment; 3) No-fines concrete - this type of concrete is created by omitting the fine aggregate from the mix so that a large number of interstitial voids is present; normal weight coarse aggregate is generally used [2,3]. In this research ultra-light weight concrete based on aluminium powder was created and partial of cement was replaced with granite dust waste.

In this case expansion of the concrete is obtained by adding aluminium powder to the water/cement mix. Manufacture process is without autoclave, however some external energy is required. The chemical reaction between cement and aluminium is the following: $2Al + 3Ca(OH)_2 + 6H_2O \Rightarrow 3CaO\ AL_2O_3\ 6H_2O + 3H_2$ [7]. As stated by the formula $3H_2$ is formed. The creation of these hydrogen bubbles causes the mix to expand thus density decreases. The dosage of aluminium powder typically depends on W/C ratio, cement, mixer type, used chemical composition, temperature etc.

Waste of granite dust is a by-product of crushed granite, due to very fine particles and surface area, inserts of clays, cannot be anywhere properly utilized, thus waste of granite dust typically is discarded in landfills [4-6]. Incorporating this waste in ordinary concrete is not possible, because it increases W/C ratio, decrease durability properties (mainly resistance to salt-scaling), increases permeability, decrease strength and etc. However granite waste dust utilisation in ultra-light weight concrete is possible, because ULWC has completely different structure, it has high porosity concertation, high W/C ratio, and usually lower compressive strength in this type of concrete is expected. Utilisation of granite dust in ULWC could help reduce the amount of waste discharged in landfills.

The main goal of this research is to determine effect of waste granite dust on main properties of ultra-light weight concrete and create non-autoclaved ultra-light weight concrete.

2. Materials

Cement. Portland cement CEM I 52.5 R was used in the experiments. Main properties: paste of normal consistency – 28.5%; specific surface (by Blaine) – 4840 cm^2/kg; soundness (by Le Chatielier) – 1.0 mm; setting time (initial/final) – 110/210 min; compressive strength (after 2/28 days) – 32.3/63.1 MPa. Mineral composition: C_3S – 68.70; C_2S – 8.70; C_3A – 0.20; C_4AF – 15.90. The particle size distribution is shown in Fig. 1.

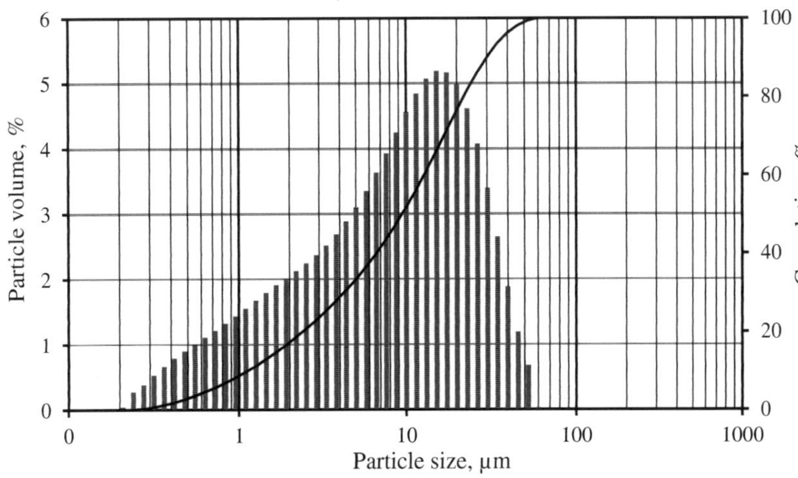

Figure1. Particle size distribution of Portland cement

Waste of granite dust. Granite dust is a by-product of crushed granite produced by JSC "Granitas" (Lithuania). Main properties: density – 2612 kg/m^3, bulk density – 1600 kg/m^3, amount of clay impurities – 32%. In experiment waste of granite dust was additionally milled, thus specific surface area varied: 4532 cm^2/g, 5805 cm^2/g and 6822 cm^2/g, but for ultra-light weight concrete properties research was used dust with specific surface area 6822 cm^2/g.

Aluminum powder. In the research was used aluminium powder (Al) covered with paraffin and average size of particles - 30 μm.

Gypsum (β-calcium sulphate hemihydrate). The main properties of gypsum: specific surface area (by Blaine) – 5000 cm^2/g, specific density – 2318 kg/m^3, the average particle size – 34.42 μm. Chemical composition: CaO – 38.2 %, SO_3 – 50.1%, insolubles – 4.2 %, loss of ignition – 6.8 %.

Polypropylene fibers. Main properties: raw material – polypropylene; shape – rounded; diameter -16 μm; fiber length -12 mm, fiber quantity - 275 million pcs/kg; flexibility modulus - 3500-

3900 N/mm²; tensile strength - 400 N/mm²; melting temperature - 160-170°C; flammable temperature - >320°C; chemical material resistance - good.

3. Methods

Mix preparation. Ultra-light weight concrete was prepared by using dry aggregates. Cement, gypsum, granite dust aluminium powder, NaOH and water were dosed by weight. Cubes (100x100x100 mm) were formed for the research to determine main properties of the concrete. Composition were mixed with laboratory mixer (G-250). Homogeneous mixes were cast in moulds and kept for 24 h in laboratory environment (20±2 °C and 60 RH). After demoulding, all samples were kept under laboratory condition (20±2 and RH 40-60 %) till age of 28 days. Samples were not kept underwater. In the experiment (Table 1), notation of created mixes: C100G0 – pure Portland cement without any substitution of granite dust, C85G15, C70G30, C55G45 – when certain percentage (15%, 30% and 45% accordingly) of Portland cement were substituted to granite dust (by mass).

Table 1. Compositions of ultra-light weight concrete

<table>
<tr><th>No.</th><th>Composition</th><th>Water, l</th><th>Cement, kg/m³</th><th>Granite dust, kg/m³</th><th>Al powder, kg/m³</th><th>Gypsum, kg/m³</th><th>Polypropylene fibers, kg/m³</th></tr>
<tr><td>1</td><td>C100G0</td><td>205.5</td><td>290.8</td><td>-</td><td>0.8</td><td>1.8</td><td>0.81</td></tr>
<tr><td>2</td><td>C85G15</td><td rowspan="3">205.5</td><td>247.2</td><td>43.62</td><td rowspan="3">0.8</td><td rowspan="3">1.8</td><td rowspan="3">0.81</td></tr>
<tr><td>3</td><td>C70G30</td><td>203.6</td><td>87.62</td></tr>
<tr><td>4</td><td>C55G45</td><td>159.9</td><td>130.9</td></tr>
</table>

Specific surface and particle size distribution. The specific surface (fineness) was measured by using Blaine instrument, according to the EN 196-6:2010 standard [8]. The particle size distribution was measured with a ''Mastersize 2000'' instrument.

Density and compressive strength. The compressive strength test was performed according to EN 12390-4:2000 standard [9]. For compressive strength, there were used 3 cubes (100x100x100 mm). Density was calculated as average value mass dividing by volume from 3 cubes (100x100x100 mm).

Thermal conductivity. Thermal conductivity was calculated according to the LST EN 1745:2012 standard [10].

Water absorption. Water absorption was performed according to EN 772-11:2011 standard [11].

4. Results

4.1. Elemental analysis of granite dust

Elemental analysis of granite dust waste was performed. Main results are given in table 2 and Figure 2. As raw material granite dust in the research cannot be used, because there were a lot of conglomerations, thus additional milling was performed, and powder with specific surface areas 4532 cm²/g (without milling), 5805 cm²/g (10 min of milling) and 6822 cm²/g (15 min of milling) was researched. The idea of elemental analysis was to estimate, that prolonged milling will not put additional impurities of Fe, analyse, to examine how chemical elements can effect structure of concrete, and compare with clay.

SiO_2 and Al_2O_3 are two main chemical compounds can be found in granite rock. The same compounds also could be found and in clays, the main different is, that in granite rock these compounds usually are found in crystalline form, however in chemical reaction with cement are more preferable amorphous form. If compounds SiO_2 and Al_2O_3 are in crystalline form they under normal condition they will act as inert fillers and will not affect properties of concrete in any way. Amorphous structure of SiO_2 has positive effect in reaction with hydrated lime ($Ca(OH)_2$) and low basicity C-S-H are formed, which has positive effect on mechanical and durability properties. The Al_2O_3 compound in crystalline form is mainly found in various clays. This compound has negative effect on concrete properties: increases W/C ratio, increases required amount of superplasticizer, increases shrinkage, increases water absorption, decreases resistance to salt-scaling, increases porosity. Negative effect of this compound can be eliminated by heating and changing structure from crystalline to amorphous. Founded compounds as MgO increases inhomogeneity of the mix, K_2O and Na_2O can increase pH value of pore solution, which increases possibility to appear alkali silica reaction with certain coarse aggregate. When pore solution has very high concentration of K^+ and Na^+ solubility of Ca^+ tends to decrease, thus overall amount of C-S-H will decrease. Higher amounts of FeO and Fe_2O_3 can mainly effect the colour of concrete, other compounds do not have any negative effect or it is insignificant.

Table 2. Amount of element of granite dust waste after different milling times

Element	Amount of elements after milling		
	4532 cm²/g	5805 cm²/g	6822 cm²/g
O	63.10	64.62	63.82
Si	18.02	16.65	17.14
Al	7.04	6.51	7.44
Fe	2.80	2.18	2.59
K	2.35	1.80	2.25
Ca	2.02	2.09	2.11
Mg	1.91	1.75	1.75
Na	1.89	2.20	1.96
S	0.54	0.55	0.63
Ti	0.24	0.19	0.24
Mn	0.08	-	0.08
P	-	0.21	-

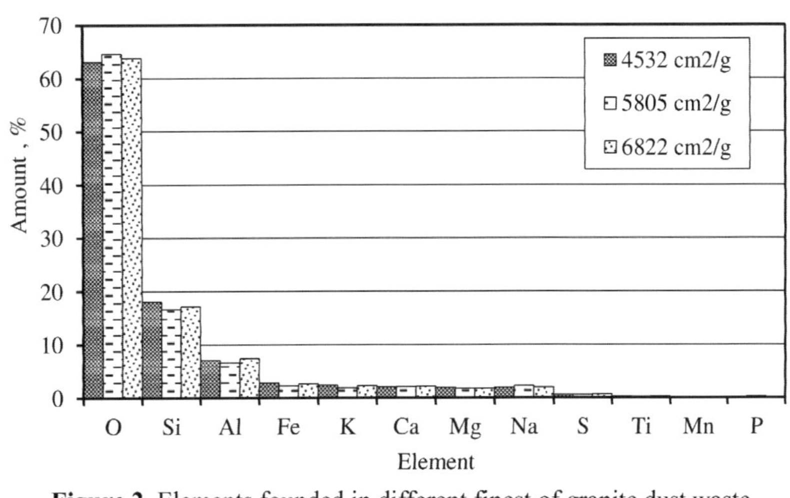

Figure 2. Elements founded in different finest of granite dust waste

4.2. SEM analysis of granite dust

SEM micrographs of granite dust can be found on Figure 3. It could be seen various forms particles with sharp angles, before milling particles size were between 10 μm and 100 μm (Fig.3-a an Fig.3-c), also were a lot of elongated particles, surfaces discontinuities, which has negative effect on proper particle size distribution, and increases amount of water for normal consistency mix. After milling particles became more spherical which have positive effect on fresh mix rheology.

Figure 3. SEM micrograph of waste granite dust a) and b) without milling (acordingly x1000 and x4000 magnification); c) and d) after milling (acordingly x1000 and x4000 magnification)

4.3. Density and compressive strength

Main results of density and compressive strength could be found on Table 3 and Fig. 4. It was noticed, that when Portland cement was substituted to granite dust from 0 % up to 45 % (by mass), density decreased about 15 % from 697 kg/m^3 (C100G0) down to 595 kg/m^3 (C55G45), however compressive strength decreased about 50 % times, from 3,33 MPa (C100G0) down to 1,66 MPa (C55G45). In order to build one storage individual houses, minimum compressive strength of blocks, should not be less than 3 MPa. According to the results could be stated, that 15 % substation of cement to granite dust is the maximum amount possible, to keep specification.

Table 3. Properties of ultra-light weight concrete

Composition	Density, kg/m^3	Compressive strength, MPa	Water absorption after 48 h.	Thermal conductivity, W/m·K
C100G0	697	3.33	42	0.1794
C85G15	677	3.08	44	0.1754
C70G30	629	2.13	46	0.1658
C55G45	595	1.66	53	0.1585

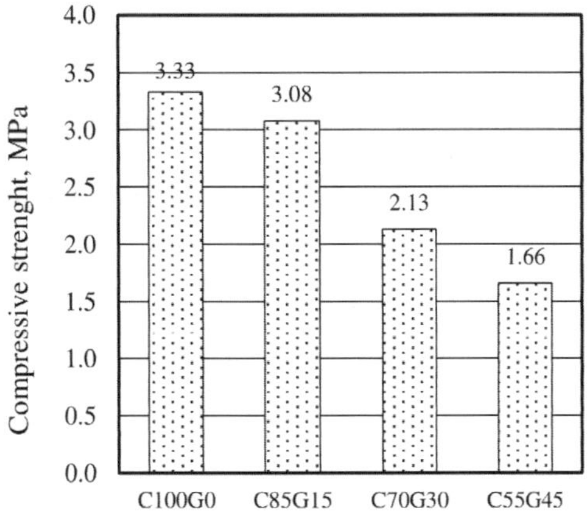

Figure 4. Relationship between amount of granite dust waste and compressive strength

4.4. Water absorption, thermal conductivity and microstructure

Water absorption and thermal conductivity testing/calculations results could be found on Table 3. The highest water absorption was observed in mixture (C55G45), where 45 % of Portland cement was substituted to waste of granite dust. Water absorption varied from 42 % (C100G0) up to 53 % (C55G45). Increase of water absorption probably related more due to decreased density than due to utilized higher amount of granite dust waste. Theoretically calculated thermal conductivity coefficient varied from 0.1794 (W/m·K) up to 0.1585 (W/m·K). Value of thermal conductivity is very similar to autoclave aerate concrete, from which most ultra-light weight concrete blocks are made. Microstructure of created ULWC sampled could be found on Figure 5. It could be observed, that there is no significant difference in air pore structure, between reference mixture (C100G0) (Figure 5-a) without waste of granite dust, and comparing with composition (C85G15), where 15 % of Portland cement was substituted to granite dust (Figure 5-b).

Figure 5. Structure of ultra-light weight concrete: a) without waste of granite dust and b) with waste of granite dust (replaced 15 % by mass of Portland cement to granite dust), when magnification x80

5. Conclusions

1. Waste of granite dust does not have any negative impact on micro and macro structure of ultra-light weight concrete.
2. Waste of granite dust can be properly utilized in ultra-light weight concrete without significant compressive strength loss.
3. Ultra-light weight concrete with density of 677 kg/m3, compressive strength of 3.08 MPa and thermal conductivity of 0.1754 W/m·K can be created when 15 % of Portland cement is substituted to waste of granite dust.

References

[1] Q.L. Yu, P. Spiesz, H.J.H. Brouwers. "Ultra-lightweight concrete: Conceptual design and performance evaluation". Cement & Concrete Composites 61 (2015) 18–28. DOI: https://doi.org/10.1016/j.cemconcomp.2015.04.012
[2] F.Roberz, R.C.G.M.Loonen, P.Hoes, J.L.M.Hensen. "Ultra-lightweight concrete: Energy and comfort performance evaluation in relation to buildings with low and high thermal mass". Energy and Buildings, Volume 138, 1 March 2017, Pages 432-442. DOI: https://doi.org/10.1016/j.enbuild.2016.12.049
[3] D.M.A.Huiskes, A.Keulena, Q.L.Yu, H.J.H.Brouwers. "Design and performance evaluation of ultra-lightweight geopolymer concrete". Materials & Design Volume 89, 5 January 2016, Pages 516-526. DOI: https://doi.org/10.1016/j.matdes.2015.09.167
[4] Narendra Kumar Sharma, Praveen Kumara, Sanjeev Kumar, Blessen Skariah Thomas, Ramesh Chandra Gupta. " Properties of concrete containing polished granite waste as partial substitution of coarse aggregate". Construction and Building Materials Volume 151, 1 October 2017, Pages 158-163. DOI: https://doi.org/10.1016/j.conbuildmat.2017.06.081
[5] Sarbjeet Singh, Shahrukh Khan, Ravindra Khandelwal, Arun Chugh, Ravindra Nagar. "Performance of sustainable concrete containing granite cutting waste". Journal of Cleaner Production Volume 119, 15 April 2016, Pages 86-98. DOI: https://doi.org/10.1016/j.jclepro.2016.02.008
[6] Sarbjeet Singh, Ravindra Nagar, Vinay Agrawal, Aditya Rana, Anshuman Tiwari. "Sustainable utilization of granite cutting waste in high strength concrete". Journal of Cleaner Production Volume 116, 10 March 2016, Pages 223-235. DOI: https://doi.org/10.1016/j.jclepro.2015.12.110
[7] Maciej Szumigała, Łukasz Polus. "Applications of Aluminium and Concrete Composite Structures". Procedia Engineering Volume 108, 2015, Pages 544-549. DOI: https://doi.org/10.1016/j.proeng.2015.06.176
[8] EN 196-6:2010. Methods of testing cement – Part 6: Determination of fineness.
[9] EN 12390-4:2000. Testing hardened concrete –Part 4: Compressive strenght – Specification for testing machines.
[10] LST EN 1745:2012. Masonry and masonry products –Methods for determining thermal properties.
[11] EN 772-11:2011. Methods of test for masonry units – Part 11: Determination of water absorption of aggregate concrete, autoclaved aerated concrete, manufactured stone and natural stone masonry units due to capillary action and the initial rate of water absorption of clay masonry units.

Characterization of Libyan metakaolin and its effects on the mechanical properties of mortar

Ruben Paul Borg[1], Ahmed M M Hamed[1], Ramadan Edreis[2], Ali M Mansor[3]

[1] Faculty for the Built Environment, University of Malta, Malta
[2] Industrial Research Center, Tripoli, Libya
[3] Faculty of Engineering, University of Tripoli, Libya

E-mail: ruben.p.borg@um.edu.mt

Abstract. Environmental concerns, stemming from high-energy demands and CO_2 emission associated with cement manufacture, have brought about pressures to reduce cement consumption through the use of supplementary cementitious materials (SCMs). Besides addressing environmental concerns, the incorporation of SCMs in cement bound materials and concrete can modify and improve specific concrete properties. Metakaolin (MK) in an important SCM which can enhance the performance of cementitious composites through its high pozzolanic reactivity. This study was carried out to characterize the materials and to assess the effect of Libyan metakaolin (LMK) on the mechanical properties including the compressive strength of cement mortar. LMK was produced by calcining kaolinite clay at 700°C for 2 h. X-Ray Diffraction (XRD), Scanning Electron Microscopy (SEM), Differential Thermal Analysis / Thermo-Gravimetric analysis (DTA/TG) and Fourier Transform Infrared Spectroscopy (FTIR) s were performed on the raw and calcined kaolinite powders. Seven mixes were prepared with different LMK replacement percentages (0.0 to 30%), by weight of cement, and a constant water binder ratio (w/b) of 0.5. The specimens were cured for 3, 7, 28, 56 and 90 days. At the end of each curing period, the specimens were tested for compressive strength. The results confirm the transformation of kaolinite clay into metakaolin and the pozzolanic reactivity of the produced LMK and conforms to ASTM requirements in this respect. The study confirms that LMK could be effectively used in reducing cement content up to 30% by weight without compromising compressive strength of the cement mortar.

Keywords: Libyan clay, metakaolin, cement mortar

1. Introduction

The production of Portland cement is not only costly and energy intensive, but it also requires large quantities of natural raw materials and produces large amount of carbon emission. The production of one ton of Portland cement produces approximately one ton of CO_2 in the atmosphere [1]. Supplementary cementitious materials (SCMs) are often used to reduce cement content in concrete and to improve some desired property. The addition of SCMs in cement based materials effectively addresses environmental concerns associated with cement production but also leads to improved durability performance of Portland cement based materials, in particular when exposed to aggressive environments.

Alternative SCMs such as metakaolin, silica fume, and quarry dust are effectively used in concrete. Specific SCMs constitute silica or silica-alumina that react with portlandite released during cement hydration to produce additional CSH, enhancing the properties and pore structure of concrete or cement mortar.

There is a growing interest in utilizing calcined clays rather than industrial by-products as SCMs, as these materials are highly available all over the world and present more homogeneous physico-chemical properties compared with by-products, simplifying the process of quality control of the finished cement [2]. Many studies have been conducted regarding the characterization and the use of different types of SCMs, and different results have been introduced.

A. Tironia, et al. 2012 [3], studied the influence of different thermal treatments on the pozzolanic activity of raw kaolin with 98 % kaolinite and ordered structure. The results showed that for the production of metakaolin to be used as pozzolanic material, the calcination temperature and the residence time are variables to be considered. The used kaolin treated at 700 °C showed that its pozzolanic activity were directly proportional to the residence time. On the other hand, when it was treated at 800 °C the pozzolanic activity decreases when the residence time increases. The peak area of bands assigned to the amorphous phase in FTIR spectrum of kaolin calcined at 700 ºC during 30 min is higher than that obtained at 800 ºC. For 800 °C and 750 °C, a low residence time (10 minutes) reduces the structural rearrangement and increases the pozzolanic activity.

Biljana R. Ilic, et al. [4], investigated the ability of the metakaolin to be used as a supplementary cementitious material and they reported that the metakaolin can be produced by thermal treatment (calcination) of the starting high-quality kaolin clay from Serbia. They stated that the optimal calcination parameters, for which nearly complete dehydroxylation of the material was achieved, are: temperature 650 °C and heating time of 90 min. The conversion of the kaolinite to metakaolin was confirmed by XRD and IR analyses of the starting and thermally treated kaolin samples.

He et al. 1995 [5], measured compressive strength of mortars on mini-RILEM prisms cured in water at 40°C for 28 days. The mortars had a ratio of cement, clays and sand of 70, 30 and 300 by weight, respectively. Kaolin compared with other clays and Ordinary Portland Cement (OPC) increases the compressive strength the most, up to 113 % of pure OPC when calcined at 550°C. And it remains in this range during further calcination up to 950°C.

This study is based on Libyan metakaolin (LMK) as partial replacement of cement in cement based materials. The research reported in this paper is based on the material characterization and experimental investigation with respect to mechanical properties of the concrete. LMK was produced by calcining Libyan kaolinite clay at 700°C for 2 h. X-Ray diffraction (XRD), scanning electron microscopy (SEM), differential thermal analysis / thermo-gravimetric analysis (DTA/TG) and Fourier transform infrared spectroscopy (FTIR) techniques were implemented for material characterization. The Strength Activity Index (SAI) was also determined for the pozzolanic activity. All of the characterization tests confirmed the transformation of the Libyan kaolinite clay into metakaolin. Seven cement mortar mixes with different cement-metakaolin replacement levels (0.0 – 30 %) by weight were prepared, cured and tested for compressive strength at 3, 7, 28, 56 and 90 days. The results showed that up to 30% replacement level, the blended mixes had higher compressive strengths than the control mix, at all testing ages, but for the 3 days strength.

2. Materials

Metakaolin which was obtained by calcining Libyan kaolinite clay at a temperature of 700ºC for two hours was used. The metakaolin was used as a partial replacement of cement at 0 to 30%) by weight. Ordinary Portland cement CEM II/42.5 R and certified natural standard sand to ISO 679:2009 [6] were used in the study. Cement mortar strength was determined using EN 196-1:2005 [7].

3. Experimental programme

3.1 Material characterization

Metakaolin was obtained by calcining Libyan natural kaolinite clay at temperatures of 700°C for 2 hours. The material was collected from the village of Samnu located at the South part of Libya. The material preparation, calcination process and material characterization and testing were conducted in the University of Malta laboratories. The DTA/TGA analysis was carried out at University of Kaunas, Lithuania. The material was characterized using X-Ray Diffraction (XRD), Scanning Electron Microscopy (SEM), Differential Thermal Analysis / Thermo-Gravimetric (DTA/TG) and Fourier Transform Infrared Spectroscopy (FTIR) techniques for raw clay and calcined materials.

The X-Ray Diffraction (XRD) results of raw and calcined materials are presented in figures 1 and 2 respectively. As indicated in the figures, the kaolinite peaks disappeared during the calcining process, indicating complete transformation of kaolinite to metakaolin. The XRD pattern revealed the presence of quartz, which could be distinguished from XRD pattern before and after calcinations (peak position and intensities). The characteristic peaks (2Θ= 12.18°, 20.26° and 24.72°), which look to be in principle Smectite-Kaolinite disappeared while the peaks representing Quartz (2Θ= 20.72° and 26.46°) remained after calcination. Figures 1 and 2 also show SEM images of raw (crystalline plates) and calcined (amorphous) materials respectively.

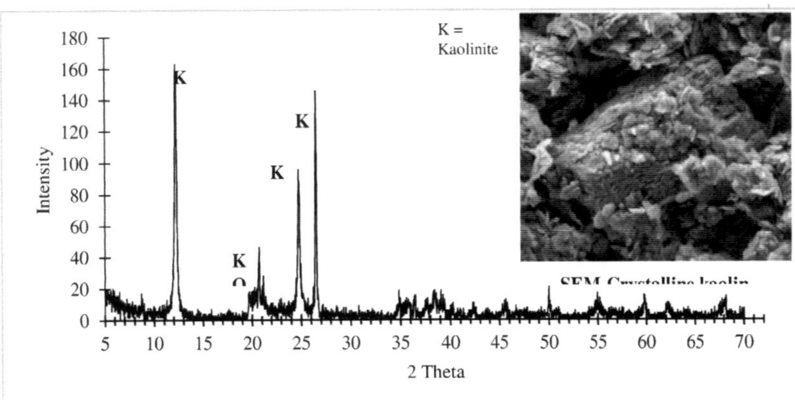

Figure 1. XRD pattern and SEM image of the raw kaolin

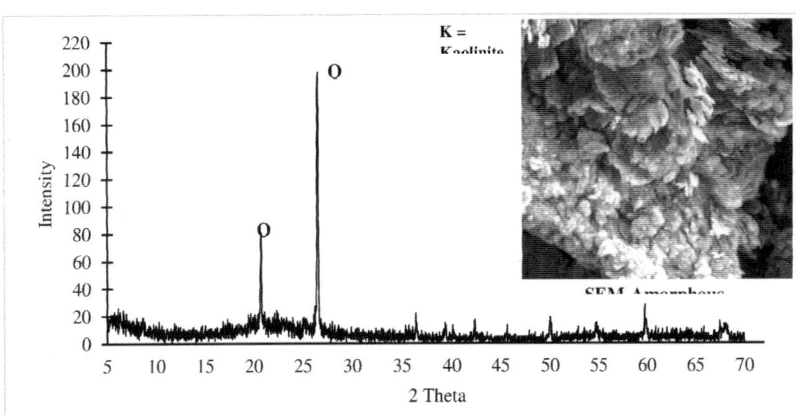

Figure 2. XRD pattern and SEM image of the metakaolin

IOP Conf. Series: Materials Science and Engineering **442** (2018) 012005 doi:10.1088/1757-899X/442/1/012005

Figure 3 represents Differential Thermal Analysis / Thermo-Gravimetric analysis DTA/TG test results obtained. It shows that the sample exhibits a total loss of 10.87 % of weight. The curve shows peaks between 38-179°C, 250-283°C and 460-582°C normally associated with loss of surface and absorbed water, pre-dehydroxylation and dehydroxylation processes of the clay material respectively. Pure kaolinite has a theoretical total loss of (±14 %) [9&10]. The difference noted between experimental result and those reported in literature confirm that this clay is associated with microcline and quartz.

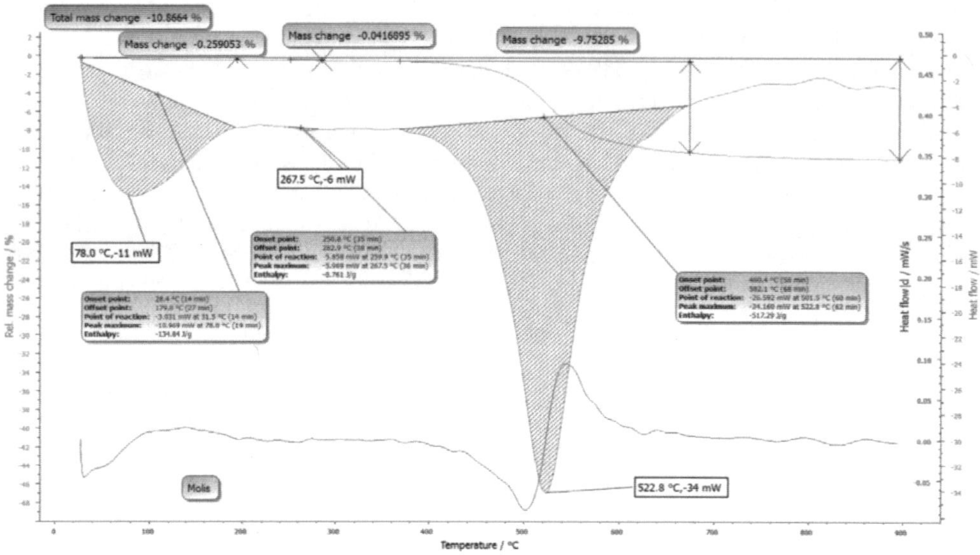

Figure 3. DTA/TG for the Libyan kaolin clay

The Fourier Transform Infrared Spectroscopy (FTIR) technique was used to distinguish between different types of clay minerals and to derive information concerning their structure, composition and structural changes upon chemical modification. A KBr pressed disk technique was used for the material characterization. Figures 4 and 5 show the FTIR spectrum of raw and calcined clay respectively.

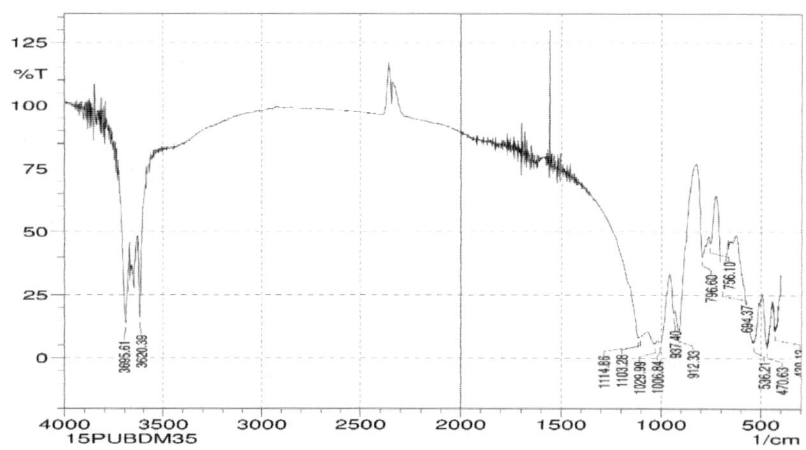

Figure 4. FTIR for the raw kaolin clay

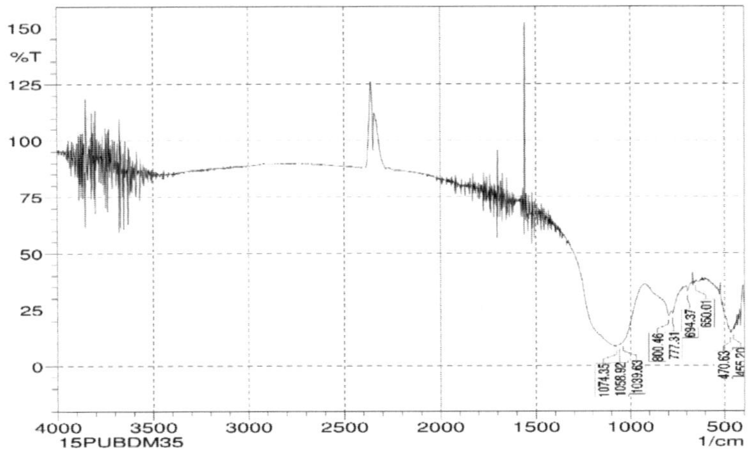

Figure 5. FTIR for the calcined kaolin clay (LMK)

The absorption band at 3695 cm-1 represents free OH stretching (hydroxyl sheet) which is very close to the 3699 cm^{-1} value obtained by Yleana [12], while the 3620 cm^{-1} frequency band represents the inner layer OH (Al-O…H) which falls at 3626 and 3622cm^{-1} obtained by Davarcioglu et al [13] and Yleana [12] respectively. The absorption band at 1114cm^{-1} is assigned to Si-O normal to the plane stretching which is very close to the band at 1120cm^{-1} obtained by Aroke et al [11]. The bands placed at 1029cm^{-1} and 1006cm^{-1} correspond to Si-O planar stretching which agree closely with 1027 and 1009cm^{-1} obtained by Davarcioglu et al [13]. The peaks at 937cm-1 and 912cm^{-1} correspond to inner-surface Al-OH deformation which agree to 940cm-1 and 913cm-1 obtained by Aroke et al [11] and Ilic et al [14] respectively. After calcinations, the absorption bands in the 3695 cm^{-1} and 3620.39cm-1 region as well as 1114 cm^{-1} disappeared or merged into a broad profile. The appearance of a new band at 800.46 cm^{-1} as obtained by Ilic et al [14].

3.2 Compressive strength

3.2.1 Strength activity index (SAI). The pozzolanic activity of the Libyan kaolinite clay before and after calcination was determined through the strength activity index test (SAI). Three sets of test specimens were prepared with respect to the standard mix design: one with 100 % Cement (control), and the other two were prepared with part of the Cement replaced by a corresponding mass (20%) of the materials to be tested. After curing, the specimens were tested for compressive strength. The effect of the pozzolanic material is expressed as strength activity index SAI = stress (tested mix) / stress (control). ASTM C618 [15] requires a SAI greater than 0.75 after 7 and 28 days for fly ash (FA) and natural pozzolans at a cement replacement of 20% [8]. Table 1 shows the result of SAI for the raw and calcined (metakaolin) Libyan clay. The test results showed that the calcined clay was active and could be considered as a pozzolanic material, while the raw clay was not active.

Table 1, Result of SAI for the raw and calcined Libyan clay.

Material	7 day Stress (MPa)	28 day Stress (MPa)	SAI	
			7 d	28 d
OPC	33	39		
Raw clay	18	22.3	0.55	0.57
Calcined clay (LMK)	32	44	0.97	1.13

3.2.2 Compressive strength of LMK mortar. After confirming the pozzolanic activity of LMK, specimens were cast for the determination of compressive strength, using mortar containing one part by mass of cement and three parts by mass of standard sand with a water/cement ratio of 0,50. The cement was partially replaced with LMK at different levels (0 - 30%) of the cement weight. The mortar was mixed mechanically n a mortar mixer, cast in 50 mm cube moulds and then compacted with the aid of a vibrating table. The specimens in the moulds were stored for 24 hours under laboratory conditions before demoulding. All samples were cured in a water bath until the time of testing (3, 7, 28, 56 and 90 days). The compressive strength test was conducted based on the EN 196-1:2005.

4. Results and Discussion

Results of the compressive strength test for mortar containing LMK at different replacement levels (0.0 – 30%) by weight of cement and at various curing ages (3, 7, 28, 56 and 90 days) are presented in figure 6. It can be seen that the incorporation of LMK improves the compressive strengths at early and later stages and for all replacement levels. The exception was for 3 days curing age where the control mix showed better compressive strength with respect to the higher replacement values. It is worth mentioning that mixes with LMK replacement level of 15% and higher were less workable than the mixes with lower replacement values. To overcome the workability issue, a separate set of additional mixes for those replacement levels were prepared with the use of superplasticizer (sp) type Master Glenium SKY 698, at a dosages of 1%, 1.25%, 1.5% and 1.5% of the total powder for 15%, 20%, 25% and 30% LMK replacement levels, respectively. The specimens were cast, cured and tested as indicated above. Figure 7 presents the results of 28 days compressive strength of the specimens with and without sp. Slightly higher compressive strength is reported for the specimens with sp when compared to specimen without sp. This can be attributed to the, improved workability and compaction achieved for samples with sp. The superior performance of specimens containing metakaolin could be due to both the pozzolanic reaction and micro-filler effect of metakaolin.

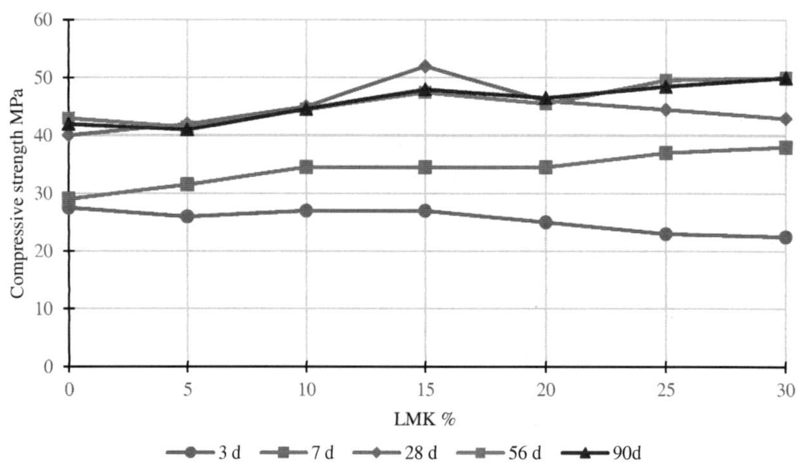

Figure 6. Compressive strength with different LMK levels and different ages

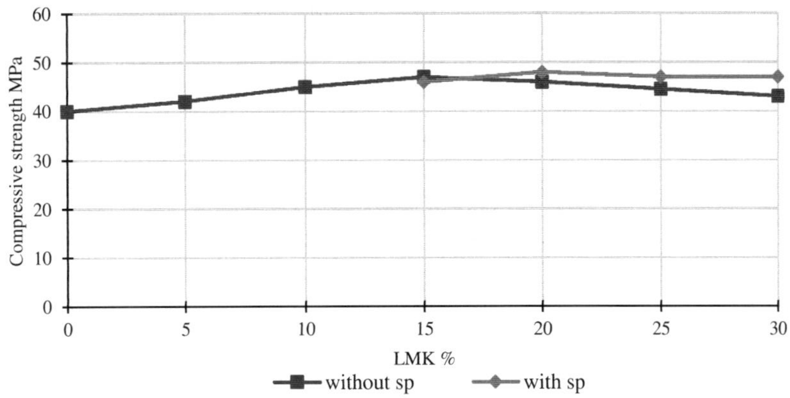

Figure 7. 28 day compressive strength of cement mortar with and without superplasticizer

5. Conclusions

From the study of the Libyan kaolinite clay and its effect on the strength of cement mortar, the following conclusions are drawn:

1. The utilization of different characterization techniques allows one to distinguish between different clay minerals and to provide fundamental information on their chemical composition and structural changes that occur as a consequence of their chemical modifications.
2. The two main components of the tested clay are kaolinite and quartz.
3. The Libyan kaolinite clay can be effectively transformed into an active metakaolin.
4. The compressive strengths of blended mortar specimens containing up to 30% replacement of LMK by weight of cement were superior to the control, both fort early age and also longer term.
5. LMK can be effectively used as a cement replacement, resulting in sufficiently high strength and a reduction in the consumption of cement.
6. The incorporation of higher quantities of LMK in concrete leads to a reduction in workability, Therefore, the use of chemical admixtures is necessary for higher cement replacement levels.

Acknowledgements

The authors would like to acknowledge the support of the following: Mr.N. Azzopardi (Faculty for the Built Environment, University of Malta), Ing. J. Camilleri (Department of Metallurgy and Materials Engineering, University of Malta), Mr. M. Zerafa (Department of Chemistry, University of Malta), staff at the Kaunas University of Technology, XRD laboratory staff at the Petroleum Institute of Libya; ERDF (Malta) for the financing of the testing equipment through the project: ''Developing an Interdisciplinary Material Testing and Rapid Prototyping R&D Facility (Ref. no. 012)''

References

[1] Yogesh R. Suryawanshi, et al. 2015, Experimental study on compressive strength of concrete by using metakaolin, International Research Journal of Engineering and Technology (IRJET), Vol. 02, pp. 235-9.

[2] Susan A, et al. 2017, Characterization of supplementary cementitious materials by thermal analysis, Materials and Structures 50:26, DOI 10.1617/s11527-016-0909-2

[3] A. Tironia, et al. 2012, 11th Int. Congress on metallurgy & materials SAM/CONAMET, Thermal treatment of kaolin: effect on the pozzolanic activity, Procedia Materials Science 1, pp. 343-50

[4] Biljana R. Ilic, et al. 2010, Thermal treatment of kaolin clay to obtain metakaolin, Institute for Testing of Materials, Belgrade, Serbia.
[5] Changling He, et al. 1995, Pozzolanic reactions of six principal clay minerals: activation, reactivity assessments and technological effects, Cement and Concrete Research 25, pp 1691-702·
[6] ISO 679: 2009, Cement test methods – Determination of strength.
[7] EN 196-1: 2005 Methods of testing cement – Part 1: Determination of strength.
[8] S. Donatello, et al.2010, Comparison of test methods to assess pozzolanic, Cement and Concrete Composites, Vol. 32, Issue 2, pp 121-7
[9] Newman, A.D.C; 1987, Chemistry of clays and clay minerals; Longman scientific and technical; USA.
[10] Ekosse, G. 2001, Provenance of the Kgakgwe kaolin deposit in southeastern Botswana and its possible utilization; Applied Clay Science; 20:137-52.
[11] Aroke U. O., El-Nafaty U. A. and Osha O. A. 2013, Properties and characterization of kaolin clay from alkaleri, Nortth-Eastern Nigeria, International Journal of Emerging Technology and Advanced Engineering, Vol.3.
[12] Yleana, M. C. L. 2005, FTIR Micro-spectroscopy of RDX Interactions with Clay Minerals, Unpublished M. Sc Thesis, University of Puerto Rico Mayaguez Campus.
[13] Davarcioglu B and Ciftci E 2009 Investigation of central anatolian clays by FTIR spectroscopy (Arapli-Yesilhisar-Kayseri, Turkey), Int. Journal of Natural and Engineering Sciences 3, pp.154-61.
[14] B. R. ILIC, A. A. METROVIC, L. R MILICIC: THERMAL 2010, treatment of Kaolinite clay, Institute for testing of materials, Belgrad, Serbia.
[15] ASTM C618, Standard specification for coal fly ash and raw or calcined natural pozzolan for use in concrete.

Self-compacting concrete produced with limestone waste

K Kopecskó

Budapest University of Technology and Economics, Hungary

E-mail: kopecsko.katalin@epito.bme.hu

Abstract. The aim of the research was to design powder type self-compacting concrete containing high amount of limestone quarry waste as aggregate and fines. The light colour aggregate mixed with white cement resulted in esthetical surfaces of exposed concrete for architectural purposes. Large amount waste originated from the limestone mining process requires deposition with further treatment. In our study we tried to replace also the limestone powder fines by the slurry produced during the cut of limestone blocks. Seven recipes were examined during the research: three mixtures of quartz aggregate with different particle size distributions were used as reference mixtures, three mixtures composed of crushed limestone aggregate and one with mixed quartz-limestone composition, respectively. In the recipe of mixed limestone-quartz aggregate dried quarry waste slurry was successfully applied instead of direct ground limestone powder. In the research different test methods were used to determine the fresh properties as well as the hardened properties of concretes.

Keywords: self-compacting concrete, limestone waste, chloride ion migration

1. Introduction

The nature of mining processes creates a potential negative impact on the environment both during the mining operations and after the mine is closed. The limestone mining in Hungary dating back to Roman times, has been to present a number of operating mines and surface engineering to pursue stone processing. The mining and processing of limestone results in large amount of unusable stone blocks and also wet limestone cutting slurry, which are considered as mining waste and requires deposition with further treatment. It is important to use the excavated limestone to the greatest extent minimizing the debris have permanently disposed of, and positioned as a mining waste, thereby further reducing the environmental impact, the landscape damaging effect [1]. Reuse of these materials may reduce the environmental impacts.

Self-compacting concrete (SCC) is a type of structural concrete which strongly deviates from ordinary concrete in the fresh state, as it de-aerates and flows without the application of additional compaction energy [2]. The definition of self-compacting concrete (SCC) was described by Ozawa et al. [3], then Okamura and Ozawa [4]. We can distinguish the powder type and the stabilizer type self-compacting concretes. The main difference in composition between normal concrete and powder type SCC is the use of higher amount of fines in SCC to reach the suitable, approximately 30% higher amount of paste content which ensures the high fluidity in fresh state. Application of fines is necessary in order to prevent bleeding and segregation. Direct ground limestone powder is well suited additional material for SCCs. Quarry waste limestone slurry could replace the limestone powder products in concrete decreasing the material and energy consumption.

The aim of the present research was to design powder type SCC containing high amount of limestone waste for architectural purposes (e.g. facades, street furniture).

2. Experimental programme

In the mix design we have followed the Okamura method [5] which is based on a three-fold approach considering the paste, mortar and concrete phases [2]. In figure 1 the volume fractions are given for these phases.

In the Okamura method paste, mortar and concrete phases are distinguished and the desired volume ratios are given for these phases:

- estimation of air content, usually 2-3 vol% (V_{air});
- limitation of the coarse aggregate content (4-16 mm) in 50 vol%, then to determine the mortar volume ($V_{mortar} = V_{concrete} - V_{air} - V_{aggr,4/16}$);
- limitation of fine aggregate content (0-4 mm) in the mortar phase in 40 vol%, then to determine the paste volume ($V_{paste} = V_{mortar} - V_{aggr,0/4} = V_{water} + V_{powder}$);
- determine the composition of the paste volume by defining the volume ratio of filler and cement (V_{filler}/V_{cement}), experimentally measuring the water demand of the powder (cement-filler), mix β_{powder}, calculating the necessary water content ($V_{water} = V_{paste} \cdot \beta_{powder}$) and powder content ($V_{powder} = V_{paste} - V_{water}$);
- experimental optimization of the resulting phases (paste, mortar and concrete).

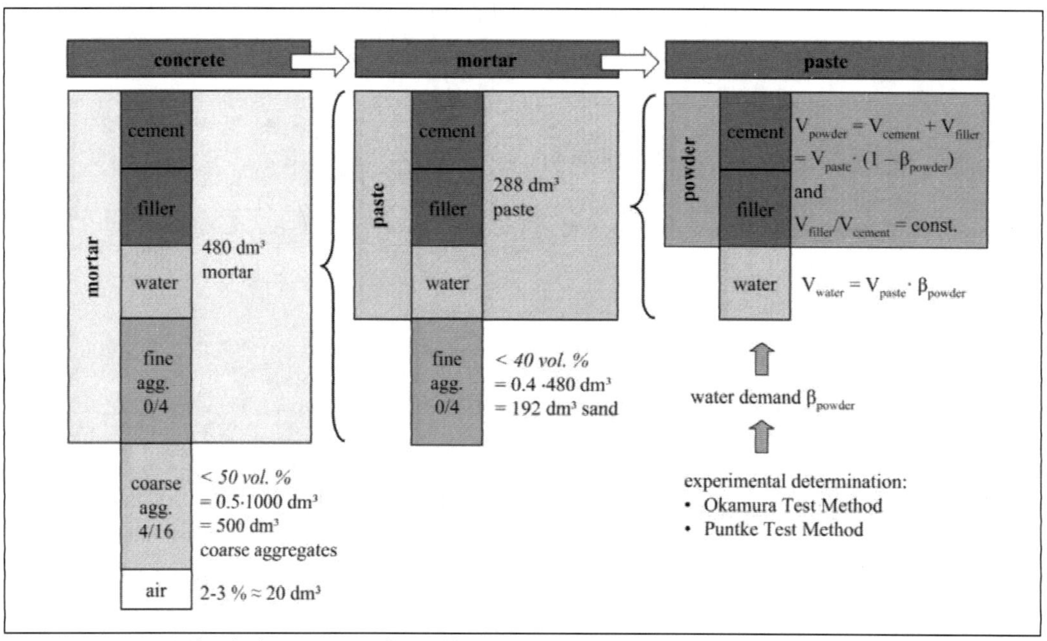

Figure 1. Mix-design concept for self-compacting concrete according to Okamura et al. [5] (source of figure: *fib* Bulletin No. 51 [2], courtesy of *fib*).

In our mix design the limitations for concrete volume fractions (coarse aggregate 4/16 < 50 vol %) were not taken into account because in the mixtures 4 mm maximum grain size of aggregate (d_{max}) was chosen. It means that in our case the fine aggregate content was equal with the total aggregate content. Instead of the difficult determination of water demand (β_{powder}) the water/cement ratio was chosen 0.45 based on the result of the trial mixing.

Seven mixtures of self-compacting concrete were studied during the research. Three basic recipes were designed with different particle size distributions of the coarser aggregate, where the fractions of 1/2 and 2/4 mm varied in different ratios (25:75, 35:65 and 45:55). Three mixtures were made with quartz aggregate (K1-K3, used as reference), further three mixtures were made with crushed limestone

aggregate fractions of 1/2 and 2/4 mm (M1-M3). Based on the results one recipe was selected for mixed quartz-limestone composition (MK). The origin of the limestone is the freshwater limestone quarry Süttő (Hungary). Crushed aggregate fractions were produced in the laboratory conditions.
Proposed by Daczko [6] in the mix design the 0/0,125 fraction of fine quartz sand was added to the paste phase (i.e. paste volume).

To keep the consistency and workability high performance superplasticizer admixture was applied. The amount of cement, limestone powder, fine sand (0/0.125 mm) and the total amount of the aggregate was constant for all recipes. The type of cement used in the experiment is CEM I 52,5 N white cement. The type and amount of polypropylene microfiber was the same in all recipes (0.1 vol%).

In the studied mixtures the aggregate fractions of 1/2 and 2/4 mm varied in different mass ratios:

- 25:75 in the mixtures K1 and M1;
- 35:65 in the mixtures K2, M2 and MK;
- 45:55 in the mixtures K3 and M3.

The volume fraction of fine aggregate (1/4 mm) for all mixtures was approximately 42 vol% (table 1).

Table 1. Recipe of the mixture K1 (aggregate 1/2 and 2/4 mm fractions in the ratios of 25:75).

Material	Type or fraction	Density (kg/m^3)	Mass (kg)	Volume (l)	Mass/mix (kg/60 l)
Cement	CEM I 52,5 N white	3100	400	129.0	24.0
Limestone powder	commercial	2700	450	166.7	27.0
Fine sand	0/0.125 mm		216	81.5	13.0
	1/2 mm (25%)	2650	284	107.2	17.0
Sand	2/4 mm (75%)		850	320.8	51.0
	Sum of sand (100%)		1134	427.9	68.0
Microfiber	Fibrofor Multi	910	0,9	1.0	0.054
Admixture	ViscoCrete 20HE (2.5%)	1090	10	9.2	0.600
Water	m_w/m_c=0.45	1000	180	180	10.8
Air		-	-	5	-
Sum			2391	1000	

Commercial ground limestone powder was applied as fines (filler), except in the recipe of mixed quartz-limestone aggregate (MK), where dried quarry waste slurry was utilized. The quarry waste slurry was produced during the cut of the large limestone blocks, and it is available in large amount. The specific surface of the fines was measured by Blaine method. The specific surface of limestone waste slurry was close to the specific surface of the white cement, and higher than for the commercial limestone powder (table 2).

Table 2. Specific surface of fines

	CEM I 52.5 N	Limestone powder	Limestone slurry
Specific surface (m^2/kg)	417	343	407

The trial mixtures were made with quartz aggregate. In case of limestone aggregate the consistency and the workability was not appropriate with the same amount of mixing water because of the porosity resulting water uptake. To reach the same consistency first we measured the 30 minutes water uptake

(4.8 m%) of the porous crushed aggregate. Before mixing the different concrete mixtures the limestone aggregate was mixed with the necessary water demand and kept to saturate for 30 minutes in the mixer. Without this treatment exclusively the increase of amount of admixture did not result adequate consistency and workability.

3. Results and Discussion

In the study various test methods were used to determine both the fresh properties (e.g. slump-flow test, T50 time, J-ring test, loss of consistency) of concrete mixtures as well as the hardened properties of concrete specimens at different ages (compressive strength, flexural tensile strength, modulus of elasticity, abrasion resistance, freeze-thaw resistance).

3.1. Fresh properties of concrete mixtures

Studying the fresh properties of self-compacting concrete has a great importance. All recipes resulted in self-consolidating, self-compacting concrete (table 3).

3.1.1. Slump-flow test, Visual Stability Index and J-ring test

As we expected specimens with limestone aggregate reached lower values of compressive and flexural Slump-flow test according to the MSZ EN 12350-8 standard [7] was used to assess flowability and flow rate. The flow values of concretes K (quartz gravel) exceeded the size of the flowtable (80 cm), so these values were fixed to 80 cm uniformly. The flow values of limestone additive were slightly less, between 74 and 80 cm. This means that the shear-tension in the concrete mixtures made with quartz aggregate is smaller. According to the MSZ EN 206-9:2010 standard [8], all the tested concrete can be classified into the SF3 (760-850 mm) slump-flow class.

T_{50} times were uniformly above 2 seconds, which are to be classified into VS2. Comparing the T_{50} times, it can be seen that concrete mixtures made with quartz aggregate reached slower the 50 cm diameter than concrete mixtures made with limestone, which means that their plastic viscosity is higher. According to Daczko [6], the smaller viscosity is better for mixtures with smaller slump-flow value, so that they can fill the formwork and the space between the reinforcements properly. Otherwise, the filling ability of concrete is inadequate. However, in the case of concretes with higher slump-flow value, the higher plastic viscosity is appropriate.

The J-ring test can be used to determine the passing ability of self-consolidating concrete. The results of J-ring tests were significantly affected by the shelf life of the concrete. The passing abilities of limestone concretes were much better, with less than 5 cm for all three recipes. For the other recipes, the maximum difference in spread is the highest for concrete K3 (24 cm), which indicates a poor passing ability. The passing ability of MK (mixed quartz-limestone aggregate) recipe represents about the average with the 9 cm difference in spreading.

The stability of self-consolidating concrete can be observed visually by examining the concrete mass and therefore can be used for quality control of self-consolidating mixtures. Visual rating criteria were used to classify the ability of the self-compacting concrete mixtures to resist segregation (VSI, visual stability index). According to ASTM C 1611, after spreading of the concrete has stopped, visually inspect the concrete mixture by observing the distribution of the coarse aggregate within the concrete mass, the distribution of the mortar fraction particularly along the perimeter, and the bleeding characteristics. VSI is rating from 0 (highly stable: no evidence of segregation or bleeding) up to 3 (unstable: mixture without stability and bleeding) [9]. The results of measurements of fresh properties are summarized in table 3. Figures 2-5 shows the VSI of different mixtures.

3.1.2. Shelf life of concrete mixtures

The shelf life of concrete mixtures was also studied . At the age of 30 minutes the mixtures lost the self-compatible property: they showed rather slump than flow. The reason is the relative early setting of the CEM I 52,5 N white cement. However the drop of the efficacy of the admixture could also influence the shelf life and change the workability (table 4).

Table 3. Fresh properties of mixtures

	Q1	Q2	Q3	L1	L2	L3	QL
slump-flow test, k_t (cm)	>80	>80	>80	75	80	74	77
J-ring test, k_{tJ} (cm)	45	71	56	72	78	70	68
T_{50} time (s)	3.5	4.5	5.0	2.5	2.5	3.0	2.0
Visual stability index (VSI)	0	0	0	1.5	1	2	1

Figure 2. Spread of the Q2 mixture, VSI 0

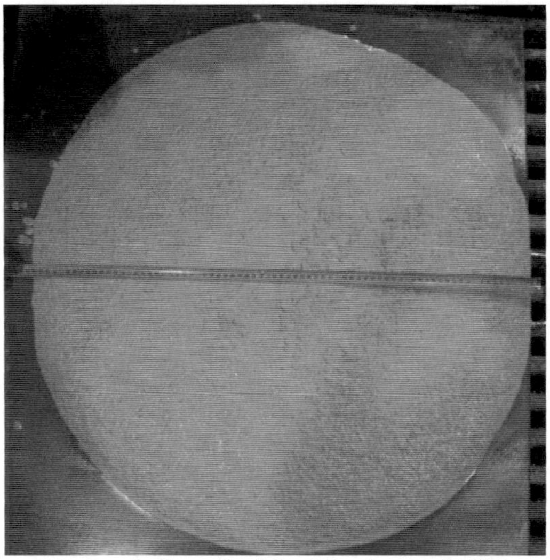

Figure 3. Spread of the L3 mixture, VSI 0

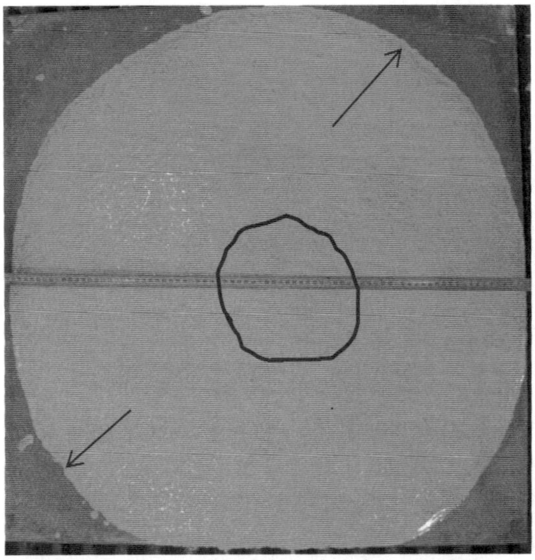

Figure 4. Spread of the QL mixture, VSI 1

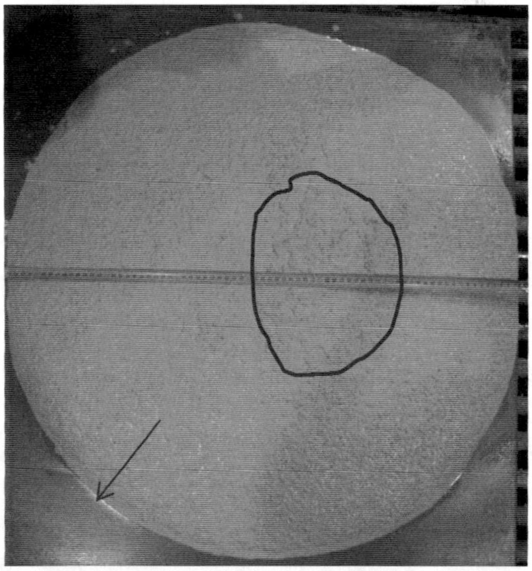

Figure 5. Spread of the L3 mixture, VSI 2

Table 4. Shelf life of the concrete mixtures

Haegermann spread	Q1	Q2	Q3	L1	M2	M3	QL
0 min. (cm)	30	28	28	28	26	27	25
15 min. (cm)	28	25	26	20	24	20	22
30 min. (cm)	10	11	11	n.a.	n.a.	n.a.	n.a.

3.1.3. Air content and density of concrete mixtures

The air content in the recipes of concrete with quartz aggregate was estimated to be 0.5 vol% (5 l/m^3), and for concretes made with limestone aggregate to be1.0 vol% (10 l/m^3). This was different from the proposed mix design where the air-void content estimated to be 2-3 vol %. Based on the test results of the air content, the Okamura's suggestion (2-3 vol% air content) can be realistically designed in SCC recipes. It is difficult to achieve lower air content with self-compacting concrete. Limestone additive concretes due to the angular shape and the coarser surface of the grains are able to keep more air content. Due to the round, smooth surface of quartz aggregates, the admixture is effective to drive out the adsorbed gas layer.

The densities of the fresh concrete mixtures varied between 2250 and 2370 kg/m^3, the prepared concretes can be considered as concretes with ordinary density, however slightly less than for conventional concretes with quartz aggregate (without addition of large amount of limestone filler). The density of hardened concrete was measured at the age of 28 days (table 5).

Table 5. Air contents and densities of fresh and hardened concrete

	K1	K2	K3	M1	M2	M3	MK
Air content (vol%)	1.42	1.58	2.04	2.89	2.88	4.36	2.66
Density of fresh concrete (kg/m^3)	2368	2365	2354	2275	2280	2250	2282
Density of hardened concrete (kg/m^3)	2365	2363	2346	2235	2246	2227	2275

3.2. Hardened properties of concrete mixtures

3.2.1. Strength and modulus of elasticity

The compressive and flexural tensile strength values were determined on 40·40·160 mm prisms (figures 6 and 7) as well as the compressive strength was determined on cubes of 150 mm (figure 8). As we expected specimens with limestone aggregate reached lower values of compressive and flexural tensile strength than concrete with quartz aggregate. The average compressive strengths of quartz containing mixtures (K1-K3) at the age of 28 days were above 80 MPa; mixtures with limestone aggregate (M1-M3) shown about 35% smaller compressive strength as well as mixture with mixed aggregate (MK) about 20% smaller.

The moduli of elasticity of quartz containing mixtures (K1-K3) are greater than 35,000 MPa; in case of mixtures with limestone aggregate (M1-M3) are below 30,000 MPa and for mixed aggregate concrete (MK) is 33,400 MPa.

3.2.2. Abrasion resistance

The abrasion resistance was determined according to the standard MSZ EN 14157:2005 [10]. For the test the Böhme abrasion wheel tester was used. The aggregate properties affect the performance of concrete. Based on the test results mixtures containing limestone aggregate (M1-M3) are not considered abrasion resistant. This is the consequence of the different hardness of the minerals. Surface hardness of aggregates is important where aggregates are likely to wear in service. The Mohs' relative hardness of calcite ($CaCO_3$) is 3 while the Mohs' relative hardness of α-quartz (α-SiO_2) is 7 [11]. The abrasion resistance of the mixture MK improved by the application mixed quartz-limestone aggregate (figure 9).

3.2.3. Freeze-thaw resistance

Mixtures containing limestone aggregate (M1-M3) in terms of freeze-thaw resistance also performed worse than concretes containing quartz aggregate. The frost resistance test carried out in deionised water at 50 cycles. After removing the specimens the residual compressive strength was determined. The most significant decrease in compressive strength was measured in case of concrete containing mixed aggregate (MK), with an average decrease of 15 MPa (21%); the decrease in average compressive strength of concrete made with aggregate quartz aggregate is 5.00 MPa (5.8%) and for concrete made with limestone aggregate is 3.40 MPa (5.2%). Based on the results of the salt scaling tests concretes without quartz aggregate (M1-M3) are not suggested for applications required use of de-icing salt in winter.

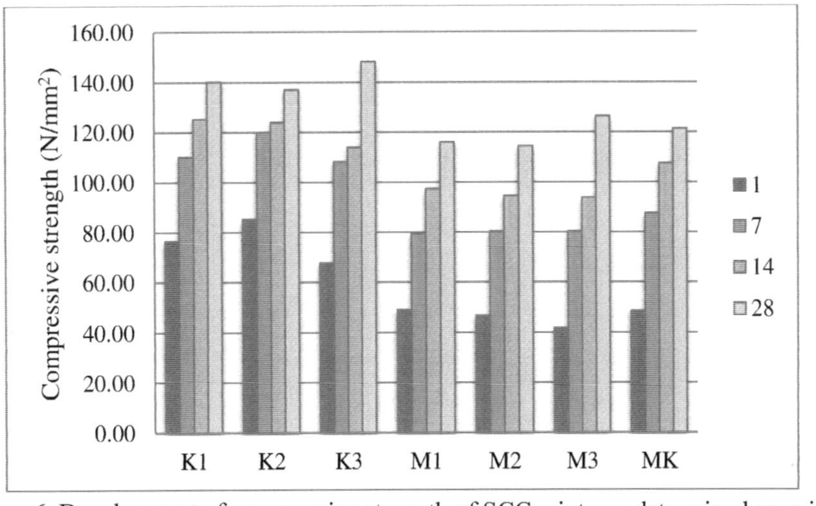

Figure 6. Development of compressive strength of SCC mixtures determined on prisms

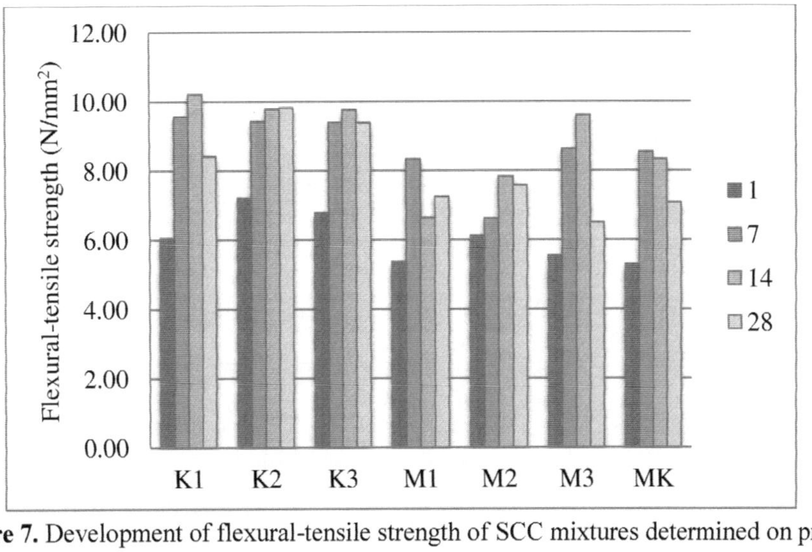

Figure 7. Development of flexural-tensile strength of SCC mixtures determined on prisms

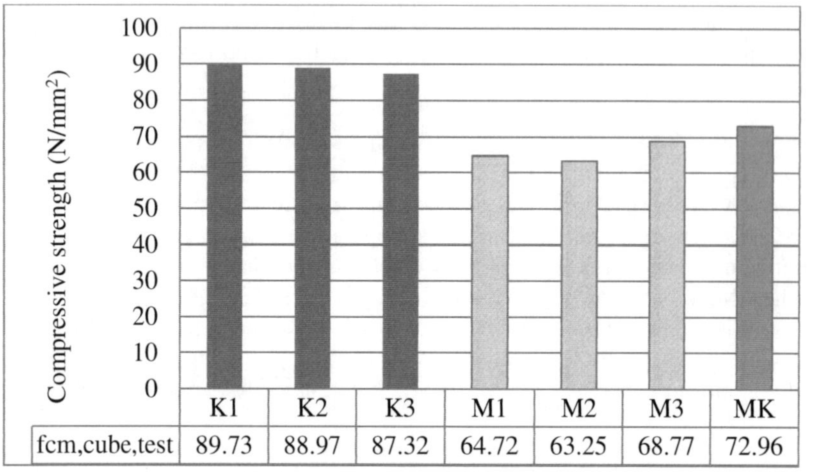

Figure 8. Compressive strength of SCC mixtures at the age of 28 days, determined on cubes

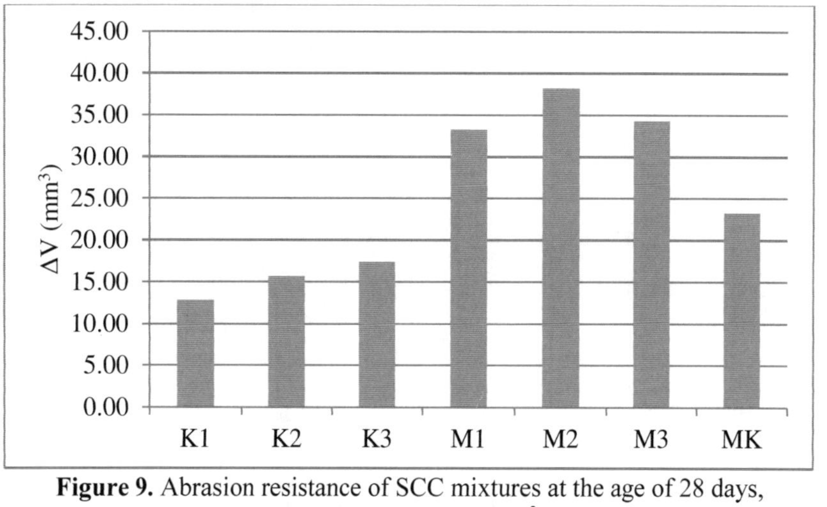

Figure 9. Abrasion resistance of SCC mixtures at the age of 28 days, decrease of volume (mm^3)

3.2.4. Chloride ion migration

Tang and Nilsson [12] developed a rapid electrical migration (HTC) method to determine the diffusion coefficient of chloride ions and the degree of chloride penetration. The method was further developed in NT Build 492 [13]. For the experiment, discs with 50±2 mm height were cut from the Ø100x200 mm cylinders (figure 10). The test started at the age of 28 days. For the comparative test all specimens were subjected to the 30 V DC for the same time (8 hours). After the measurement, the specimen has to be splitted and then the split surface is treated with silver nitrate ($AgNO_3$) solution for the colorimetric determination of the depth of chloride penetration. Figures 11 and 12 shows pictures after the whitish precipitation (AgCl) occurred during the reaction between $AgNO_3$ and Cl^- ions.

In specimens made with quartz aggregate, the chloride penetration depth as well as the nonsteady-state migration coefficient was much smaller than in specimens made of limestone additives. The depth of chloride penetration depends on the permeability concrete (i.e. pore structure). The porosity of the limestone aggregate increases the permeability. It can be noticed that among the three discs of one specimen, the disc number 2 (i.e. the middle of the cylindrical specimen) has the largest value of penetration depths or migration coefficients. This can be explained by the self-compacting properties of concrete. In case of SCC there is no vibration, so the upper third is richer in the paste phase, while the lower third is denser due to the more aggregate grains.

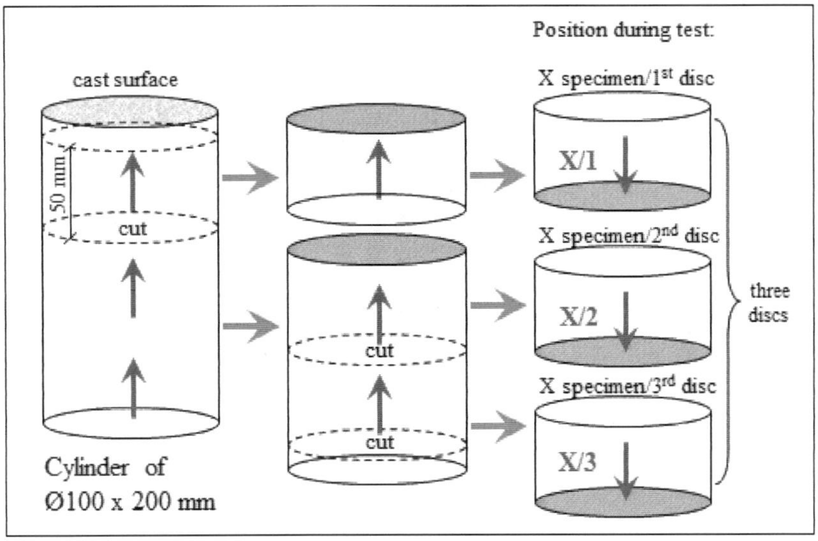

Figure 10. Preparation of specimens (discs) for testing chloride migration

Figure 11. White precipitation of AgCl shows the depth of chloride ingress, mixture K1

Figure 12. White precipitation of AgCl shows the depth of chloride ingress, mixture M2

4. Conclusions

The aim of the present research was to design powder type SCC containing high amount of limestone waste for architectural purposes (urban furniture, cladding, etc.). The origin of the limestone is the freshwater limestone quarry Süttő (Hungary). Aggregate fractions were made crushing of the freshwater limestone in laboratory scale. Seven recipes were examined during the research: three mixtures of quartz aggregate with different particle size distributions were used as reference mixtures, three mixtures composed of crushed limestone aggregate and one further with mixed quartz-limestone composition. Based on the test results the recipe of self-compacting concrete containing combination of quartz-limestone aggregate (MK) was suggested to use in precast concrete industry. In the recipes of K1-K3 and M1-M3 commercial ground limestone powder was used as fines. In the recipe of mixed quartz-limestone aggregate (MK), instead of commercial ground limestone powder the waste limestone slurry was successfully applied. Limestone slurry is a waste material, which is produced during cut the large limestone blocks. This concrete has increased resistance against abrasion comparing with concretes containing only limestone aggregate. With the proposed concrete mixture the purpose of containing the largest possible amount of limestone waste in a self-compacting concrete with appropriate fresh and hardened properties was achieved.

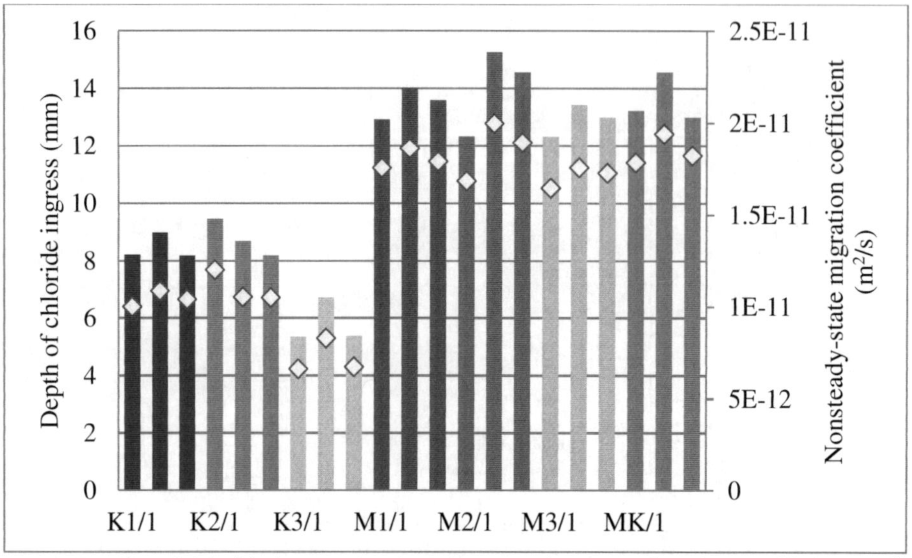

Figure 13. Depths of chloride ingress (mm) and nonsteady-state migration coefficients (m^2/s) of SCC mixtures at the age of 28 days (migration test by 30 V DC, 8 hours).

Acknowledgements
Author acknowledges the support by the Hungarian Research Grant NVKP_16-1-2016-0019 "Development of concrete products with improved resistance to chemical corrosion, fire or freeze-thaw".

References

[1] Mang B 2004 Az ásványi nyersanyagtermelés főbb hatásai (in English: The impact of mineral raw material production), *Publication of the University of Miskolc*, Series A, Mining, Vol. 66, pp. 15-21.
[2] *fib* Bulletin No. 51. 2009 Structural concrete. *Textbook on behaviour, design and performance*, Vol. 1. pp. 96-99.
[3] Ozawa K, Maekawa K and Okamura H 1990 High performance concrete with high filling capacity, *Admixtures for concrete, improvement of properties*, Vasquez E (Ed.), Chapman & Hall, London, Great Britain, pp. 51-62.
[4] Okamura H and Ozawa K 1996 Self compactable high performance concrete in Japan, *High performance concrete*, Zia P (Ed.), SP-159, ACI, Farmington Hills, USA, pp. 31-44.
[5] Okamura H and Ozawa K 1995 Mix design for self-compacting concrete, *Concrete Library of JSCE*, Vol. 25, No. 6, pp. 107-120.
[6] Daczko J A 2012 Self-consolidating concrete – Applying what we know, *Spoon Press*, Abingdon, pp. 118-137.
[7] MSZ EN 12350-8:2010. Testing fresh concrete. Part 8: Self-compacting concrete. Slump-flow test.
[8] MSZ EN 206-9:2010. Concrete. Additional rules for self-compacting concrete (SCC).
[9] ASTM C1611 / C1611M – 14 Standard Test Method for Slump Flow of Self-Consolidating Concrete.
[10] MSZ EN 14157:2005. Natural stone test methods. Determination of the abrasion resistance.
[11] http://webmineral.com/help/Hardness.shtml#.WqAo5nxG2Cg. Downloading date: 01/02/2018, search engine: google.hu, search terms: Mineral hardness.
[12] Tang L and Nilsson L 1992 Chloride diffusivity in high strength concrete, *Nordic Concrete Research*, Vol. 11, pp. 162-170.
[13] Nordtest NT Build 492. Concrete, mortar and cement-based repair materials: Chloride migration coefficient from non-steady-state migration experiments, Approved 1999-11, Finland.

Self-healing cement-based materials: an asset for sustainable construction industry

L Ferrara

Department of Civil and Environmental Engineering, Politecnico di Milano, Italy

E-mail: liberato.ferrara@polimi.it

Abstract. Worldwide increasing consciousness for sustainable use of natural resources has made "overcoming the apparent contradictory requirements of cost and performance effectiveness a challenging task" as well as a major concern. Self-healing cement-based materials, by controlling and repairing cracks, could prevent "permeation of driving factors for deterioration", thus extending the structure service life, and even provide partial recovery of the engineering properties relevant to the application. This paper will outline the current state of art on self-healing cement-based materials and experimental methods for the assessment of the self-healing capacity. Moreover, it will also critically analyse the current hindrances which challenge the engineering community in paving the way towards the reliable and consistent incorporation of self-healing concepts and effects into a durability-based design framework for buildings and structures made of or retrofitted with self- healing concrete and cementitious composites.

Keywords: self -healing; advanced cement based materials; experimental methods; durability-based design

1. Introduction

According to a survey recently published by the WEF, the construction industry accounts for about 6% of global GDP (from 5% in developed countries up to 8% in developing ones) with total annual revenues of about 10 trillion USD and 3.6 trillion USD added values, such figures are expected to increase to an estimated 15 trillion USD revenues by 2025 [1]. Construction industry, which currently directly employs more than 100 million people worldwide, serves all the industry verticals, with 38% of global construction volume accounted for by residential housing, 32% by transport, energy and water infrastructures, 18% by institutional and commercial buildings and 13% by industrial sites. The importance of the construction industry, for countries to enjoy an inclusive and sustainable growth, has been further highlighted by a quite recent (2014) estimate, that for an extra 1% GDP investment in infrastructures, in advanced economies, a 1.5% GDP increase after four years can be expected [1,2].

The construction industry is also the single largest global consumer of resources and raw materials and generator of solid waste as well. For example, about 40% of solid waste in the US and 35% in EU28 countries (Eurostat 2014 estimate) derives from construction and demolition activities [1].

In this global framework, concrete plays a role of paramount importance, with about 10 billion tons (corresponding to about 4 billion cubic meters) produced each year. This makes it not only the second largest used material worldwide, the first being water, but also correspond to twice as much than the total of all other building materials, including steel, wood, aluminum and plastic. The production and consumption of such large quantities of concrete feeds the deman0d on the one hand for a likewise huge production of cement, which has overcome 4 billion tons/year, with continuous increase even in the worst years of the financial crisis, and, on the other hand, for the consumption of aggregates, with a demand of about 48 billion tons in 2015, as well as of water.

The production of cement as well requires large quantities of raw materials (approximately about 2 tons of limestone and shale per ton of cement), and high energy (about 4 GJ per ton of cement) and produces about 1 ton CO_2 per ton of cement, which makes the cement industry responsible of about 5% not energy related greenhouse gas emissions worldwide.

These figures anyway have not to be deceptive and misleading about the carbon footprint of concrete as a construction material, since "the reason concrete has a high footprint as a whole is that there are just such huge concrete quantities used", and the reason for which it is used in such large amounts and "has been used in [...] pioneering architectural feats for millennia" is that "it is, simply, a remarkably goof building material" and "is in fact a very low impact material" [3]. Besides the well-known and likewise appreciated good performance of concrete in terms of, among the others, compressive strength, thermal inertia and cost effectiveness, what makes it a low impact material is the worldwide local availability of its raw constituents, including cement, which can be nowadays regarded as a global commodity. This implies that "if you replace concrete with another material, it would have a bigger carbon footprint" [3].

In this framework it well worth remarking that both cement and concrete industry are really making giant leaps forward in improving their sustainability signature. The adopted measures range from optimization of cement production processes, including, *e.g.*, reuse of waste heat or use of secondary fuels such as waste tires in cement kilns, to peculiar and nowadays consolidated practices in concrete technology which encompass the use of supplementary cementitious materials, as replacement of cement in concrete and in the production of multi-blended cements, to the use of recycled aggregates, originated from construction and demolition wastes, as a replacement of natural ones, up to the use of treated waste water and even sea water in the mix-design [4].

Concepts and figures exposed above do actually refer to the sustainability signature of the material "per se" and to its production process. A thorough evaluation of the sustainability of a construction work, in the framework of the built environment as a whole, has anyway to consider the performance of the construction in service conditions all along its life cycle, including its durability.

The durability of (reinforced) concrete structures is, as of today, one of the major concerns, if not the most dramatically serious one, of the construction industry. Reinforced concrete structures more and more frequently exhibit earlier and faster and more and more severe decay of their level of performance in the service scenarios and conditions they have been supposedly designed for.

Though a comprehensive analysis of the causes of the aforementioned problem is so far lacking, several issues can be called for to justify the somewhat premature decay of structural performance and anyway their higher durability sensitiveness, including, *e.g.*:

- the fact that current construction technologies, and in some cases the same structure concept, and the qualification of the workmanship have not adequately paced up with the development in material concept and technology of concrete and cement-based materials;
- climate change and environment pollution issues, together with the fact that the material and design concept together have pushed ahead the service stress boundaries, are going to threaten more severely the material, when in service in the structure.

It has also always to be borne in mind that, because of its inherent brittleness and low tensile strength, concrete is expected to crack and is generally used in combination with, prevalently steel, reinforcement, which takes the tensile forces generated by the applied actions. Cracks open an ingress pathway for aggressive agents to penetrate inside a structural element, reaching the reinforcement and the inside bulk concrete, and activating complex degradation mechanisms, among which the corrosion of the same reinforcement is surely among the most threatening ones. Unexpected and/or not correctly predicted decay of the structural performance results into unpredicted maintenance needs which, besides being costlier than if correctly predicted and planned, hardly can restore the pristine level of performance, thus implying also an increased frequency of the subsequent following maintenance actions and an uncontrolled growth of the life cycle costs. Recent estimates have shown that, *e.g.*, the cost of repairing corrosion damages (including also automotive and aircraft industry) sums up to about 3.5% of the world GDP [5] and that currently a significant share of the year budget of construction industry in developed countries accounts for maintenance and repair of existing structures. At the same time, the lifespan of the same maintenance and repair works is dramatically shortening, as from a case history analysis recently provided by the CON-REP-NET project [6], which has shown that 50% of the repaired concrete works failed again, 25% of which in the first 5 years after repair. A percentage share which increases to 75% and 95% respectively if the time observation frame is extended to 10 and 15 years after the repair.

A true "durability based" design approach is actually far from being formulated in current design codes, though it is appropriately recognized that the achievement of the required durability is the complex outcome of the suitable choice of structure concept and shape, material selection, as well as of the enforcement of "operational" design criteria, which limit the crack width under the anticipated actions to suitable scenario-based threshold values [7,8].

Limitation of crack width being hence recognized as the major requisite for the intended durability, concrete technology has developed, over the past fifty years, Fiber Reinforced Concrete (FRC). Thanks to the dispersed fiber reinforcement, an effective control of crack width can be achieved throughout the entire structure and starting from the very early ages. As a matter of fact, because of their wire-like features, fibers are able to interact with cracks much fines than what obtainable with the smallest commercially available bar diameters [9]. The boundaries of this concept have been pushed forward up to the formulation of the so-called High Performance Fiber Reinforced Cementitious Composites (HPFRCCs), whose composition is designed through micro-mechanical concepts based on fiber pull-out and crack tip toughness balance. Thanks to this, after the formation of a first crack, fibers effectively provide a through crack stress redistribution which enables new multiple cracks to be formed while controlling the opening of the previously formed ones, which are basically stopped from further widening, up to the unstable localization of one major crack. This makes the material able to "spread" the entity of a single damage (crack) into a series of tightly spaces and narrowly opened multiple cracks, whose single width is hence much less detrimental to the structural durability [10].

The composition of this category of advanced cementitious composites features high contents of cement and supplementary binders, with either pozzolanic (fly ash, silica fume) or delayed cementitious activity (slag) and low water content. The resulting high amount of reactive material which remains un-hydrated and entrapped inside the bulk volume of a structural element may be, upon cracking exposed to outdoor environment, in case featuring presence of liquid or vapor water. These can both activate delayed hydration reactions as well as carbonation ones, whose products precipitate onto the crack surfaces sealing it. The reconstruction of the through crack matrix continuity and, in case, matrix densification at the interface with the fibers, which also yields improved fiber-matrix bond, may also result into "proprie dictum" material healing, i.e. a recovery of the post-cracking mechanical performance, in terms of load bearing capacity, stiffness, toughness and ductility etc [11-13].

Self-healing materials are well known in the field of biology. Blood clotting, skin cicatrization and bone reconstruction are "sparkling" examples of self-healing functionalities inborn in biological materials which material science has successfully attempted to incorporate also in man-made ones, including, among the others, polymers, metals, asphalts, paintings and cement-based construction materials as well [14]. It is blatantly undeniable that the possibility of engineering or stimulating the self-repairing functionalities in cement-based construction materials would enhance the material and structural durability also resulting into reduced maintenance needs over an extended structure service life. In the whole framework herein outlined self-healing cement-based materials would hence represent an exceptional asset for shaping the sustainability of the cement, concrete and construction industry [15].

This paper after, after having outlined the current state of art on self-healing cement-based materials and experimental methods for the assessment of the self-healing capacity, will also critically analyze the current hindrances which challenge the engineering and research community in the "breakthrough in understanding and technology" path which could only lead to a consistent and reliable "design code incorporation" of the self-healing cement-based material concepts.

2. What do we know: autogenous and autonomous (engineered) self-healing

Reportedly known for about two centuries, as from a 1836 "citation" by the French Academy of Sciences with reference to the "cicatrization" of cracks in concrete pipes [16], the self-healing capacity of concrete has been heuristically investigated throughout the last century, studies having literally boomed in the last decade [17,18].

Besides reversible crack closure, due to the swelling of cement paste in the crack flanks, or mechanical clogging, due to debris produced by the same cracking process or by water transported impurities, the main causes of autogenous healing of cement-based materials are the formation of calcium silicate hydrates (C-S-H) due to delayed hydration of binder particles remained anhydrous and suddenly exposed to environment moisture/water upon cracking and/or the precipitation of calcium carbonate ($CaCO_3$) due to the reaction of calcium ions/calcium hydroxide with carbon dioxide either present in the atmosphere or dissolved in water. The first mechanism, which is likely to yield products reconstructing the through crack matrix continuity as strong as the original ones, is likely to prevail in concretes undergoing cracking at earlier ages and/or in concrete containing pozzolanic or latent cementitious additions. The second mechanism, which in terms of effectiveness is likely to prevail over the former though its products are weaker, more frequently happens in old-cracked concrete, the use of supplementary cementitious materials with pozzolanic activity, which favoring, as said above, the former mechanism, counteracts with the latter because of the consumption of calcium hydroxide by the same pozzolanic reactions, which reduces the potential availability of calcium ions [19]. Different techniques have been proposed and validated to stimulate and enhance the aforementioned autogenous healing capacity, making its effectiveness less scattered and more reliable. The most consolidated ones consist in the use of crystalline admixtures [16, 20] and superabsorbent polymers (SAPs) [21].

The latter are characterized by a high water absorption capacity, as high as several times their initial weight. Such a characteristic being reversible, the water initially absorbed by SAPs when added into concrete can constitute a long term reservoir able to feed the self-healing reactions over time. Moreover, if coming into contact with water at a cracked site, the swelling of SAPs can contribute to an initial physical blocking of the cracks, while the chemical healing starts developing. The technique has been validated in [22-26], the obvious negative effects of SAP addition of the concrete strength remaining the major hindrance to its wider use, mainly in load bearing structural applications.

The formers, on their hand, generally consist of a proprietary mix of active chemicals, carried out in a carrier of cement and sand, which, because of their highly hydrophyllic nature, are able to react with water, cement particles but also with the soluble phase of cement hydration products and form CSH and other pore blocking precipitates. These products increase the density of the CSH phase and deposit in the existing capillaries and micro-cracks, activating the self-repairing process. The mechanism is analogous to the formation of CSH and the resulting crystalline deposits become integrally bound with the hydrated cement paste, thus contributing to a significantly increased resistance to water penetration under pressure. Research on crystalline admixtures as stimulators of the autogenous healing capacity has been conducted in [16, 20], confirming the effectiveness of the healing products not only in sealing the cracks, even under aggressive environmental conditions, such as chloride exposure [27] but also in restoring the through-crack stress transfer capacity, both in the case of plain and fiber reinforced concrete [28-30] and under repeated cracking/healing actions [30]. In the case of HPFRCCs an interesting synergy has been highlighted at the fiber-matrix bond level between the dispersed fiber reinforcement and the healing products depositing in the crack, most likely resulting in a sort of distributed "chemical self-prestressing" from which the recovery of the material performance can greatly benefit [31].

The presence of a homogenously dispersed fiber reinforcement, effectively controlling and limiting the crack width, is helpful to self-healing, since narrower cracks are definitely much easier to be healed. Moreover, the technology of fiber reinforced cementitious composites has in the last decade or so pushed ahead the boundary of the "crack propagation control concept" through the developing a signature category of materials broadly known, as above said, as HPFRCCs. The composition of these materials is designed through micromechanics concepts based on the balance between crack tip toughness and fiber pull out work. Once a crack is formed, the crack-sewing and stress redistribution effect provided by the fibers makes it "easier" to form a new crack at another location rather than continuing to open the existing one, which would require to pull out the fibers. The iteration of this concept results into a multiple cracking, with each single crack remaining as narrowly opens as a few tens of microns, and a strain hardening behavior, up until saturation of cracking spacing and unstable localization of one major crack. As a matter of fact, the matrix composition which makes it possible the aforementioned behavior, is characterized by a high content of binders (cement and cement substitutes) and low water content, well below the stoichiometric ratio, which, actually leaving large amounts of un-hydrated and hence potentially reactive binder material, is highly conducive to self-healing. The very narrow width of the cracks guaranteed by the fibers favors a more effective and quicker healing [32-47].

As a matter of fact, the effectiveness of autogenous healing depends not only on the mix composition but on several other interacting factors characterizing the structure service scenario. These parameters range from the environmental conditions, to the age and width of the cracks and also encompass other important, and so far not deeply enough investigated, issues, such as the existence of a sustained through crack stress state and/or repeated actions and their frequency, which may affect the degree and kinetics of healing. The need of guaranteeing not only the effectiveness of the healing mechanism, but also its versatility and repeatability in front of the variety of the structure service scenarios, has led to fruitful development of self-healing engineering techniques, able to provide the required functionality through the incorporation of materials and systems other than the "usual" concrete constituents. These techniques can be grouped into two major categories: encapsulation of polymers and use of bacteria.

In the former, the material which is going to provide the healing functionality in enclosed into capsules which are dispersed into the concrete matrix at the mixing stage. The capsule material and geometry shall be designed so that the capsules do not break during the mixing but do so once they are intercepted by a propagating crack [48-50]. In this respect, it is also of the utmost importance that capsules do homogeneously distribute within the concrete matrix, due to the randomness of crack formation sites. Once capsules intercepted by the crack rupture, the material they contain is released and flows into the crack, and, once hardened, because of its proper polymerization reactions, provides the crack sealing action. A huge variety of materials has been tested, both for the "cargo" and the shell, optimizing in the former case, *e.g.*, the viscosity for a better filling of hairline cracks and, in both cases, the bond with the cementitious matrix, which, in the case of the cargo material is likely to guarantee a stronger and longer persistence of the sealing action and, for the shell, to increase the probability of rupturing upon interception by a propagating crack. Both micro-encapsulation (capsules with a diameter lower than 1 mm) and macro-encapsulation have been investigated, with a large variety of capsule shapes ranging from spherical to tubular [51]. An evolution of the concept is represented by vascular systems [52,53], implementing a biomimetic approach to self-healing, which, besides the evident topological advantage, would also benefit from the further possibility of a supply of the healing material which could be continuous and in case tailored to the nature and amount of the damage to be repaired.

The bacterial approach to self-healing [54-49], on its hand, relies upon the production of calcium carbonate from the microbiological activity of bacteria and other different types of microorganisms, all featuring the ability to survive into the highly alkaline environment of a cementitious matrix, though contributing to the precipitation of calcium carbonates through different metabolic pathways, including:

- Ureolytic bacteria, which though urease produce, from urea and water, ammonia and carbonate ions, which then form calcium carbonate with the calcium available from the matrix; the bacterial activity requires oxygen to be initiated, which may not be easily available deep inside the cracks;
- Denitrifying bacteria, mutuated from solid consolidation techniques, which produce carbonate and bi-carbonate ions from a nutrient source in the form of calcium formates and nitrates;
- Aerobic heterotrophic bacteria, which produce calcium carbonate as a result of aerobic metabolic conversion of organic compounds, such as calcium lactate.

Protection of micro-organisms from the quite harsh cementitious matrix environment and up to the tome of cracking, when their activity will be requested, has been so far successfully attempted through micro-encapsulation, encapsulation in hydrogels and impregnation in porous granules, including lightweight aggregates already commonly employed in (light-weight) concrete technology [60,61].

3. What do we know: experimental characterization of self-healing

Quantifying the effectiveness of a healing technique, be it based either on stimulated autogenous mechanisms or on an engineered process, is a quite comprehensive experimental task which has to encompass a multifold set of variables, ranging from the creation of a "controlled" crack to the implementation of curing conditions representative of the intended service scenarios, as described in the previous section, to the evaluation of the sealing of the crack and/or of the recovery of the engineering property of interest, either mechanical or durability related or both, to the comprehension of the mechanisms underlying the aforementioned macroscopic observation, through characterization of the healing products. Such a thorough assessment is not only needed for the validation of any proposed self-healing technology but has also to constitute the logical framework for the development of self-healing testing standards, which stands as the first step towards the incorporation of self-healing concepts into code-wise durability based design approaches.

A critical review of the testing methodologies for the assessment of self-healing capacity of cement-based materials has been recently provided in [62]. The choice of the methodology to produce/induce a crack/damage to be healed depends, on the one hand, on the property whose healing-related recovery has to be measured but, on the other, also on the type of cement-based material under investigation and on its fracture toughness, i.e. on the stability of its post-cracking response.

In the case of plain concrete preferably three-point bending tests have been used, the determinedness of the crack position making the procedure easier to implement in case feedback signal closed loop test control systems are not available. Moreover, the geometry of the specimen and of the produced crack allows tests based on the post-healing recovery of flexural performance as well as of durability based properties such as sorptivity to be consistently performed.

In the case of fiber reinforced concrete and high performance fiber reinforced cementitious composites, the need of reproducing, also at the lab-characterization scale, the signature multiple cracking, has led four-point bending tests to be preferred or even direct tensile tests. Recovery in terms of strength, stiffness, ductility and deformation capacity can thus be directly quantifiable, as of interest also in the sight of the intended applications of this category of materials.

Splitting tensile tests have also been used in the case of fiber reinforced concrete, fibers appropriately improving the stability of the post-cracking response, and because of the suitability of the test specimens to verify healing induced recovery of permeability or chloride penetration resistance.

In the case of weak materials, such as lime mortars, whose self-healing properties can be of high interest in restoration of building heritage, compression tests have been used to induce a damage by loading to a prescribed fraction of the compressive strength and assessing the healing induced recovery of the same strength as well as of physical properties, related to elastic stress wave propagation [63,64].

As from the cited literature survey, healing related recovery of the material performance has been evaluated with reference to both mechanical and durability properties, depending on the application.

Visual observation of crack closure, and in case its quantification by means of image analysis, has often complemented the investigation, also in an attempt to establish a correlation with the measured recovery of durability related and mechanical properties. In this framework it is of the utmost importance to highlight that, since the properties of sound concrete do evolve with age, as a function of the scenario, the healing induced recovery of both durability and mechanical properties has to be made considering not only the "self-comparison" between the performance of the same specimen, as measured immediately after cracking and after scheduled healing, but also considering the evolution of that same property for sound concrete exposed to the same exposure conditions as cracked/healing ones [65].

The complete comprehension of the problems finally requires the characterization of the healing products, as well as of the mechanisms underlying their formation, which has been so far performed by means of conventional microstructure investigation techniques, including Scanning Electron Microscopy (SEM), in case combined with energy dispersive X-ray analysis (EDX) or X-ray diffraction (XRD), Fourier Transform Infrared Spectroscopy (FTIR), Thermogravimetric Analysis (TGA), though other less conventional techniques, such as X-ray Computed Tomography, Nuclear Magnetic Resonance (NRM) and neutron radiography have also been used [62].

4. What do we need? Paving the way towards "design-code" integration of self-healing concepts.

Any innovative construction technology, for a consistent and widespread adoption in the practice, requires its main concept to be "formulated" in a code-wise framework, which also has to rely upon standardization of experimental techniques to identify the material parameters which have to be employed in the design method equations. In spite of the tremendous developments in the concept and validation of healing stimulating/engineering techniques, as summarized in Section 2 of this paper, the "healing assessment" experimental techniques are far from achieving a standard recognition, their critical reappraisal having been only recently attempted. Such review has appropriately allowed to highlight the hindrances which have so far limited the market penetrability of self-healing technologies, at the same time allowing the clearly identify the main topics which still need and deserve intense research work in order to fill the knowledge gap which still prevents not only a widespread market penetration of self-healing materials and concept but also their recognition in testing and product standards as well as in design guidelines and recommendations.

As a first instance, it has to be remarked that, with the exception of autogenous healing and healing stimulating techniques based on the use of supplementary cementitious materials, crystalline admixtures and, in case, superabsorbent polymers, most of the healing engineering techniques have been often tested with reference to cement pastes and mortars. Scaling up to concrete implies issues such as interaction with coarse aggregates, even since the mixing stage, to be taken into account, with reference to the dispersion of the capsules into a medium featuring larger inclusions as well as to the "dilution" effect in the concrete matrix, where the paste or mortar only represent a limited volume fraction.

Besides this issue, which would call for further technological improvement of the capsule shells and encapsulated cargo materials, including bacterial cultures and feeding sources, what is really required to proof the robustness of any self-healing technology and pave the way for its conscious acceptance by the construction industry is the investigation of the healing effectiveness in real structure service scenarios. To this purpose, a surely not exhaustive list of research needs is hereafter provided.

Since cracks in a structure may form at any time, the "rate of survival" of the healing functionality, whether autogenous or autonomous, should be assessed for cracks forming at ages even significantly older than the conventional 28 or 56 days. Very few studies exist on this topic [29, 65]. Ferrara et al. [29] assessed the autogenous healing capacity of HPFRCC pre-cracked at about one year after casting. They interestingly found that the healing functionality still holds, even if at 50% lower effectiveness than for the same material pre-cracked about two months after casting (in the mix half of the required cement dosage – 1200 kg/m^3 – was replaced with an equivalent volume of slag).

In a real structure, a crack undergoes, if any, the healing process while the structure itself sustains the permanent loads and a fraction of the variable loads; this share of the target design service load can be as high as more than 60-70% in reinforced concrete structures and infrastructures. This implies, cracks in concrete being cohesive in nature and also considering that generally reinforcing bars or dispersed fiber reinforcement cross the cracks, that the same cracks heal under a sustained through-crack stress, whose influence on the kinetics and effectiveness of the healing as well as on the same nature of the healing products has to be suitably investigated. It is furthermore worth remarking that, while in most cases such stresses are tensile, whose effects on the healing may be expected to be detrimental [66, 67], few interestingly representative cases also hold in which, e.g., an early formed crack may experience in the structure service condition a through crack compressive stress state, which may be expected to benefit its healing. This may be the case, for example, of precast tunnel segments, which may experience restrained shrinkage and/or even flexural cracks during transient situations, including transportation and handling, whereas, once the annulus is completed, experience circumferential compressive stresses whose effect is likely to favor the healing [68]. The same may happen with reference to, e.g., restrained shrinkage cracks in pavements (joints) which, because of fixed or moving loads may experience bending moments whose sign is likely to result into through crack compressive stresses.

This also implies the need to include in the healing assessment scenarios moving/repeated loadings, the magnitude and frequency of the likely alternate stresses they can induce onto the crack faces surely affecting, to an extent so far hardly investigated by anyone, the healing mechanisms and kinetics.

In a similar framework the effect of the load rate application is also worth being considered, the effectiveness in the release of the healing agent from capsules of vascular systems being most likely affected by it, which would result, e.g. in case of high strain rates, in a likewise affected effectiveness of the related healing functionality. It is also worth remarking that, even in the case of load events whose frequency may be very low and hence resembling the quasi static loads so far normally investigated in laboratories and in a few pilot real scale studies, the issue of healing/cracking repeatability has to be investigated, also with reference to the rate of "survival" of the healing functionality across the same cracked sites, upon repeated cracking and healing events. Moreover, the possibility as to be evaluated of "engineering" such a good healing, and material stress redistribution capacity, that, upon a repeated loading event the pristinely formed and finally healed crack is so strong that it does not reopen but a new crack is formed elsewhere, through which the same level of healing should be guaranteed [31, 69].

The assessment of the healing capacity in real structure scenarios is quite a challenging task, involving non-conventional experimental capabilities and most of all requiring the investigation to span along a time frame whose extension could hardly match with the needs of an effective and market oriented technology transfer (healing periods up to a few years have been considered in research so far). In this respect the availability of sound and reliable predictive modelling tools becomes of the utmost importance, since only they would allow to extrapolate the results of performed experimental characterization of self-healing capacity of cement-based materials to real structure service scenarios. A review of the currently existing modelling approaches for self-healing of cement-based materials has been recently provided by Jefferson et al. [70], highlighting on the one hand the complex multifold set of phenomena whose appropriate comprehension and description is required. These range from coupled chemical hydration and mechanical behavior and transport properties to simulation of embedded microcapsules, capsule breakage to release and flow of healing agents. On the other hand, such a review has also pointed out the out the paucity of experimental data suitable for proper validation of numerical models and the scattering not only among the values of the experimentally identified material parameters but also among the concept which has informed the experimental investigation, the way in which the parameters have been identified and the same nature of the thus garnered material properties.

This first of all highlights the compelling need, already remarked in the previous section, of standardizing not only the testing methods for the measurement of concrete self-healing capacity, but, hopefully, of achieving a consensus about the philosophy underlying the testing approaches aimed at identifying material parameters which also have to serve as the basis for a design oriented modelling. In this respect, aiming at the formulation of design concepts/approaches incorporating self-healing in a semi-probabilistic format, it has to be clearly kept in mind that the "statistical" processing of experimental data, and their modelling through physical based models, is a fundamental and necessary but surely not sufficient step. The randomness of structure service scenarios, e.g. in terms of environmental conditions, load patterns etc., requires a likewise mathematically sound treatment, playing a role of equal importance in the design concept formulation.

The goal towards which such a "design oriented" predictive modelling activity has to aim, should be, in the author's opinion, the formulation of a "healable crack width concept", which would replace the current crack width limits in serviceability limit state design. The healable crack width threshold value, which would form the core of a durability based design approach, calibrated on the basis of the randomness of scenarios and materials, would actually represent the crack width limits that the material would be able to heal, with a given assigned/accepted probability, under the intended environmental conditions and under an assigned and accepted probability that the anticipated cracks will form under the anticipated load combinations and over the intended target design service life.

In order the benefits of self-healing materials to be fully appreciated into an engineering framework but also from the economic point of view, the outcomes of the whole aforementioned activity should be framed into a life cycle analysis approach which together with life cycle cost and social life cycle analysis will quantify the benefits as above in terms of extended material/product/structure service life and reduced maintenance and life cycle cost, including indirect benefits in terms of public awareness of the contribution of the construction industry to the whole sustainability goal.

5. Conclusions

In this paper the topic of self-healing cement-based materials has been dealt with, presenting it also with the aid of state of art reviews recently completed and published in the framework of transnational research collaboration initiative, in the challenging socio-economic context of the construction sector worldwide. Besides reviewing the current state of development of healing technologies and healing assessment testing methods, the paper has also highlighted the most challenging research needs which have to be tackled aiming at fostering the market penetration of self-healing cement-based materials, through incorporation of self-healing concepts into durability based and life cycle design approaches.

References

[1] European Commission. The European construction sector. A global partner. 11/03/2016. 16 pp. (accessed through https://ec.europa.eu/growth/content/european-construction-sector-global-partner-0_en on March 1, 2018)
[2] WEF (2016). Shaping the future of construction. A breakthrough in mindset and technology. 61p. (http://www3.weforum.org/docs/WEF_Shaping_the_Future_of_Construction_full_report__.pdf)
[3] The concrete conundrum. Chemistry World, March 2008, 62-66. Accessed on March 1, 2018 through http://www.rsc.org/images/Construction_tcm18-114530.pdf.
[4] Barros, J.A.O., Ferrara, L. and Martinelli, E., eds. (2017). Recent advances on green concrete for structural purposes. The contribution of the EU-FP7 project EnCoRe. Springer, pp. 427+x, ISBN 978-3-319-56797-6.
[5] https://inspectioneering.com/news/2016-03-08/5202/nace-study-estimates-global-cost-of-corrosion-at-25-trillion-ann (accessed on March 1, 2018).
[6] Matthews S. (2007). CONREPNET: Performance-based approach to the remediation of reinforced concrete structures: Achieving durable repaired concrete structures. *Journal of Building Appraisal,* 3(1), 6-20.
[7] EN 1992-1-1. Eurocode 2: Design of concrete structures - Part 1-1. December 2004.
[8] *fib* Model Code for Concrete Structures 2010. DOI: 10.1002/9783433604090.
[9] di Prisco, M., Plizzari, G. and Vandewalle, L. (2009). Fiber reinforced concrete. New design perspectives.

Materials and Structures, 42 (9), 1261-1281.

[10] Naaman, A.E., & Reinhardt, H.W. (2006). Proposed classification of HPFRC composites based on their tensile response. *Materials and Structures*, 39 (5), 547-555.

[11] Li, V., Lim, Y., and Chan, Y. (1998). Feasibility of a passive smart self-healing cementitious composite. *Composites Part B: Engineering*, 29, 819-827.

[12] Cuenca, E.A. and Ferrara, L. (2017). Self-healing capability of FRC. State of the art and perspectives. *Journal of the Korean Society of Civil Engineers*, 21(7), 2777-2789.

[13] Snoeck, D. and De Belie, N. (2015). From straw in bricks to modern use of microfibers in cementitious composites for improved autogenous healing. A review. *Construction and Building Materials*, 95, 774-787.

[14] Gosh, S.K. (ed.) (2009). Self-Healing Materials: Fundamentals, Design Strategies, and Applications. Wiley. ISBN 9783527318292.

[15] De Rooij, M.R., Van Tittelboom, K., De Belie, N., Schlangen, E. (eds.) (2013). Self-Healing phenomena in Cement-based materials. State-of-the-Art Report of RILEM Technical Committee 221-SHC. Springer, 266 pp. ISBN: 978-94-007-6623-5.

[16] Ferrara, L., Krelani, V. and Carsana, M. (2014). A fracture testing based approach to assess crack healing of concrete with and without crystalline admixtures. *Construction and Building Materials*, 68, 515-531.

[17] Mihashi, H. and Nishiwaki, T. (2012). Development of engineered self-healing and self-repairing concrete – State-of-the-art report. Journal of Advanced Concrete Technology, 10, 170-184.

[18] Van Tittelboom, K. and De Belie, N. (2013a). Self-healing in cementitious materials – a review. *Materials*, 6, 2182-2217.

[19] Sisomphon, K., Copuroglu O. and Koenders, E. A. B. (2012). Self-healing of surface cracks in mortars with expansive additive and crystalline additive. *Cement and Concrete Composites*, 34, 566–574.

[20] Van Tittelboom, K., Gruyaert, E., Rahier, H. and De Belie, N. (2012). Influence of mix composition on the extent of autogenous crack healing by continued hydration or calcium carbonate formation. *Construction and Building Materials*, 37, 349-349.

[21] Snoeck, D. and De Belie, N. (2016). Repeated autogenous healing in strain-hardening cementitious composites by using superabsorbent polymers. *ASCE Journal of Materials in Civil Engineering*, 28(1), 1-11.

[22] Snoeck, D., Steuperaert, S., Van Tittelboom, K., Dubruel, P. and De Belie, N. (2012b). Visualization of water penetration in cementitious materials with superabsorbent polymers by means of neutron radiography. *Cement and Concrete Research*, 42(8), 1113-1121.

[23] Snoeck, D., Van Tittelboom, K., Steuperaert, S., Dubruel, P. and De Belie, N. (2014). Self-healing cementitious materials by the combination of microfibers and superabsorbent polymers. *Journal of Intelligent Material Systems and Structures*, 25(1), 13-24.

[24] Gruyaert, E., Debbaut, D., Snoeck, D., Díaz, P., Arizo, A., Tziviloglou, E., Schlangen, E., De Belie, N. (2016). Self-healing mortar with pH-sensitive superabsorbent polymers: testing of the sealing efficiency by water flow tests. Smart Materials and Structures, 25 (084007), 11 p. DOI: 10.1088/0964-1726/25/8/084007.

[25] Snoeck, D., Dewanckele, J., Cnudde, V., and De Belie, N. (2016). X-ray computed microtomography to study autogenous healing of cementitious materials promoted by superabsorbent polymers. *Cement and Concrete Composites*, 65, 83-93.

[26] Lee, H.X.D., Wong, H.S., and Buenfeld, N.R. (2016). Self-sealing of cracks in concrete using superabsorbent polymers. *Cement and Concrete Research*, 79, 194-208.

[27] Borg, R.P., Cuenca, E., Gastaldo Brac, E.M. and Ferrara, L. (2018): "Crack sealing capacity in chloride rich environments of mortars containing different cement substitutes and crystalline admixtures", *Journal of Sustainable Cement-based Materials,* 7(3), 141-159.

[28] Roig-Flores, M., Moscato, S., Serna. P. and Ferrara, L. (2015). Self-healing capability of concrete with crystalline admixtures in different environments. *Construction and Building Materials*, 86, 1-11.

[29] Roig Flores, M., Pirritano, F., Serna Ros, P. and Ferrara, L.. (2016). Effect of crystalline admixtures on the self-healing capability of early-age concrete studied by means of permeability and crack closing tests. *Construction and Building Materials*, 114, 447-457.

[30] Ferrara, L., Krelani, V. and Moretti, F. (2016). On the use of crystalline admixtures in cement-based construction materials: from porosity reducers to promoters of self healing. *Smart Materials and Structures*, 25 084002 (17pp) DOI:10.1088/0964-1726/25/8/084002.

[31] Cuenca, E., Tejedor, A. and Ferrara, L. (2018). A methodology to assess crack sealing effectiveness of crystalline admixtures under repeated cracking-healing cycles. Submitted to *Construction and Building Materials*.

[32] Qian, S., Zhou, J., De Rooij, M., Schlangen, E., Ye, G., and Van Breugel, K. (2009). Self-healing behavior of strain hardening cementitious composites incorporating local waste materials. *Cement and Concrete Composites*, 31, 613-621.

[33] Kan, L., Shi, H., Sakulich, A. and Li, V. (2010). Self-healing Characterization of Engineered Cementitious Composite Materials. *ACI Materials Journal*, 107(6), 617-624.

[34] Li, M., and Li, V. (2011). Cracking and healing of Engineered Cementitious Composites under chloride environment. *ACI Materials Journal*, 108(3), 333-340.

[35] Qian, S.X, Zhou, J. and Schlangen, E. (2010). Influence of curing condition and precracking time on the self-healing behavior of Engineered Cementitious Composites. *Cement and Concrete Composites*, 32, 686–693.

[36] Yang, Y., Yang, E.-H- and Li, V.C. (2011b). Autogenous healing of engineered cementitious composites at early age. *Cement and Concrete Research*, 41, 176-183.

[37] Nishiwaki, T., Koda, M., Yamada, M., Mihashi, H. and Kikuta, T. (2012). Experimental Study on Self-Healing Capability of FRCC using different types of synthetic fibers. *Journal of Advanced Concrete Technology*, 10, 195-206.

[38] Wu, M., Johannesson, B. and Geiker, M. (2012). A review: self-healing in cementitious materials and engineered cementitious composite as a self-healing material. *Construction and Building Materials*, 28, 571-583.

[39] Kim, D., Kang, S., and Ahn, T. (2014). Mechanical Characterization of High-Performance Steel-Fiber Reinforced Cement Composites with Self-Healing Effect. *Materials*, 7(1), 508-526.

[40] Nishiwaki, T., Kwon, S., Homma, D., Yamada, M. and Mihashi, H. (2014). Self-healing capability of Fiber-Reinforced Cementitious Composites for recovery of watertightness and mechanical properties. *Materials*, 7, 2141-2154.

[41] Snoeck, D., Van Tittelboom, K., Steuperaert, S., Dubruel, P. and De Belie, N. (2014). Self-healing cementitious materials by the combination of microfibers and superabsorbent polymers. *Journal of Intelligent Material Systems and Structures*, 25(1), 13-24.

[42] Şahmaran, M., Yildirim, G., Noori, R., Ozbay, E. and Lachemi, M. (2015). Repeatability and pervasiveness of Self-Healing in Engineered Cementitious Composites. *ACI Materials Journal*, 112(4), 513-522.

[43] Snoeck, D., Smetryns, P., and De Belie, N. (2015). Improved multiple cracking and autogenous healing in cementitious materials by means of chemically-treated natural fibres. *Biosystems Engineering*, 139, 87-99.

[44] Yildirim, G., Kasap Keskin, Ö, Bahadir Keskin, S., Şahmaran, S. and Lachemi, M. (2015). A review of intrinsic self-healing capability of engineered cementitious composites: recovery of transport and mechanical properties. *Construction and Building Materials*, 101, 10-21.

[45] Ferrara, L., Krelani, V., and Moretti, F. (2016b). Autogenous healing on the recovery of mechanical performance of High Performance Fibre Reinforced Cementitious Composites (HPFRCCs): part 2 - correlation between healing of mechanical performance and crack sealing. *Cement and Concrete Composites*, 73, 299-315.

[46] Ferrara, L., Krelani, V., Moretti, F., Roig-Flores, M., and Serna, P. (2017a). Effects of autogenous healing on the recovery of mechanical performance of High Performance Fibre Reinforced Cementitious Composites (HPFRCCs): part 1. *Cement and Concrete Composites*, 83, 76-100.

[47] Ferrara, L., Ferreira, S.R., Krelani, V., Nunes da Cunha Moreira, T., De Andrade Silva, F. and Toledo Filho, R.D. (2017b). Self-healing capacity of hybrid sisal-steel fiber reinforced cementitious composites and comparison with steel only HPFRCCs. Submitted to *Construction and Building Materials*.

[48] Li, W, Jiang, Z., Yang, Z., Zhao, N., and Yuan, W. (2013). Self-healing efficiency of cementitious materials containing microcapsules filled with healing adhesive: Mechanical restoration and healing process monitored by water absorption. *PloS ONE*, 8(11), article ID e81616.

[49] Souradeep, G., and Kua, H.W. (2016). Encapsulation Technology and Techniques in Self-Healing Concrete. *ASCE Journal of Materials in Civil Engineering*, 28(12), 1-15.

[50] Gilabert, F.A., Van Tittelboom, K., Van Stappen, J., Cnudde, V., De Belie, N. and Van Paepegem, W. (2017). Integral procedure to assess crack filling and mechanical contribution of polymer-based

healing agent in encapsulation-based self-healing concrete. *Cement and Concrete Composites*, 77, 68-80.

[51] Formia, A., Irico, S., Bertola, F., Canonico, F., Antonaci, P., Pugno, N.M. and Tulliani, J.M. (2016). Experimental analysis of self-healing cement-based materials incorporating extruded cementitious hollow tubes. *Journal of Intelligent Materials Systems and Structures*, 27, 2633-2652.

[52] Teall, O., Davies, R., Pilegis, M., Kanellopoulos, A., Sharma, T., Paine, K., Jefferson, A., Lark, R., Gardner, D., Al-Tabbaa, A. (2016) Self healing concrete full scale site trials. In: Proceedings of the 11th fib International PhD Symposium in Civil Engineering, 639-646.

[53] Teall, O; Davies, R; Pilegis, M; Kanellopoulos A; Sharma, T; Gardner D; Paine, K; Jefferson, A; Abir Al-Tabbaa,, A; Lark, R. (2017). Large Scale Application of Self-Healing Concrete: Design, Construction and Testing. Submitted to *Case Studies in Construction Materials*.

[54] Jonkers, H., Thijssen, A., Muyzer, G., Copuroglu O. and Schlangen, E. (2010). Application of bacteria as self-healing agent for the development of sustainable concrete. *Ecological engineering*, 36(2), 230-235.

[55] Van Tittelboom, K., De Belie, N., De Muynck, W. and Verstraete, W. (2010). Use of bacteria to repair cracks in concrete. *Cement and Concrete Research*, 40, 157–66.

[56] Jonkers, H. (2011). Bacteria-based self-healing concrete. Heron 56,1/2:1–12.

[57] Luo, M., Qian. C.X. and Li, R.X. (2015). Factors affecting crack repairing capacity of bacteria-based self-healing concrete. *Construction and Building Materials*, 87, 1-7.

[58] Wang, J., Mignon, A., Snoeck, D., Wiktor, V., Van Vlierberghe, S., Boon, N., & De Belie, N. (2015). Application of modified-alginate encapsulated carbonate producing bacteria in concrete: a promising strategy for crack self-healing. *Frontiers in Microbiology*, 6.

[59] Khaliq, W. and Ehsan, M. B. (2016). Crack healing in concrete using various bio influenced self-healing techniques. *Construction and Building Materials*, 102, 349-357.

[60] Wang, J.Y., Snoeck, D., Van Blierberghe, S., Verstraete, W. and De Belie, N. (2014a). Application of hydrogel encapsulated carbonate precipitating bacteria for approaching a realistic self-healing in concrete. *Construction and Building Materials*, 68, 110-119.

[61] Wang, J.Y., Soens, H., Verstraete, W., and De Belie, N. (2014c). Self-healing concrete by use of microencapsulated bacterial spores. Cement and Concrete Research 56 (2014) 139-152.

[62] Ferrara, L., Van Mullem, T., Alonso, M.C., Antonaci, P., Borg, R.P., Cuenca, E., Jefferson, A., Ng, P.L., Peled, A., Roig, M., Sanchez, M., Schroefl, C., Serna, P., Snoeck D., Tulliani, J.M. and De Belie, N. (2018). Experimental characterization of the self-healing capacity of cement-based materials and its effects on the material performance: a state of the art report by COST Action SARCOS WG2. *Construction and Building Materials*, 167, 115-142.

[63] De Nardi, C., Bullo, S., Ferrara, L., Ronchin, L. and Vavasori, A. (2017). Effectiveness of crystalline admixtures and lime/cement microcapsules in engineered self-healing capacity of lime mortars. *Materials and Structures,*50 (4), 191.1-191.12

[64] De Nardi, C., Cecchi, A., Ferrara, L., Benedetti, A. and Cristofori, D. (2017). Effect of age and level of damage on the autogenous healing of lime mortars. *Composites-Part B: Engineering,* 124, 144-157.

[65] Hearn, N. (1998). Self-sealing, autogenous healing and continued hydration: what is the difference?. *Materials and Structures,* 31(8), 563-567.

[66] Yildirim, G., Hamidzadeh, Kiavi, A., Yeşilmen, S. and Şahmaran, M. (2018). Self-healing per-formance of aged cementitious composites. *Cement and Concrete Composites*, 87, 172-186.

[67] Özbay, E., Şahmaran, M., Yücel, H., Erdem, T., Lachemi, M., and Li, V. (2013a). Effect of sustained flexural loading on Self-Healing of Engineered Cementitious Composites. *Journal of Advanced Concrete Technology*, 11, 167-179.

[68] Yildirim, G., Alyousif, A., Şahmaran, M., and Lachemi, M. (2015b). Assessing the self-healing capability of cementitious composites under increasing sustained loading. *Advances in Cement Research*, 27(10), 581-592.

[69] Ferrara, L., Albertini, I., Gettu, R., Krelani, V., Moscato, S., Pirritano, F. and Theeda, S. (2015a). Self-healing of cement-based materials engineered through crystalline admixtures: experimental results form a multinational university network. in M.A. Chiorino et al., eds., Durability and Sustainability of Concrete Structures –ACI SP 305, 13.1-13.10.

[70] Jefferson, A.D., Javierre, E., Freeman, B., Zaoui, A., Keoners, E. and Ferrara, L. (2018). *Research* progress on numerical models for self-healing cementitious materials. Accepted for publication in *Materials Interface Science.*

A multi-criteria decision-making based approach to assess the sustainability of concrete structures

Albert de la Fuente[1], **David Fernández-Ordóñez**[2]

[1] Department of Civil and Environmental Engineering, Polytechnic University of Catalonia, Spain
[2] International Federation for Structural Concrete (*fib*), Switzerland

E-mail: albert.de.la.fuente@upc.edu

Abstract. The use of sustainability assessment tools is gaining importance in the construction sector. There exist several methods with different approaches and scopes; however, there is still no consensus about which method should be used to deal with the sustainability assessment of concrete structures. Among these, the multi-criteria decision-making based approach called MIVES seems to be a suitable and flexible model that allows taking into account all those indicators and parameters (of economic, social and environmental nature) involved in the sustainability assessment of concrete structures. The objective of this research document is two-fold: (1) to expose the basis and concepts related with the MIVES model as a sustainability assessment tool and (2) to present 3 real study cases (wind precast concrete towers; steel fibre reinforced precast concrete tunnel linings and reinforced concrete pile-supported slabs) in which this model has been used to make decisions. The authors of this research consider that similar approaches should be included in future national and international structural concrete guidelines (as the Spanish Structural Concrete Code does) to perform sustainability analysis of new designed concrete structures.

Keywords: AHP; decision-making; fibre reinforced concrete; MIVES; structures; sustainability

1. Introduction

There exist several tools, database and methods available to assess sustainability and environmental aspects within the architectural and civil engineering area. However, no consensus has been yet reached respect to which is the most suitable tool to assess the sustainability in building construction. Table 1 presents 8 of these certification tools for building sector.

The acceptance of these methods has shown to be variable, as only two have been internationally applied for decades, while the others are mainly used in the country of origin. These methods also differ in being credits or percentage based rating tools, the application's complexity and the outcomes resulting from each method, which is in most of the cases, either a certification with a qualification of satisfaction or a graphic sustainability index. All these tools have contributed to advance towards a more sustainable construction sector and to raise awareness of this issue within the sector. Only a few methods are capable of quantifying all the different social, economic and environmental requirements that permit researchers to derive a global sustainability index. The main drawback arises, however, when trying to assess the sustainability of independent structural elements or products (e.g., piles, beams, slabs, pipes, walls) for which these methods are scarcely representative.

Table 1. Different sustainability assessment tools for buildings.

Name	Institution	Origin	Use	C	E	S	CR	PR	C	R
BEAM	BEAM	Hong Kong 1996	N	L	X	L	X	-	L	CQ
BREEAM	BRE	UK 1990	I	-	X	L	X	-	L	CQ
DGNB	DGNB	Germany 2008	N	X	X	X	-	X	H	CQ
EcoEffect	KTH	Sweden 2000	N	X	X	-	X	-	M	GI
Green Star	GBCA	Australia 2003	N	-	X	X	X	-	M	CQ
HQE	AssoHQE	France 1996	N	X	X	X	-	X	M	CQ
LEED	USGBC	USA 2000	I	L	X	L	X	-	L	CQ
VERDE	GBCE	Spain 2010	N	X	X	X	-	X	M	CQ

I. internationally consolidated, N. nationally consolidated

C. Economic requirements (cost, time…). X means included, L means low consideration.

E. Environmental requirements (energy consumption, CO_2 emissions…). X means included.

S. Social requirements (health, safety, quality…). X means included, L means low consideration.

CR. X means credits based rating tool that gives credits to carry out the assessment.

PR. X means percentage based rating tool that assesses the percentage of satisfaction of each indicator.

C. Tool complexity of application. H means high, M means medium, L means low.

R. Result. CQ means certification with a qualification of satisfaction; I means graphic index.

As a potential solution, the MIVES method (from the Spanish Integrated Value Model for the Sustainability Assessment) is herein proposed to assess the sustainability index of concrete products and systems. MIVES is a Multi-Criteria Decision Making (MCDM) method capable of defining specialized and holistic sustainability assessment models to obtain global sustainability indexes. The MIVES method is a MCDM based on the use of value functions to assess the satisfaction of those stakeholders involved in the decision-making process. The use of these functions allows minimizing the subjectivity in the assessment.

A description of the MIVES method is firstly presented and, posteriorly, the method is applied to measure the sustainability index of different alternatives within the field of tunnelling construction and supports for wind turbines.

2. MIVES Method

MIVES is a multi-criteria decision-making method capable of defining specialized and holistic sustainability assessment models to obtain global sustainability indexes of structures or products. MIVES combines: a) a specific holistic discriminatory tree of requirements (Figure 1); b) the assignation of weights for each requirement (α_i), criteria (β_i) and indicator (γ_i); c) the value function concept [1] to obtain particular and global indexes and d) seminars with experts using Analytic Hierarchy Process (AHP) [2-3] to define the aforementioned parts. The sustainability index (SI) is assessed by means of Equation 1.

$$SI = \sum_{i=1}^{i=N} \alpha_i \cdot \beta_i \cdot \gamma_i \cdot I_{ind}(X_i) \quad (1)$$

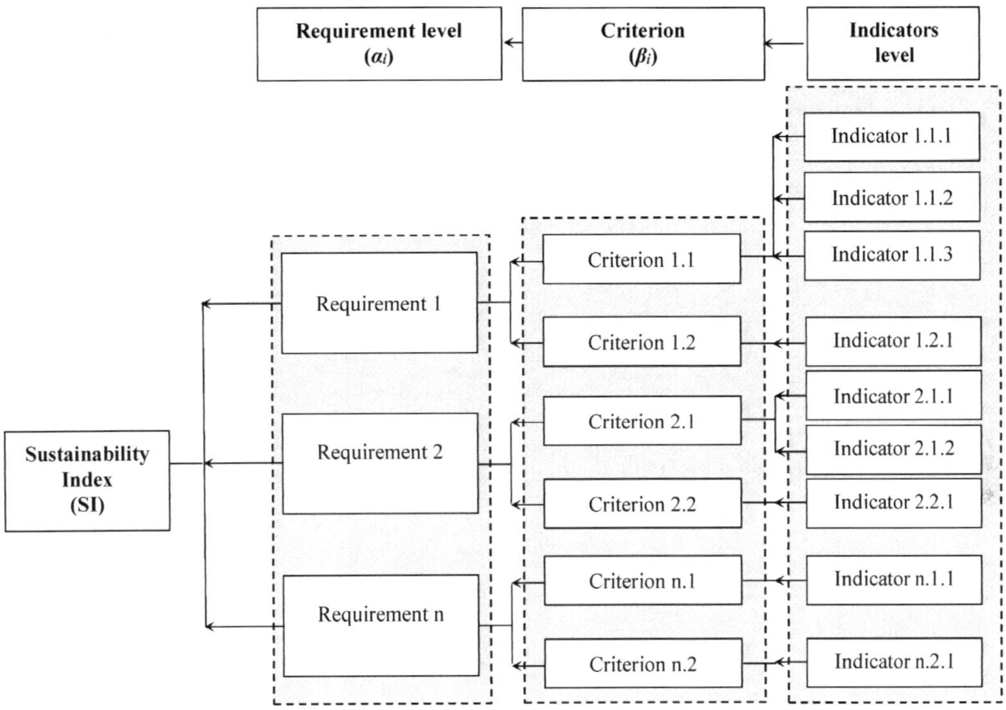

Figure 1. General requirements tree

Value functions (V_{ind}) are assigned to the previously described indicators. These functions transform physical units of each indicator (e.g., €/m³, Ton/m³, dB) into dimensionless values ranging from 0 to 1. These values represent the sustainability or satisfaction of each indicator. Equation 2 shows the general form of a value function.

$$V_{ind}(X) = A + B\left[1 - e^{-K_i\left(\frac{|X_{ind} - X_{min}|}{C_i}\right)^{P_i}}\right] \quad (2)$$

In Equation 2, B is the value of V_{ind} for X_{min}; X_{min} is the minimum abscissa value of the indicator interval assessed; X is the abscissa value for the indicator assessed; P_i is a shape factor which defines whether the curve is concave ($P_i<1$), convex ($P_i>1$), linear ($P_i=1$) or S-shaped ($P_i>1$) (Figure 2); C_i approximates the abscissa at the inflexion point; K_i tends towards V_{ind} at the inflexion point; B, the factor that prevents the function from exceeding the range (0, 1), is obtained by Equation 3, X_{max} being the abscissa value of the indicator that gives a response value of 1 for increasing value functions.

$$B = \left[1 - e^{-K_i\left(\frac{|X_{max} - X_{min}|}{C_i}\right)^{P_i}}\right]^{-1} \quad (3)$$

There have already been numerous applications of MCDM in engineering [4], most focusing on economic aspects and fewer about environmental issues or social aspects. The MIVES method is a unique MCDM based on the use of value functions [5] to assess the satisfaction of the different stakeholders involved in the decision-making process. The use of these functions allows minimizing the subjectivity in the assessment. So far, MIVES has already been used for industrial buildings [6-8], underground infrastructures [9], hydraulic structures [10-11], wind towers [12], sewage systems [13], post-disaster sites and housing selection [14-15] and construction projects [16-17]. It should be highlighted that in the current Spanish Structural Concrete Code [18], MIVES method is proposed for assessing the sustainability of concrete structures [19]. Finally, it must be added that the MIVES method has even been expanded to include the uncertainties involved in the process of analysis [2].

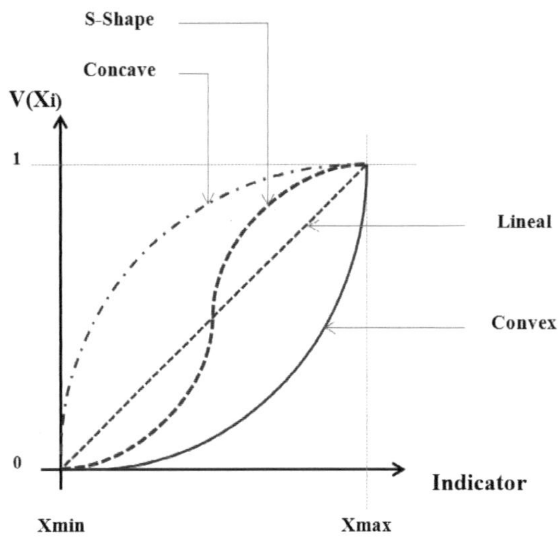

Figure 2. Shapes of the value function

3. Example of application 1. Segmental linings for tunnels

3.1. Introduction

The use of precast concrete linings (Figure 3) to resist the earth actions in TBM-constructed tunnels is a widespread practice. These elements are generally reinforced with steel curved-cages. However, the replacement of this traditional reinforcement for structural fibres is increasing due to diverse reported technical and economic advantages as well as the acceptance of the fibre reinforced concrete (FRC, hereinafter) as structural material in several standards.

Figure 3. Precast concrete segments: (a) Traditional reinforcement (Metro L9 of Barcelona) and (b) FRC full – scale bending test carried out at the UPC of Barcelona

Besides, it should be born in mind that these segments are, usually, subjected to reduced bending moments and the likelihood of cracking is relative low (especially during service). In this regard, the higher bending moment basically occur during transient loading stages (see Figure 4) for which the segments are designed no to crack; thus, minimum reinforcement is required. As a consequence, the competitive amount of structural to be used as a replacement of the rebars makes the FRC an attractive material for this application.

Even though different current code already permits the use of FRC in structural elements and the solution has proven to be both technically and economically attractive in the segmental linings used in over 50 TBM tunnels built to date, some restrictions still persist concerning the use of FRC in this particular application. Among these, the main factors that designers and contractors compare traditional and FRC solutions are based solely on direct material costs without taking into account either indirect costs or social and environmental factors, that is, without considering the sustainability of possible solutions.

The aim of this practical example of application of MIVES consist in proposing a multi-criteria decision-making method based on the MIVES to assess the different viable solutions for reinforcement of precast concrete segments.

Figure 4. Some of the transient load situations in which the segments are subjected to bending moments: (a) demoulding and (b) stacking at the yard

IOP Conf. Series: Materials Science and Engineering **442** (2018) 012008 doi:10.1088/1757-899X/442/1/012008

3.2. Sustainability assessment MIVES model for precast concrete segments

The requirements tree defined for the sustainability assessment of precast concrete segments is presented in Table 2. In this case, although a cradle to cradle LCA could be carried out, a cradle to the placement of the segments inside the tunnel has been considered. This can be assumed since the type of reinforcement is not a variable that affects the rest of operations after the placement of the segments. Based on the results of the seminars, 1 km tunnel was considered to be enough representative to integrate all those factors involved in assessing the sustainability of the segment, omitting consideration of infrastructure and other elements not crucial to the analysis, such as vertical shafts and stations.

It must be emphasized that the reinforcement alternatives for a certain segment, either traditional reinforcement, FRC or hybrid configurations (rebars + fibres), considered for a certain boundary conditions (lining thickness, internal diameter and loads) should complain with the structural and project requirements. Otherwise, these should not be considerate as alternatives to be compared.

3.3. Case study: segmental lining of the L9 Extension to the Barcelona Airport.

As an example of application of this MIVES model, the segmental lining of the L9 Tunnel Extension to the Barcelona Airport has been considered. This consist in a 2.84 km long TBM-constructed tunnel in service since 2016. The lining (Fig. 5) was made up with a universal ring with a mean length of 1.60 m and an internal diameter of 9.60 m. The ring is 0.32 m thick and is composed of 6 segments and 1 key.

Table 2. Requirements' tree proposed to assess the sustainability of precast concrete linings

Requirement	Criteria	Indicator	Units	Function
R_1 Economic (40%)	C_1 Direct costs (90%)	I_1 Total costs (100%)	M€/km	DS
	C_2 Cost of repairs (10%)	I_2 Probability of repair (100%)	Attributes	
R_2 Environmental (45%)	C_3 Resources consumption (30%)	I_3 Cement and aggregates (50%)	Ton/km	DCx
		I_4 Water (20%)		
		I_5 Reinforcing steel (30%)		
	C_4 Emissions (40%)	I_6 CO_2 emissions (100%)	TonCO_2-eq/km	DS
	C_5 Energy (30%)	I_7 Embodied energy (100%)	MWh/km	
R_3 Social (15%)	C_6 Labour conditions (100%)	I_8 Noise pollution (70%)	Db	DCx
		I_9 Risks during handling (30%)	Attributes	

For the former project, conventional reinforced concrete (CRC) segments with 110 kg/m^3 of steel rebars and concrete with a characteristic compressive strength value f_{ck} of 45 N/mm^2 were designed. The designers also verified that the design forces do not exceed the crack resistance of the segment in any of the loading stages and fixed a minimum reinforcement to ensure adequate ductile behaviour in a hypothetical cracking situation. However, two new solutions for the segments using only FRC have been proposed: (1) using conventionally vibrated FRC concrete and (2) using self-compacting fibre-reinforced concrete (SC-FRC).

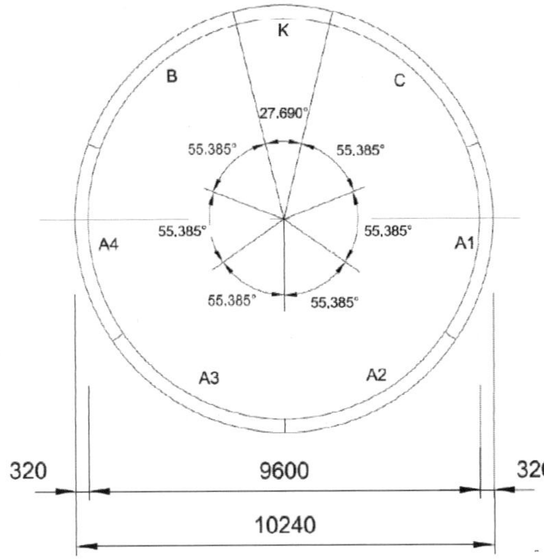

Figure 5. Dimensions (in mm) of the tunnel lining (L9 Extension to the Barcelona Airport)

The different concrete dosages are presented in Table 3.

Table 3. Dosages (in kg/m^3) considered for the different concrete mixes.

MATERIALS	CRC	FRC	SC-FRC
CEM I 52.5	315	315	381
Sand 0/5	817	817	1,200
Fine aggregate 5/12	404	404	500
Coarse aggregate 12/20	810	810	200
Water	150	156	165
Superplasticiser	2.80	2.80	4.60
Steel fibres	0	50	45

Hooked – end steel fibres with a yielding strength of 1000 N/mm^2, length of 50 ± 5 mm and a diameter of 1.0 ± 0.1 mm were used for both FRCs. In this regard, the experimental results on notched prismatic specimens according to the testing procedure EN 14651:2005 have confirmed that 50 kg/m^3 (FRC) and 45 kg/m^3 were sufficient to reach the required 4d (f_{R1k} = 4.0 N/mm^2 and 1.1≤f_{R3k}/f_{R1k}<1.3) strength class to replace all the rebars while guaranteeing the ductile behaviour of the segments. It is worth to note that SC-FRC requires 10% less fibre material than FRC because of the better orientation of the fibres in the pouring process of the self-compacting concrete due the flow forces and boundary conditions imposed by the walls of the mould.

The construction of the tunnel lining involves 12,425 segments (1,775 rings), requiring 28,322 m^3 of concrete. The segments will be fabricated in an existing plant specifically designed for the purpose. The distance from the plant to the TBM access shaft is 110 km. The plant is expected to be in operation for a period of 16 months between the start of preparations and final shutdown. It is estimated that the fabrication of all segments will take nine months with two 8-hour work shifts a day.

IOP Conf. Series: Materials Science and Engineering **442** (2018) 012008 doi:10.1088/1757-899X/442/1/012008

The data gathered in Table 4 must be considered to assess each of the 9 indicators fixed in the requirements' tree (Table 2). Likewise, the constitutive parameters required to define the specific value functions are presented in Table 5.

Table 4. Values of the main features of the alternative reinforcement configurations

Indicator	CRCS	FRCS	SC-FRCS
I_1 Direct costs (M€/km)	2.89	2.60	2.61
I_2 Probability of repair	Moderate	Low	Low
I_3 Cement and aggregates (Ton/km)	66,444	66,444	64,603
I_4 Water (Ton/km)	15,590	10,863	11,668
I_5 Reinforcing steel (Ton/km)	1,097	499	449
I_6 CO_2 emissions (TonCO_2-eq/km)	5,305	4,601	5,083
I_7 Embodied energy (MWh/km)	12,411	9,375	9,904
I_8 Noise pollution (Db)	90	90	60
I_9 Risk during handling	Reduced	High	High

By applying the additive formula shown in Equation (1), the requirements' satisfaction degrees and the SI of each reinforcement alternative can be estimated (Table 6).

Table 5. Value function parameters for each indicator

Indicator	X_{max}	X_{min}	C	K	P
I_1 Direct costs (M€/km)	4,00	2,24	1,00	1,00	2,50
I_2 Probability of repair	Steel: 0.00 – 0.25 (very high); low fibre content: 0.25 – 0.50 (high); steel + low fibre content: 0.50 - 0.75 (moderate); High fibre content: 075 - 1.00 (low)				
I_3 Cement and aggregates (Ton/km)	70,000	65,000	67,000	0.10	2.50
I_4 Water (Ton/km)	29,000	7,500	15,000	0.10	2.50
I_5 Reinforcing steel (Ton/km)	1,350	450	800	1.00	2.50
I_6 CO_2 emissions (TonCO_2-eq/km)	7,800	3,800	5,000	2.50	200
I_7 Embodied energy (MWh/km)	18,500	7,500	10,000	2.50	2.00
I_8 Noise pollution (Db)	150	0	80	3.00	10.00
I_9 Risks during handling	Very high: 0.00 – 0.25; High: 0.,25 – 0.50; Acceptable: 0.50 – 0.75; Reduced: 0.75 – 1.00				

Table 6. Sustainability index of each reinforcement alternative

	CRCS	FRCS	SC-FRCS
SI	**0.578**	**0.754**	**0.856**
I_{R1}	0.703	0.899	0.909
I_{R2}	0.513	0.786	0.836
I_{R3}	0.438	0.326	0.775

The results presented in Table 6 highlight that using FRC, vibrated and self-compacting concrete, leads to more sustainable solutions that using conventional reinforced concrete. Specifically, SC-FRCS (SI = 0.856) represents an increase of 48% in SI over CRCS (SI = 0.578) and an increase of 14% over FRCS (SI = 0.754).

4. Example of application 2. Precast concrete wind towers

4.1. Introduction
Precast concrete wind towers have been progressively introduced in the market, these gaining importance over other existing alternatives for heights above 100 m due to diverse technical and economic advantages. However, still there is not an objective tool that allows quantifying the sustainability of wind towers considering the three pillars (economic, environmental and social). The aim of this practical example of application of MIVES consist in presenting the whole procedure carried out to establish the components (tree of requirements, value functions, weights and analysis of the results) that permits to assess the sustainability index of wind towers. As a particular case of sustainability assessment, a precast concrete tower has been chosen.

4.2. Sustainability assessment MIVES model for wind towers
In Table 7 the requirement tree defined is presented. The LCA embraces from cradle to the deconstruction of the wind field. The unit of analysis consist of the structural elements (foundation and tower). Likewise, the maximum transport distance of the precast concrete elements from plant to site is not superior to 350 km. It must be emphasized that this model can be applied to any composition of structural materials that the tower might be made of.

The *economic requirement* (R_1) takes into account the impact of the different costs, both direct and indirect, identified during the seminars. The *environmental requirement* (R_2) is used to consider the impact of the construction process and materials involved in the tower's installation. The *social requirement* (R_3) is used to assess key factors for the social acceptance of wind farms.

Table 7. Requirements' tree proposed to assess the sustainability of wind towers

Requirement	Criteria	Indicator	Unit
R_1 Economic (50%)	C_1 Construction cost (40%)	I_1 Direct cost (50%)	€/tower
		I_2 Cost deviations (50%)	Points
	C_2 Maintenance cost (40%)	I_3 Planned works (100%)	€/tower
	C_3 Deconstruction (20%)	I_4 Deconstruction (100%)	€/tower
R_2 Environmental (35%)	C_4 Resources consumption (33.3%)	I_5 Material (100%)	Tn/MW
	C_5 Energy (33.3%)	I_6 Energy (100%)	GWh/MW
	C_6 Emissions (33.3%)	I_7 CO_2 (100%)	TnCO_2-e/MW
R_3 Social (15%)	C_7 Occupational hazards (30%)	I_8 Risk of accident (100%)	Points
	C_8 Perception (60%)	I_9 Proportions (50%)	
		I_{10} Flexibility (50%)	
	C_9 Technology integration (10%)	I_{11} New patents (100%)	

The assigned weights (λ_{R1} = 50%; λ_{R2} = 30% and λ_{R3} = 15%) have been derived from experts' seminars. It can be seen that a higher weight to the economic requirement has been established respect to the environmental aspects. In this regard, it should be mentioned that wind farms have a lower environmental impact in terms of energy than electricity-generation technologies based on fossil fuels. Furthermore, the difference between the energy produced and consumed is positive over the tower's entire life. Thus, in this specific case, a lower weight can be assigned to the environmental requirements. This tree can be used to assess the sustainability index score for towers in other scenarios (different system constraints and/or social perceptions) and from the viewpoint of other stakeholders by adjusting the weightings and boundary conditions accordingly.

4.3. Case study: precast concrete tripod
A prototype precast concrete tripod (Spanish patent No. 7,123,455, see Figure 6) is used as an example of sustainability assessment of wind towers. This support is capable of bearing and resisting the forces

transmitted by large turbines (P ≥ 3.0 MW) installed at height ranging in 100 – 120 m. This structural system consist of a three-legged tower of 20 m – length precast prestressed concrete segments joined in by means of post-tensioned tendons. These three legs are reinforced transversely with double-T structural steel profiles (Figure 6). In Figure 7 the geometric details of the foundation designed for this precast concrete tower are presented.

The data gathered in Table 8 must be considered to assess each of the 11 indicators fixed in the requirements' tree (Table 7). Likewise, the constitutive parameters required to define the specific value functions are presented in Table 9.

Table 8. Values of the main features of the proposed tripod tower

Feature	Value	Unit
Height	100	M
Power output of supported turbine	3.5	MW
Foundation weight	698	t/tower
Tower weight	1,263	t/tower
Construction cost	1,022,000	€/tower
Maintenance cost	6,545	€/tower·year
Deconstruction cost	120,200	€/tower
Energy consumption (LCA)	0.68	MW/tower
CO_2 emissions (LCA)	299	TnCO_2-e/tower

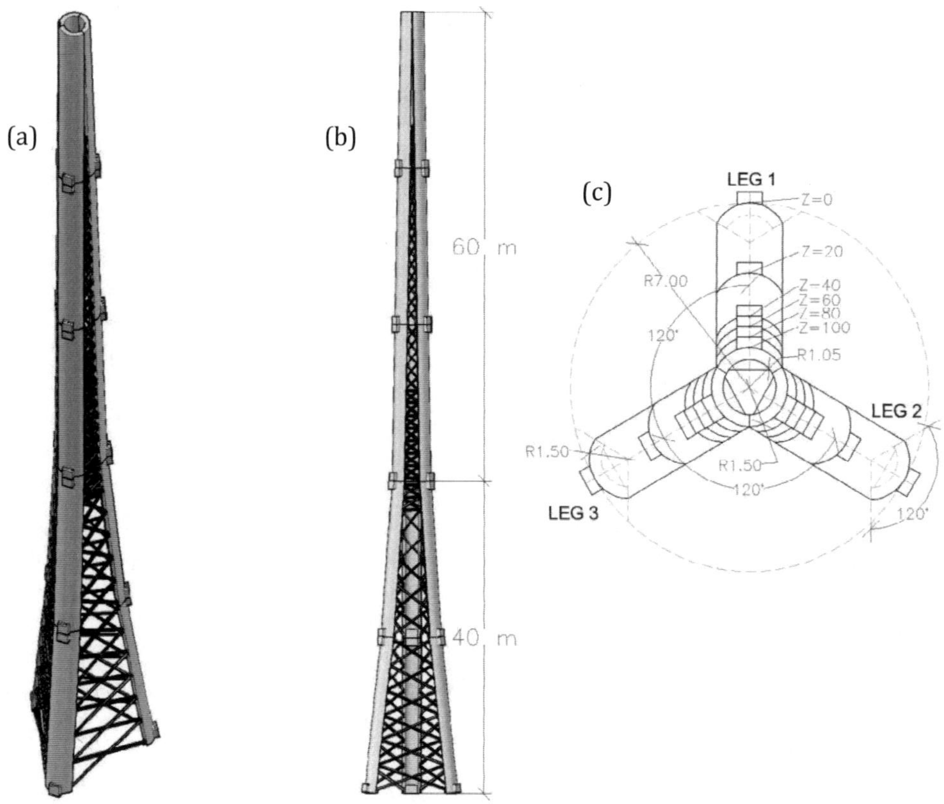

Figure 6. (a) 3D, (b) frontal, and (c) upper views of the precast concrete tripod for wind tower

IOP Conf. Series: Materials Science and Engineering **442** (2018) 012008 doi:10.1088/1757-899X/442/1/012008

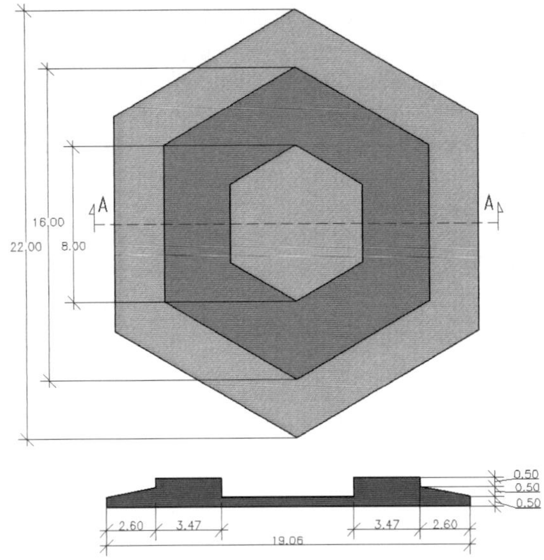

Figure 7. Upper and frontal views of the foundation for the precast concrete tripod for wind tower

Table 10 shows the values for each indicator and requirement for the precast concrete tripod. In this regard, considering the weights established for the three requirements (Table 7), the sustainability index derived from applying the Equation (1) is **SI = 0.68**.

Table 9. Value function parameters for each indicator

Indicator	Unit	x_{max}	x_{min}	C	K	P	Shape
I_1. Direct cost	€/tower	2,000,000	900,000	1,100,000	1.00	2.5	DCv
I_2. Cost deviations	points	90	40	50	1.00	2.5	DCv
I_3. Maintenance work	€/tower·year	10,000	4,000	5,000	0.05	2.5	DCv
I_4. Deconstruction	€/tower	250,000	20,000	60,000	0.05	2.5	DCv
I_5. Material	Tn/MW	2,000	200	500	0.01	1.0	DL
I_6. Energy	GWh/MW	1.5	0	0.75	1.00	1.0	DCx
I_7. Emissions	ton CO_2-e/MW	1,500	0	750	1.00	1.0	DCx
I_8. Occupational hazards	points	2.5	1.5	2.5	0.01	3.0	DCv
I_9. Proportions	points	100	0	100	0.01	1.0	DL
I_{10}. Flexibility	points	100	0	100	0.01	1.0	DL
I_{11}. New patents	points	1	0	1	0.01	1.0	DCx

Table 10. Satisfaction values for each indicator and requirement

Indicator	R_1	I_1	I_2	I_3	I_4	R_2	I_5	I_6	I_7	R_3	I_8	I_9	I_{10}	I_{11}	**Total**
Index V_i	0.57	0.83	1.00	0.33	0.38	0.86	0.80	0.86	0.98	0.64	0.31	0.90	0.60	1.00	0.68

Had other stakeholders' preferences wanted to be taken into account, the requirement weights could be calibrated according the situation. For instance, on the one hand, in case of a private owner (or investor) or in a general economic recession panorama, the economic requirement would have higher relative importance. A possible weights' distribution would be λ_{R1} = 75%, λ_{R2} = 15%, λ_{R3} = 15%, being the final SI = 0.65. On the other hand, in case of economic goodness and/or a country with strong environmental and social sensitivity, a potential weights' distribution would be λ_{R1} = 35%, λ_{R2} = 45%, λ_{R3} = 20%. In this case, the total SI = 0.71.

In view of the results, it is evident that the model allows the decision-maker contemplate different scenarios taking into account different preferences. This model can be applied independently of the structural material of the tower, height, turbine power and transport distance. So, it can be stated that this MIVES model is general for assessing the sustainability of wind towers.

5. Conclusions

The MIVES model has been presented in this paper as a potential tool to assess the sustainability of concrete products and systems. The model allows taking into account the three pillars of the sustainability objectively by means of the satisfaction function concept.

Case studies within the field of fibre reinforced concrete tunnel linings and precast concrete wind turbine supports were included aiming at emphasizing the versatility and applicability of the model to deal with the sustainability analyses of concrete structures.

References

[1] B. Alarcón, A. Aguado, R. Manga y A. Josa, A. Value Function for Assessing Sustainability:
[2] T. Saaty. How to make a decision: The analytic hierarchy process. Eur J Oper Res, vol. 48, pp. 9-26, 1990.
[3] T. Saaty. Fundamentals of decision making and priority theory with the Analytic Hierarchy Process. Vol. VI, Pittsburgh: RWS Publications, 2006.
[4] E. Kazimieras, J. Antucheviciene, Z. Turskis y H. Adeli. Hybrid multiple-criteria decision-making methods: A review of applications in engineering. Sci Iran, vol. 23, nº 1, pp. 1-20, 2016.
[5] J. San-Jose y I. Garrucho. A system approach to the environmental analysis industry of buildings. Build Environ, vol. 45, nº 3, pp. 673-683, 2010.
[6] J. San-Jose y J. Cuadrado. Industrial building design stage based on a system approach to their environmental sustainability. Constr Build Mater, vol. 24, nº 4, pp. 438-447, 2010.
[7] J. Reyes, J. San-Jose, J. Cuadrado y R. Sancibrian. Health & Safety criteria for determining the value of sustainable construction projects. Safety Sci, nº 62, pp. 221-232, 2014. J. San-José,
[8] R. Losada, J. Cuadrado y I. Garrucho. Approach to the quantification of the sustainable value in industrial buildings. Build Environ, vol. 42, nº 11, pp. 3916-3923, 2007. G. Ormazabal, B.
[9] Viñolas y A. Aguado. Enhancing value in crucial decisions: Line 9 of the Barcelona Subway. Manage Eng J, vol. 24, nº 4, p. 265–272, 2008. 2008.
[10] F. Pardo y A. Aguado. Investment priorities for the management of hydraulic structures. Struct Infrastruct E: Maint Manag Life-Cycl Design and Perform, vol. 11, pp. 1338-1351, 2014.
[11] F. Pardo y A. Aguado. Sustainability as the key to prioritize investments in public infrastructures. Environ Impact Asses. Review. Vol. 60, September 2016, 40-51.
[12] A. de la Fuente, J. Armengou, O. Pons y A. Aguado. Multi-criteria decision-making model for assessing the sustainability index of wind-turbine support systems. Application to a new precast concrete alternative. J Civ Eng Manag, In Press.
[13] A. de la Fuente, O. Pons, A. Josa y A. Aguado. Multi-Criteria Decision Making in the sustainability assessment of sewerage pipe systems. J Clean Prod, vol. 112, nº 5, pp. 4762-4770, 2016.
[14] S. Hosseini, A. de la Fuente y O. Pons. Multi-criteria decision-making method for sustainable site

location of post-disaster temporary housing in urban areas. J Constr Eng M ASCE, In Press.
[15] S. Hosseini y A. de la Fuente. Multi-criteria decision-making method for assessing the sustainability of post-disaster temporary housing units' technologies: A case study in Bam. Sustainable Cities and Society, vol. 20, pp. 38-51, 2016.
[16] O. Pons y A. Aguado. Integrated value model for sustainable assessment applied to technologies used to build schools in Catalonia. Spain. Build Environ, vol. 53, pp. 49-58, 2012.
[17] O. Pons y A. de la Fuente. Integrated sustainability assessment method applied to structural concrete columns. Constr Build Mater, vol. 49, p. 882–893, 2013.
[18] CPH 2008. Spanish Structural Concrete Standard. Annex 14: Recommendations for the use of fiber reinforced concrete, Madrid, 2008.
[19] Aguado, A. del Caño, P. de la Cruz, P. de la Cruz, P. Gómez y A. Josa. Sustainability assessment of concrete structures. The EHE approach Rev., J Constr Eng M ASCE, vol. 138, nº 2, pp. 268-276, 2012. Doi: 10.1061/(ASCE)CO.1943-7862.0000419
[20] A. Caño, D. Gómez y M. de la Cruz. Uncertainty analysis in the sustainable design of concrete structures: A probabilistic method. Constr Build Mater, vol. 37, p. 865–873, 2012.

Life Cycle Assessment Optimization of a Shotcrete Tunnel Lining

P Stepanek, I Lanikova and J Venclovsky

Brno University of Technology, Faculty of Civil Engineering, Czech Republic

E-mail: stepanek.p@fce.vutbr.cz

Abstract. The proposed contribution describes the formulation of the optimization design of a utility tunnel lining. The objective function is defined as a multicriterial problem and includes economic and environmental aspects associated not only with concrete member production but also with maintenance during utilization (life cycle assessment). The constraining conditions expressed by relations derived from equations of equilibrium (the solution of optimization calculations for a structure with a Winkler foundation using the Finite Element Method), the reliability conditions of a reinforced concrete structure and the continuity of deformations define the range of allowable solutions from the mathematical viewpoint. The thickness of the tunnel lining and areas of the top and bottom reinforcement in cross-sections are optimized. A method (algorithm) for obtaining a serviceable reinforcement design (via the implementation of reinforcement types) from an optimal solution is described. The obtained results are compared and analysed.

Keywords: optimization, life time assessment, concrete, reinforcement, tunnel lining

1. Introduction

The aim of the "sizing" optimization of reinforced concrete load-bearing structures is a design that considers the cross-sectional dimensions of individual parts of the structure, as well as its reinforcement (i.e. optimized variables). The objective function in optimization tasks (dependent on these variables) usually expresses the cost of materials or cost of the whole building structure (i.e including the cost of materials, formwork, fabrication, transportation, substructure, cladding, and erection, respectively). It means that typically one-criterion optimization task is solved to minimize construction costs [1][2][3][4][5][6]. However, nowadays it is also necessary to assess the quality of the design, in addition to the price, with regard to other aspects such as sustainability (from the environmental viewpoint), constructability, structural stiffness, etc. Therefore it is possible to solve each optimization task separately [7] and to consequently compare the obtained solutions manually (according to designer's opinion), or - and i tis better variant - to formulate and solve multicriterial optimization task and gain the Pareto set of solutions. In addition, multi-objective optimization algorithms offer a range of possibilities for the study of distinct objectives; see [8][9][10].

The former tasks of optimization of reinforced concrete structures concerned in particular to the optimization of sub-elements, i.e. beams (simply supported or continuous), columns, floor slabs, and only few publications dealt with a optimization of frame structures [1][2]. The scope of tasks, i.e. dimensions of optimized variables space, was limited by used optimization algorithm (mostly classical exact methods or iterative gradient based algorithms). With the development of metaheuristic algorithms (since the 70s of the last century), which has also been reflected in the field of engineering tasks, the scope of solved optimization tasks has expanded. These methods are often based on concept found in nature and they are characterized by the search techniques for the valid solution in design space. In particular, genetic algorithms [3][4][5][6], simulated annealing method [8][9], colliding bodies optimization [10], big bang-big crunch approach [7], harmony search [11], etc. were used to optimize framework structures.

Our research is focused on the complex evaluation of structures according to Integrated Life Cycle Assessment (ILCA) [12], which represents the multi-parametric evaluation of a structure over its whole life cycle. It integrates the main aspects of sustainability such as the total environmental impact, the total cost and the total sociocultural quality. The determination of the cost is easy, whereas the calculation of the environmental impact is complicated. Commonly, we have to take into account the following aspects: primary energy consumption (PE), global warming potential (GWP), acidification potential (AP), photochemical ozone creation potential (POCP), consumption of primary raw materials, water consumption, the amount of recyclable (reusable) materials and/or members (at the end of the lifetime of the analysed structure), and the amount of un-recyclable/un-reusable materials/members (amount of waste). The sociocultural quality is often neglected.

In order for the optimization task to be applicable not only to rectangular frames (what is usual) but also to frames which are irregular, the cross-sections for ULS assessment will need to be defined by a knowledgeable user when entering the data for a task. Therefore, the assessed cross-sections will not be automatically determined only for the frame span and joints, as is practised by authors of similar research works. The user defines the cross-sections for the individual members of the analysed structures; based on these, the structures will be discretized into finite elements. The proposed process can achieve suitable reinforcement along the length of individual members of the frame structure, i.e. it is possible to determine the lengths of the reinforced bars so that the reliability conditions are fulfilled even in places where the reinforcement area has changed, and not only in cross-sections with extreme moments.

The response structure model (which uses the FEM) considers the influence of physical non-linearity caused by the non-linear stress-strain diagram of concrete and reinforcement and by cracking in the tensile zone of concrete with consideration of the tension stiffening effect. Above mentioned approach allows the checking of deflection via calculation (not only by limiting the span/effective depth ratio [2][5][7]).

2. Formulation of optimization

In the investigated case, the objective function describes the cost and environmental impacts expressed in terms of the first four above-mentioned aspects, which are assessed for the whole lifetime of the structure. The problem is a multicriterial one, and all the aspects should be minimized. The method of weighted sums was applied. The terms of the objective function have different units and thus must be normalized by the chosen reference values. The objective function $f(\boldsymbol{x})$ can be expressed in the form

$$f(\boldsymbol{x}) = w_{\mathrm{GWP}} \frac{GWP_{\mathrm{i,tot}}(\boldsymbol{x})}{GWP_{\mathrm{i,tot}}^{\mathrm{best}}} + w_{\mathrm{AP}} \frac{AP_{\mathrm{i,tot}}(\boldsymbol{x})}{AP_{\mathrm{i,tot}}^{\mathrm{best}}} + w_{\mathrm{POCP}} \frac{POCP_{\mathrm{i,tot}}(\boldsymbol{x})}{POCP_{\mathrm{i,tot}}^{\mathrm{best}}} + \\ + w_{\mathrm{PE}} \frac{PE_{tot}(\boldsymbol{x})}{PE_{\mathrm{tot}}^{\mathrm{best}}} + w_{\mathrm{C}} \frac{C_{tot}(\boldsymbol{x})}{C_{\mathrm{tot}}^{\mathrm{best}}} \quad (1)$$

where $\boldsymbol{x}$ is the vector of optimized design variables, $GWP_{\mathrm{i,tot}}$ is the total impact of equivalent global warming potential, $AP_{\mathrm{i,tot}}$ is the total impact of acidification potential, $POCP_{\mathrm{i,tot}}$ is the total impact of photochemical ozone creation potential, PE_{tot} is the total energy consumption and C_{tot} is the total acquisition costs for the whole lifetime of the optimized structure; w_{GWP}, w_{AP}, w_{POCP}, w_{PE} and w_{C}

IOP Conf. Series: Materials Science and Engineering **442** (2018) 012009 doi:10.1088/1757-899X/442/1/012009

are weighting coefficients in the objective function. The quantity $GWP_{\text{i,tot}}^{\text{best}}$ is the impact of GWP related to a structure which has minimized GWP, i.e. the value of the monocriterial objective function:

$$f(\boldsymbol{x}) = GWP_{\text{i,tot}}(\boldsymbol{x}). \tag{2}$$

Similarly, $AP_{\text{i,tot}}^{\text{best}}$, $POCP_{\text{i,tot}}^{\text{best}}$, $PE_{\text{tot}}^{\text{best}}$ and $C_{\text{tot}}^{\text{best}}$ are the results of monocriterial optimizations with corresponding objectives. These quantities represent reference values of individual terms of objective function (1). All quantities are calculated for the whole lifetime defined by the problem designer. The other environmental aspects are not included in the objective function.

The range of allowable solutions is bounded via constraining conditions. A reinforced concrete structure must be safe and reliable, i.e. it has to fulfil the following requirements:

- reliability conditions defined in the relevant standards with regard to ultimate limit states (ULS) and serviceability limit states (SLS);
- equations of equilibrium (the solution of optimization calculations for a structure with a Winkler foundation using the Finite Element Method);
- structural provisions according to the standards;
- other requirements and limits set by the user.

ULS reliability conditions will be evaluated for a cross-section stressed with a combination of bending moment and normal force. During the calculation of deformations (i.e. displacements and rotations) of the structure for the evaluation of SLS reliability conditions, physical nonlinearity is considered (i.e. the weakening of the cross section due to the effect of cracks occurring in concrete). Constraints are more precisely described in [13].

The optimization task itself is nonlinear and we use the CONOPT solver [14], which is provided in GAMS modelling software. The task is composed of many situations where the usage of non-smooth functions is necessary (e.g. the approximation of stress-strain diagrams, and the calculation of stiffness for cross-sections, including the impact of cracks in the concrete). However, the usage of such functions is precluded by the reduced gradient algorithm. Such situations are then resolved via Hermit´s interpolation of these functions in their non-smooth parts. This is achieved in the Pascal programming language with the aid of a DLL library. The solver then calls the function values and their derivations with the aid of so-called external equations implemented in an algorithm written in the GAMS language.

However, the Winkler soil model only represents a one-way binding between the soil and the mentioned frame. It is 'one-way' in the sense that when the frame deflects into the soil, it causes a reaction, while when the frame deflects away from the soil, the reaction is non-existent. This problem is solved by repeating the optimization cycle in such a way that at first the deflections are calculated and then, according to these deflections, the reactions / non-reactions with the soil are set for the respective frame elements (part of structure).

3. Optimization of the tunnel lining

3.1. Description of the solved structure.

Our optimization problem concerns a utility tunnel, the original design geometry of which is shown in figure 1. To obtain an optimized design (defined by the objective function), a 1 m long part of the tunnel structure is considered. This part of the structure (frame) is discretized on finite elements (figure 1). The bottom of the tunnel (i.e. elements e_{21} to e_{24}) is not subject to optimization and its thickness is a constant 600 mm, as shown in figure 1.

As regards concrete, shotcrete corresponding to concrete with a strength class of C25/30 will be used, as well as B500B steel.

3.1.1. Optimized variables. The optimized design variables are the height $h^{(e)}$ of the individual rectangular cross sections (i.e. thickness of the tunnel lining), which are predefined for each of the finite elements e, and their reinforcement, which are given by the area of the reinforcement placed at the top and bottom surface ($A_{s,1}^{(e)}$ and $A_{s,2}^{(e)}$), and that of the corresponding transverse reinforcement ($A_{sr,1}^{(e)}$ and $A_{sr,2}^{(e)}$): see figure 2. It is assumed that the optimized variables are constant along the length of the finite element.

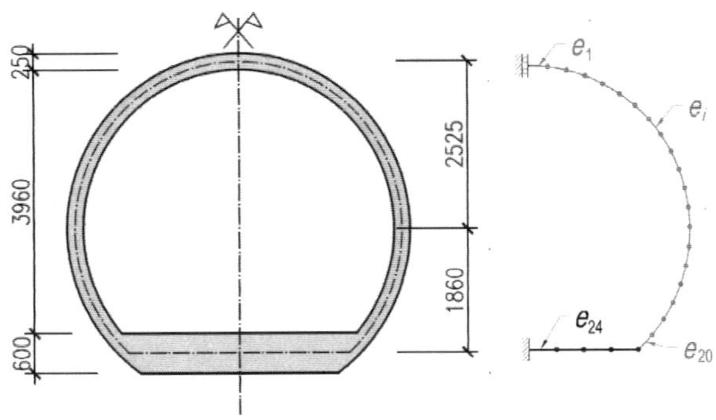

Figure 1. The shape of the tunnel lining and finite element discretization.

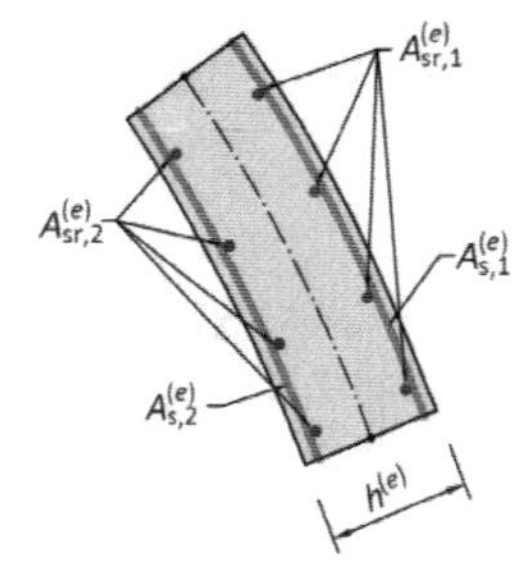

Figure 2. Optimized design variables on element *e*.

Using the optimization algorithm, different types of reinforcement can be designed for each finite element. In practice, however, such a solution is not feasible. Therefore, the next step will be to introduce reinforcement types, which is equivalent to dividing the set *E* of elements *e* that form the frame into a prescribed number *n* of subsets E_{1i}, $i = 1, ..., n$, where in each subset the value of A_{s1} must be the same (these subsets must contain consequent elements):

$$A_{s1}(e_j) = A_{s1}(e_k) \ \ \forall e_j, e_k \in E_{1i}, \ i = 1, \dots, n, \tag{3}$$

$$\bigcup_{i=1}^{n} E_{1i} = E. \tag{4}$$

This occurs analogically for A_{s2} and subsets E_{2i}, $i = 1, ..., n$.

The division into subsets will be realized via the simple 'take-the-best' method. The starting point of this method can be defined with the assumption that each subset will contain exactly one finite element. Then, every possible way of joining two consecutive subsets in both the top and the bottom reinforcement (since they are dependent) will be calculated and the one with the best objective function value will be chosen as the next step. This will be repeated until the number of subsets for both top and bottom reinforcement is reduced to the predefined value.

The anchorage length will be added to each reinforcement type. For the solution to be viable, some assumptions must be made:

- individual bars are placed at a distance of 150 mm from each other, enabling the reinforcement profile to be derived from areas A_{s1} and A_{s2};
- the anchorage length is equal to the reinforcement profile multiplied by 30 for steel reinforcement (bonding condition);
- the anchorage length is added twice to each reinforcement bar (to the beginning and end of the bar), except for the first reinforcement type, which includes element e_1, to which the anchorage length is added only once (as it is next to the axis of symmetry of the tunnel).

The amount of material used for anchoring is then added to the overall weight of reinforcement in the objective function.

3.1.2. The load. The load on the tunnel is caused by pressure from the surrounding soil and underground water. Due to complicated geo-mechanical properties, it is necessary to consider two limit values for the at-rest lateral earth pressure coefficient K_0. The first, $K_0 = 0.593$, is derived from Jaky's equation, while the second, $K_0 = 1.25$, is set manually according to the level of soil solidification (typical for neogenous clay in the Brno area, where the tunnel is situated).

The vectors of load in the equilibrium conditions are compiled for combinations of load cases (not for individual load cases) due to nonlinearity (cracking, Winkler soil model). It is predicted that the self-weight of the frame, the soil pressure and the water pressure act together, therefore the number of possible combinations for the ULS was reduced to 8. The equilibrium conditions for the SLS are solved for two load combinations (because of the two values of the coefficient K_0).

3.1.3. Life cycle assessment. The lifetime assessment of the structure was formulated as follows:

(a) The construction process phase includes the exploitation of raw materials, transport, and manufacturing techniques used to produce the materials and structure. The resulting environmental impacts and unit costs are shown in table 1; the method of their calculation and the input values are described in [15][16] and GEMIS.

(b) Utilisation and operation phase - we assume the lifetime, maintenance and repair work described in table 2 will occur. The unit costs and environmental impacts of this phase are shown in table 1.

(c) The end of life phase (i.e. demolition) – this phase was not considered.

Table 1. Unit costs and environmental impacts.

Phase	Material, activity	Unit	Energy [MJ/unit]	GWP_{equiv} [kg/unit]	AP_{equiv} [g/unit]	$POCP_{equiv}$ [g/unit]	Unit cost [€/unit]
(a)	Concrete C25/30	m^3	2966.38	362.29	1104.61	40.24	370.37
	Steel rebar	kg	27.70	2.537	15.57	0.6242	1.11
(b)	Concrete surface repair type 1[a]	$cm \cdot m^2$	35.59	4.34	13.25	0.4829	8.70
	Concrete surface repair type 2[b]	$cm \cdot m^2$	41.52	5.07	15.46	0.5634	6.11
	Replacement of reinforcement	kg	30.47	2.79	17.13	0.6867	1.85

[a] Thickness of repaired layer 30 mm.
[b] Thickness of repaired layer 70 mm.

Table 2. Maintenance and repair work conducted on inner tunnel surface.

Maintenance and repair work after:	20 years	40 years	60 years
Concrete surface repair type 1 (thickness 30 mm)	20.0 %	25.0 %	30.0 %
Concrete surface repair type 2 (thickness 70 mm)	5.0 %	7.5 %	10.0 %
Replacement of inner reinforcement	3.0 %	5.0 %	5.0 %

3.2. Optimization

3.2.1. Solved variants and requirements. A case study was carried out for two tunnel lining optimization variants. Because the structure is fabricated from shotcrete, it is possible to perform calculations for a tunnel lining of variable thickness, i.e. the optimized variable h for each element was left independent. Optimization is also performed for the lining with a constant thickness, i.e. optimized values h were bound by the following constraining condition:

$$h(e_i) = h(e_{i+1}),\ i = 1, \dots, 19. \tag{5}$$

A maximum of two types of reinforcement ($n = 2$) were used around the perimeter of the tunnel lining independently for the inner and outer surfaces of the lining.

3.2.2. Monocriterial optimization. Prior to performing the multicriterial optimization, equation (1), values were required for X^{best} (where $X = GWP_{i,tot}$, $AP_{i,tot}$, $POCP_{i,tot}$, PE_{tot} or C_{tot}). Therefore, the optimization of the structure was performed according to each sub-criterion separately, e.g. for the GWP criterion, the optimization was performed to minimize the objective function (2), the resulting optimal solution being labelled O_GWP and the value of the objective function $GWP_{i,tot}^{best}$, etc. This produced five optimal solutions according to the individual criteria (O_GWP, O_AP, O_POCP, O_PE, O_C) and values of objective functions $GWP_{i,tot}^{best}$, $AP_{i,tot}^{best}$, $POCP_{i,tot}^{best}$, PE_{tot}^{best} and C_{tot}^{best}; these are listed and displayed in red in the graph in figure 3. For the purpose of comparison, all observed criteria for all solutions (shown in blue) were then calculated.

The volume of concrete and the weight of the reinforcement of the resulting solutions are shown in table 3.

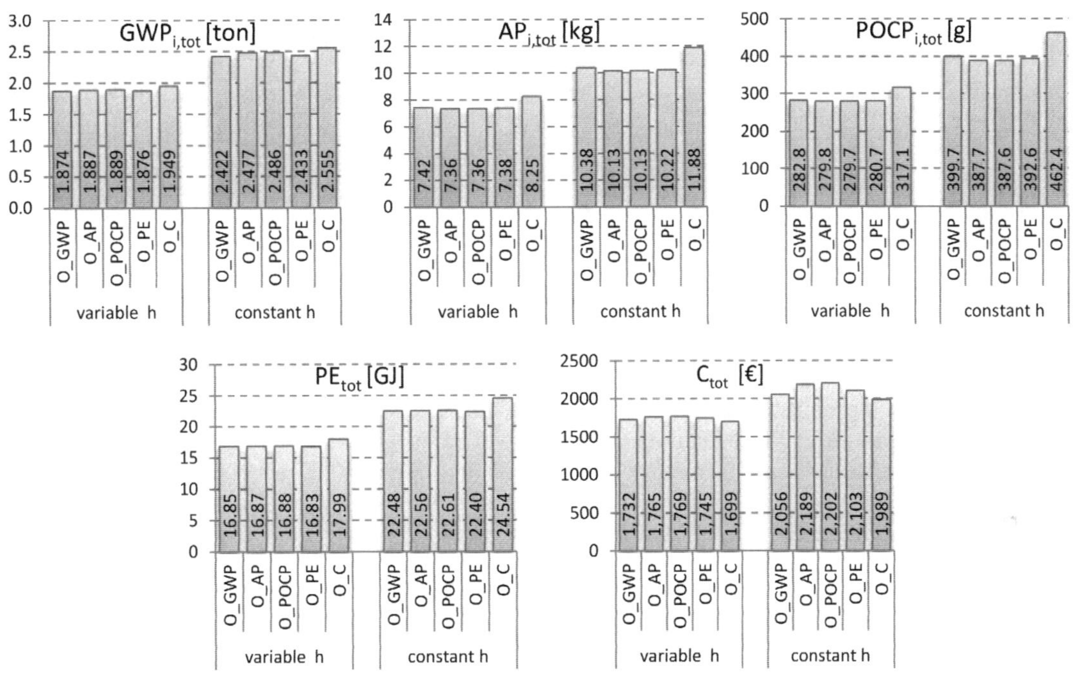

Figure 3. Amount of environmental impact, energy consumption and cost of optimal solution carried out for sub-criteria.

Table 3. The resulting volume of concrete [m^3] and amount of reinforcement [kg] in the construction process phase (monocriterial optimization)

Variant	Variable *h*		Constant *h*	
Material	Concrete [m3]	Steel [kg]	Concrete [m3]	Steel [kg]
O_GWP	3.050	207.0	3.418	360.8
O_AP	3.178	194.1	3.942	310.6
O_POCP	3.192	193.2	3.986	307.6
O_PE	3.107	200.0	3.620	337.7
O_C	2.727	278.4	2.808	492.7

From the analysis, during which optimization was carried out for the sub-criteria of the objective function, the following can be concluded:

- If the optimization criterion was cost, a thinner lining was designed using a smaller volume of concrete ($h_{O\text{-}C}$ = 232 mm for constant lining thickness, $h_{O\text{-}C}$ = from 150 to 357 mm for variable lining thickness) and a larger amount of reinforcement in comparison to the optimizations in which any of the investigated environmental impacts were the criteria.
- The greatest lining thickness and the smallest amount of reinforcement was designed for POCP minimization ($h_{O\text{-}POCP}$ = 329 mm for constant lining thickness, $h_{O\text{-}POCP}$ = from 150 to 433 mm for variable lining thickness). In other words, a thicker tunnel lining and a corresponding lower quantity of reinforcement is better from the environmental point of view, but is more expensive.

3.2.3. Multicriterial optimization. To obtain a Pareto optimal solution, it is necessary to set the weighting coefficient values so that their sum is equal to 1. By choosing their value, the user determines the importance of the accessed criteria according to his or her discretion. The optimization task was solved for the three cases of weighting coefficients presented in table 4, for which the ratios of environmental criteria (as a whole) to cost are set at 4:1, 1:1 and 1:4.

Table 4. Weighting coefficients.

Labelling of optimization results	w_{GWP}	w_{AP}	w_{POCP}	w_{PE}	w_{C}
O_W1	0.20	0.20	0.20	0.20	0.20
O_W2	0.125	0.125	0.125	0.125	0.50
O_W3	0.05	0.05	0.05	0.05	0.80

The values of the objective functions of the resulting solutions are summarized in table 5. The resulting thickness of lining and reinforcement is displayed in figure 5 (only O_W1 and O_W3); the volume of concrete and weight of reinforcement are shown in table 6. The results for environmental impacts, energy consumption and the cost of optimized solutions for both lining thickness variants (constant and variable) are displayed in the graphs in figure 4.

Table 5. Values of objective function

Variant	Variable *h*	Constant *h*
O_W1	1.0069	1.0165
O_W2	1.0123	1.0240
O_W3	1.0111	1.0185

Table 6. The resulting volume of concrete [m^3] and amount of reinforcement [kg] in the construction process phase (multicriterial optimization)

Variant	Variable *h*		Constant *h*	
Material	Concrete [m3]	Steel [kg]	Concrete [m3]	Steel [kg]
O_W1	3.083	202.5	3.555	344.5
O_W2	3.009	213.4	3.370	368.1
O_W3	2.884	237.4	3.060	426.0

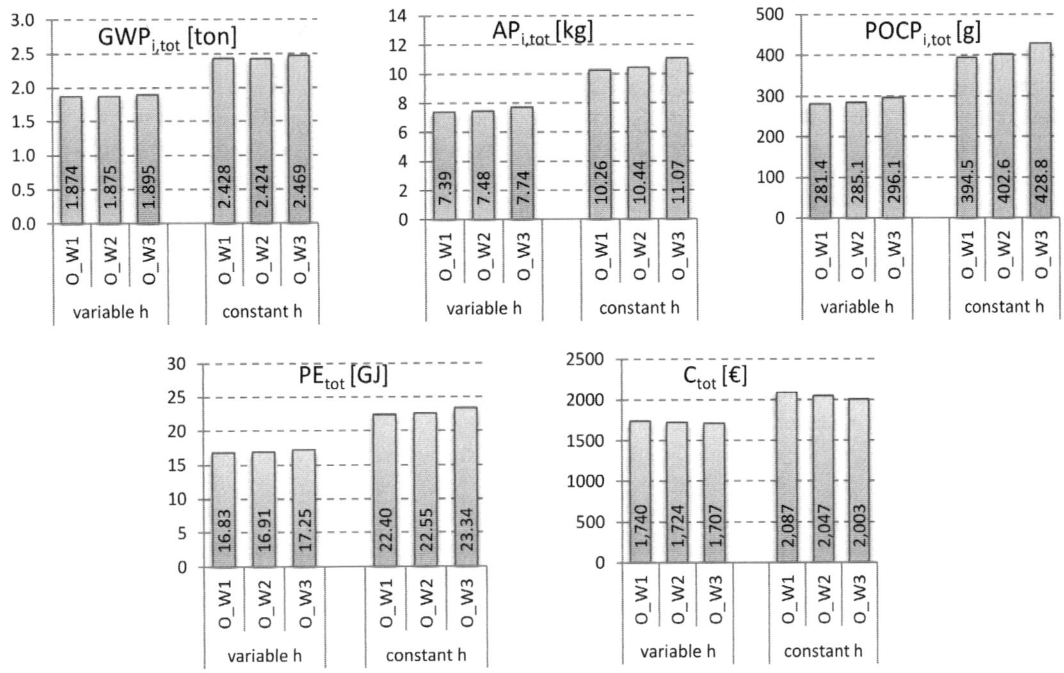

Figure 4. Amount of environmental impact, energy consumption and cost of optimal solution carried out for multiple-criteria.

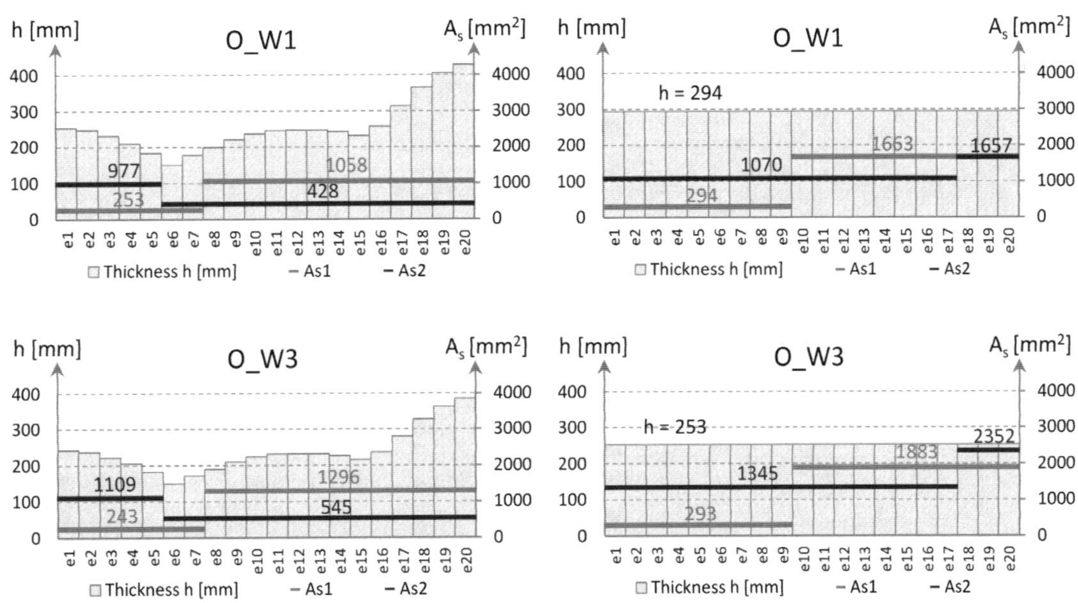

Figure 5. Thickness of lining and reinforcement of solved variants.

In the case of a lining with a constant thickness, there are greater differences in the results depending on the choice of weighted coefficients. A greater weighted coefficient lowers the value of the relevant criterion, and vice versa. In the case of tunnel lining variants with variable lining thickness, the designs for all three cases of weighted coefficients are less different.

It is clear from table 6 that the thickness of a lining, and its strengthening using steel reinforcement, are dependent on the choice of weighting coefficients in the objective function (weighting coefficients for the environmental aspects are selected in contrast with the price, see table 4, just as the results of monocriterial optimization show).

4. Conclusion

A study was conducted concerning the optimization of a structure according to life cycle environmental criteria and life cycle costs. In the investigated case, lowering the volume of concrete and raising the amount of reinforcement led to decreasing costs and increasing environmental burdens, i.e. environmental impact criteria are diametrically opposed to the criterion of cost.

This method has also been applied to the optimized design of load-bearing structures, e.g. frames or flat structures. The described mathematical algorithm is sufficiently general to allow the use of the application when designing other engineering systems.

Acknowledgements

This paper has been produced under project No. LO1408 "AdMaS UP - Advanced Materials, Structures and Technologies", supported by the Ministry of Education, Youth and Sports under "National Sustainability Program I". Some knowledge were gained within the solution under project No. TA03030851 "Rehabilitation of tunnels - technology, materials and methodology" supported by the Technology Agency of the Czech Republic has been used.

References

[1] Choi C K and Kwak H G 1990 Optimum RC member design with predetermined discrete sections *J. Struct. Eng.-ASCE* 116 2634–55
[2] Spires D and Arora J S 1990 Optimal design of tall RC-framed tube buildings *J. Struct. Eng.- ASCE* 116 877–97
[3] Lee C and Ahn J 2003 Flexural design of reinforced concrete frames by genetic algorithm *J. Struct. Eng.-ASCE* 129 762–74
[4] Camp C V and Huq F 2013 CO2 and cost optimization of reinforced concrete frames using a big bang-big crunch algorithm *Eng. Struct.* 48 363–72
[5] Govindaraj V and Ramasamy J V 2007 Optimum detailed design of reinforced concrete frames using genetic algorithms *Eng. Optimiz.* 39 471–94
[6] Kwak H G and Kim J 2009 An integrated genetic algorithm complemented with direct search for optimum design of RC frames *Comput.-Aided Des.* 41 490–500
[7] Camp C, Pezeshk S and Hansson H 2003 Flexural design of reinforced concrete frames using a genetic algorithm *J. Struct. Eng.-ASCE* 129 105–15
[8] Paya I, Yepes V, González-Vidosa F and Hospitaler A 2008 Multiobjective optimization of concrete frames by simulated annealing *Comput.-Aided Civil Infrastruct. Eng.* 23 596–610
[9] Paya-Zafortaleza I, Yepes V, Hospitaler A and González-Vidosa F 2009 CO_2-optimization of reinforced concrete frames by simulated annealing *Eng. Struct.* 31 1501–08
[10] Kaveh A and Ardalani S 2016 Cost and CO_2 emission optimization of reinforced concrete frames using enhanced colliding bodies algorithm *Asian Journal of Civil Engineering* 17 831–858
[11] Akin A and Saka M P 2014 Harmony search algorithm based optimum detailed design of reinforced concrete plane frames subject to ACI 318-05 provisions *Comput. Struct.* 147 79–95
[12] Hájek P, Fiala C, Jensen B B and Lanikova I 2013 Integrated life cycle assessment of concrete structures, *fib bulletin* 71
[13] Laníková I, Štěpánek P, Venclovský J. 2016 Optimization of a tunnel lining reinforced with FRP *Key Eng. Mat.* 691 148–59
[14] Drud S A 1996 *Conopt: a system for large scale nonlinear optimization* ARKI Consulting and Development A/S Bagsvaerd Denmark

[15] Laníková I, Štěpánek P and Simůnek P 2014 Optimized design of concrete structures considering environmental aspects *Adv. Struct. Eng.* 17 495–511
[16] Hájek P, Fiala C and, Kynčlová M 2011 Life cycle assessments of concrete structures - A step towards environmental savings *Struct. Concr.* 12 13–22

IOP Conf. Series: Materials Science and Engineering **442** (2018) 012010 doi:10.1088/1757-899X/442/1/012010

LCA-Based Comparison of the Environmental Impact of Different Structural Systems

M C Caruso, C Menna, D Asprone, A Prota

Department of Structures for Engineering and Architecture, University of Napoli Federico II

E-mail: mariachiara.caruso@yahoo.it

Abstract. Construction is one of the most impactful industrial sector because of the high consequences it generates on the society, the environment and the economy. The study presented herein aims to define a methodological framework that can be used for construction community stakeholders in order to conduct environmental sustainability comparisons among building systems at the design stage. An application of the methodology is performed by comparing structures having different building materials. Three alternative material options have been investigated: RC, steel, and wood. Each option has been designed to fulfill predefined structural, functional, and architectural requirements. Later, the environmental impacts of the structures have been assessed according to the four steps of the life-cycle assessment procedure (ISO 14040) and considering the four phases of a building life: extraction and processing of raw materials; construction; operation; end of life. LCA study is conducted for the three alternative structures with the help of SimaPro software, using both IMPACT2002+ and EPD2008 methodologies to quantify environmental impacts.

Keywords: Life Cycle Assessment methodology; reinforced concrete structure; steel structure; wood structure; construction industry sustainability.

1. Introduction

It is widely recognized that the construction and related industries have a significant global impact on the environment. Being a strategic economic sector, these industries attracts growing attention in terms of sustainability. Many assessment tools and new methodologies have been progressively developed to drive decision-making processes in the direction of achieving sustainability goals, focusing the attention on environmental issues. Life-cycle assessment (LCA) (ISO 14040, 2006) represents one of the most grounded and widespread methodology, which has the potential to analyse overall environmental factors related to the entire life-cycle of goods and processes. One of the main advantages of an LCA is that it quantifies the impacts on the environment not limited to energy or CO_2 emissions, but also use of resources, emissions and ionizing radiation. The use of the LCA in the environmental impact assessment in construction industry could be adopted for the selection of construction products and procedures. Although an LCA study of the entire building would seem to represent a system too complex to be assessed, by using a rigorous framework for life-cycle analysis, it could support the decision-making process, contributing to lower the environmental impacts. Literature offers several works dealing with the environmental assessment of buildings: most of the comparative studies, as (Peuportier, 2001), (Xing et al, 2008), (Menna et al, 2013) and (Asdrubali et al, 2013), focus on the life cycle assessment of buildings having different features, as volumes and functionality, and the "reference" unit is sometimes the building net area or volume unit. However, the required building performances should be explicitly set as basic parameters when defining the system boundary of an environmental sustainability evaluation by means of life cycle assessment. According to this, the paper presents a general framework that provides a set of requirements preliminarily defined for a comparative life-cycle assessment, and a case study where the approach is applied.

2. Proposed methodology

The main parameter driving the designer during the design phase of a buildings is its structural performance. This primary performance requirement is strictly related to many other initial choices made by a customer with reference to the final building/structure, for example, location, use, number of stories, available resources, and functional systems. Given this consideration, the present study proposes an approach for a sustainability comparative assessment among buildings, summarised in Figure 1, which defines a set of "building system requirements" for the building element/system, taking in count functional, architectural, structural, and economic performances, as well as other factors (see Figure 1, steps 1-2). The requirements are interconnected because one specific requirement can affect the others. For instance, the definition of the use of a structure/building (residential, office, strategic infra-structure etc.) primarily affects the architectural, structural, and economic features linked to the decision making at the design phase. Moreover, some of the requirements of Figure 1 (steps 1-2) are site-specific, and depend on the climatic zone, hazards, and local constraints related to the building site.

The proposed framework allows the comparison of different design options, at design stage, leading to a final comparison at the level of the construction product system. In this way, the identification of the contribution of one or more sub-assembly options in the environmental impacts of the final product is possible.

These requirements first depend on building use and location in terms of live loads, hazards, and environmental condition. Starting from these requirements, the feasible options are identified, which take into consideration several additional constraints, such as common and local construction techniques and materials, overall cost, and national standards establishing minimum performances (Figure 1, step 3). After the design parameters are computed and the considered options are designed, the sustainability assessment is performed by means of the common sustainability tools (Figure 1, step 4), allowing the examination of all the desired sustainability aspects.

According to this methodology, the comparative sustainability assessment among building system can be considered scientifically and technically reliable.

For this study, only environmental aspects of the sustainability assessment is considered (step 4). Environmental sustainability can be examined by a variety of assessment tools. LCA represents one of the most commonly used and most valuable; to calculate life-cycle environmental impacts, the LCA process is conducted by following four steps according to ISO 14040 (ISO 14040, 2006) and ISO 14044 (ISO 14044, 2006): (1) goal and scope definition; (2) life-cycle inventory (LCI) analysis; (3) life-cycle impact assessment (LCIA); and (4) results.

The general standard for an LCA is then applied to the "building system", for which a generic system boundary has to be set to properly compute the environmental impact related to the entire building life-cycle. EN 15978 (EN 15978:2011) provides indication to the definition of the system boundary for performing LCA applications in construction works.

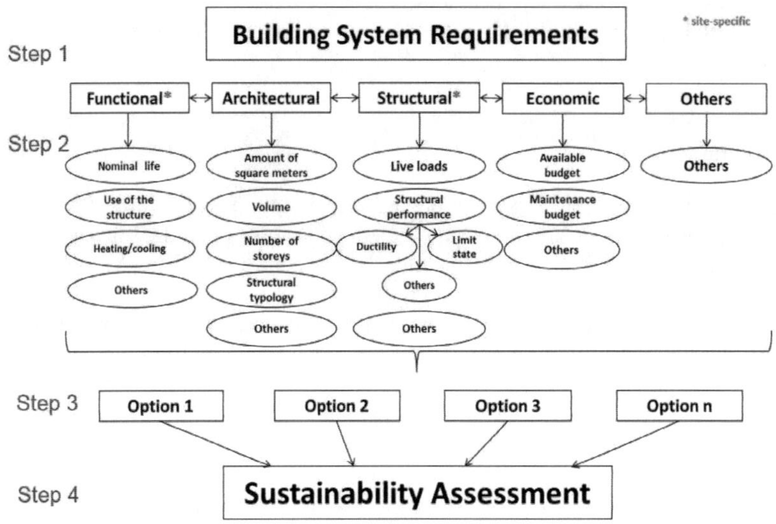

Figure 1: Proposed approach to building structures sustainability

3. Case study

The above-described methodology is applied to a case study, performing the sustainability assessment of three structural options for residential building. In particular, the study aims to the comparison of the environmental impact of materials and processes related to the complete life-cycle of a reinforced concrete (RC), steel, and wood structure, ideally located in the municipality of Rome, Italy. The LCA is only applied to the structural frame of the building because it is assumed that non-structural elements and systems do not vary within the different options. Figure 2 shows the application of the methodology for the case study. Hereafter, architectural, functional, and structural requirements, equal for the three construction options, are discussed.

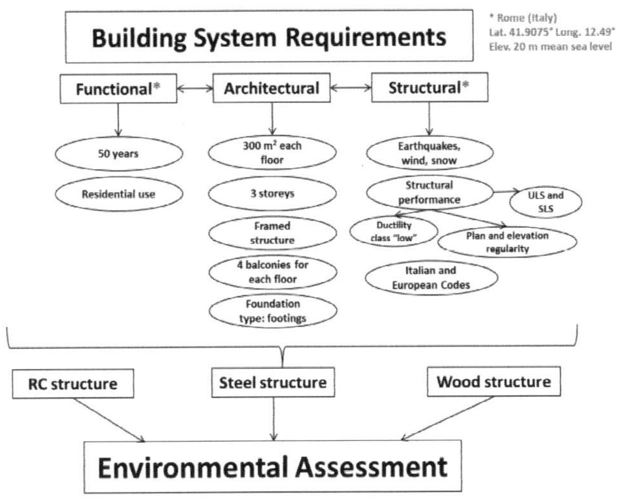

Figure 2: Proposed approach applied to the case study

3.1 Building system requirements

Three sub-categories of building system requirements are considered for the case study (Figure 2): architectural, functional and structural. Architectural requirements include: structural plan dimensions, equal to 12x25 m^2 (300 m^2 each floor); structural typology, framed structure composed by three 4 m spans in the y direction and five variable-dimension (4.7 m – 5.5 m – 4.6 m – 5.5 m – 4.7 m) spans in the x direction; number of storeys, equal to three; number of apartments, equal to two for each floor, having 2 balconies each. Functional requirements include: the use, which is residential building; and the nominal life, equal to 50 years. Structural requirements include: loads (dead loads, snow, wind, seismic actions, and live loads); structural codes (NTC 2008 (D.M. 14.01.2008) and the related Circolare n°667/2009(Ministerial Circular n.617, 02.02.2009), Eurocode 3 (EN 1993, 2005) for the steel structure and CNR-DT 206/2007 (CNR-DT 206/2007) for the wood structure); analysed limit states (ultimate limit state, ULS, and service limit state, SLS).

3.2 Design and verification of structures

The design of the RC, steel, and wood structures have been fulfilled by means of the common design software Edilus v.26.00, developed by ACCA Software S.p.A (User Manual, 2013), which is helpful for design and verification of new and existing structures according to a Eurocode-like approach (CEN (2004) Eurocode 8). The design phase of the three considered structures consists of the definition of the geometry and mechanical properties of the structural members belonging to the buildings. In detail, the RC structure consists of C25/30 concrete and B450C reinforcing steel classes, with a cross section of 30x50 cm^2 for the beams and columns and 35x50 cm^2 for the knee beams of the stairs. The cast-in-situ RC slabs are 22 cm high and the joist beams are oriented in one direction. The steel structure is comprised of S275 steel elements, with IPE270 elements for the principal beams, IPE200 for the secondary beams and HEA320 for the columns and flanged joints between the beams and columns and between the main and secondary beams. The steel-concrete slabs are made of S235 A55/P600 HI-BOND corrugated sheets and 6.5 cm high RC cover slabs. The wood structure is comprised of LL GL32h glued laminated wood class, with a cross section of 16x36 cm^2 for the beams, 20x44 cm^2 for the columns, glued S275 steel bars for the connecting beams and columns, and S275 plates for the connecting columns and the foundation elements. The wood slabs are comprised of 14x28 cm^2 wood joists with 80 cm of axle spacing, 4 cm of concrete slab, and 3.5 cm of wooden floorboards. Figure 3, 4 and 5 show the structural models.

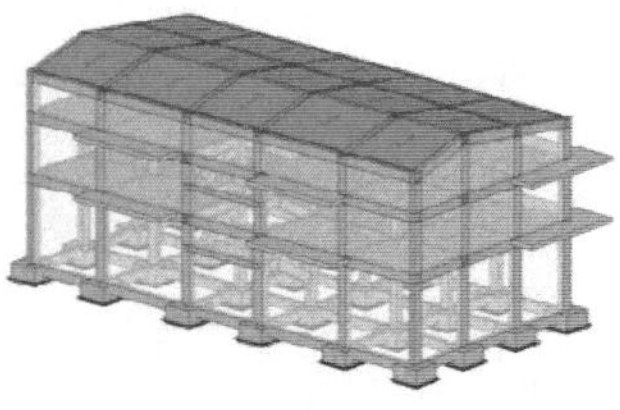

Figure 3: R C structure

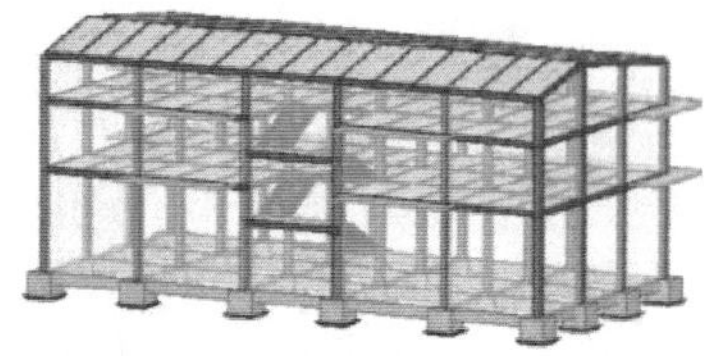

Figure 4: Steel structure

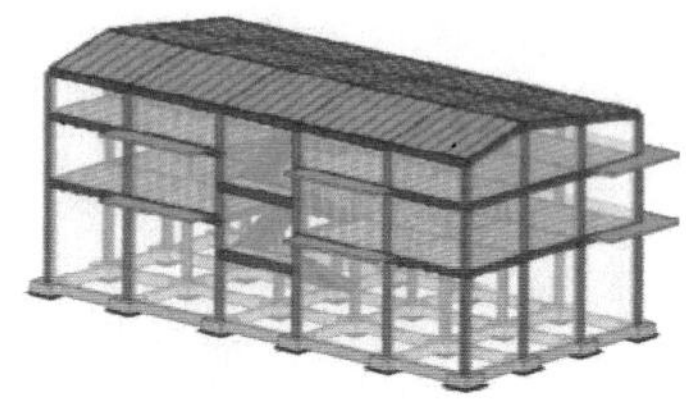

Figure 5: Wood structure

The first three vibration modes of the three structures are listed in Table 1, taking into account the x and y ULS seismic actions that indicate the period, horizontal acceleration (Sa(T)), and participating mass (in kg and in %) of each mode.

Table 1: Dynamic characteristics of the three structures

Mode	Spectrum	Period (T) s	Mode Spectrum Period (T) s	Horizontal spectral acceleration, Sa(T) m/s^2	Participating mass, N·s^2/m=kg	Participating mass, %
RC structure	1	ULS x	0.464	1.454	817784	76.5
	2	ULS y	0.416	1.454	763844	71.5
	3	ULS y	0.065	1.615	138831	13.0
Steel structure	1	ULS y	0.861	0.774	544880	67.1
	2	ULS x	0.755	0.883	496393	61.1
	3	ULS y	0.066	1.600	138186	17.0
Wood structure	1	ULS y	0.619	1.077	275093	52.6
	2	ULS x	0.573	1.164	192176	36.7
	3	ULS y	0.076	1.578	143941	27.5

3.3 Life-Cycle Assessment

The forth and last step of the methodology, set out in Figure 2, is the sustainability assessment of the three building options. To this aim, LCA models have to be built. The entire "building" is chosen as Functional Unit (FU) for this analysis. In particular, the estimated impacts are related to the materials and processes needed to build the structural system. The system boundary is shown in Figure 6, and includes: the pre-use phase, extraction and the production of materials, and construction phases; the use phase, ordinary maintenance of structural elements; and the End-of-Life (EoL) phase, building demolition, and material disposal. Mechanical demolition is chosen in the EoL phase, and it is assumed that the materials are sent for recycling, landfill, and/or to an incinerator according to the percentages declared by national institutions.

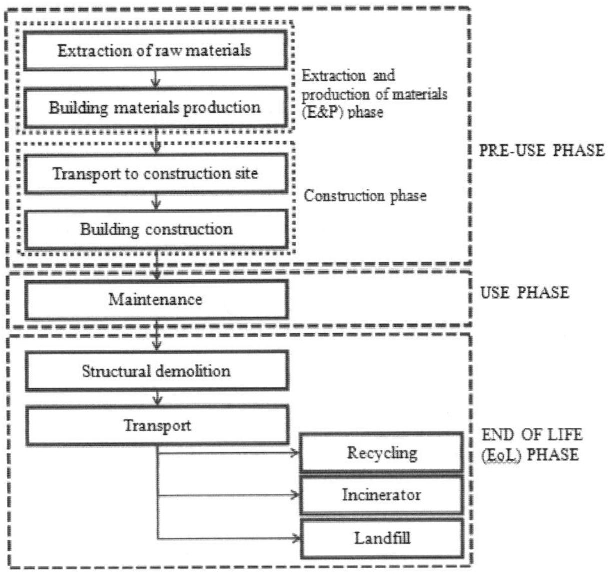

Figure 6: System Boundary of the case study

For each stage of the system boundary, the LCI analysis and LCIA are developed, and performed using the SimaPro 7.3 software (SimaPro 7, 2010), which is an efficient tool, useful to collect sustainability data and to analyse and monitor the sustainability performance of products/services. Moreover, SimaPro software allows the access to international databases and impact assessment methodologies, as Ecoinvent 3.0 (Ecoinvent Centre, 2007), the international database chosen as main source for the life-cycle inventory. Two methodologies for the impact assessment are selected: IMPACT2002+ (Humbert et al, 2012) and EPD (IEC, 2008). The IMPACT 2002+ methodology (Humbert et al, 2012) is used for the LCIA, with impacts evaluated for 15 midpoint categories, grouped into four damage categories: human health, measured in DALY (disability-adjusted life years); ecosystem quality, measured in PDF*m^2*yr (the potentially disappeared fraction of species over a certain amount of m^2 during a certain amount of year); climate changes, measured in kg-equivalents to a reference substance; and resources, measured in MJ (Humbert et al, 2012).

Each phase of the building life-cycle is described in the following section.

3.4 Case study results and discussion

3.4.1 Pre-use phase—extraction and production of materials (E&P phase)

The computation of the amounts of structural materials required for producing the beams, columns, joints, slabs, stairs, foundations, and balconies of each structure is fulfilled at E&P stage. All the processes before the construction phase are also considered, i.e. zinc coating for steel elements and wood treatments. All the materials, except concrete, are modelled according to Ecoinvent database; the data concerning concrete are modified using more refined Italian data collected by detailed environmental product declarations (EPD). According to the Ecoinvent database, structural steel is composed of 37% recycled steel (from energy-optimizing furnaces) and 63% new steel (from basic oxygen furnaces). Table 2 lists the amount of materials and processes for RC Structure, Steel Structure, Wood Structure, as well as the data sources.

Table 2: Amounts of materials and processes and related data sources

Material/Process	UdM	RC Structure	Steel Structure	Wood Structure	Data
Concrete C25/30	m^3	251.01	118.07	63.62	Average data by AITEC
Steel B450C	kg	28186.57	10547.93	4284.56	Reinforcing steel, at plant/RER U (Ecoinvent)
Steel S235-S275-8,8	kg		74265.15	24926.73	Steel, low-alloyed, at plant/RER U (Ecoinvent)
Bricks	kg	85363.2			Brick, at plant/RER U (Ecoinvent)
Glued laminated timber GL32h	m^3			138.69	Glued laminated timber, outdoor use, at plant/RER U (Ecoinvent)
Zinc coating	m^2		830		Powder coating, steel/RER U (Ecoinvent)
Powder coating	m^2		830		Zinc coating, pieces/RER U (Ecoinvent)

During this first stage, LCA results show that the wooden structure provides the highest contribution to ecosystem quality, while the steel option has the greatest impact on human health, climate change, and resources. Focusing on components, the steel material is responsible for the highest impact of a steel structure, while glued laminated timber is responsible for the greatest impact of the wooden structure on ecosystem quality.

3.4.2 Pre-Use Phase – Construction

For modeling the construction phase, transportation of materials from plants to construction sites and all the processes needed to build the structural systems are considered. Concrete and reinforcing steel plants are assumed to be available within 30 km distance from the construction site (Vitale et al, 2014). Steel elements are available from the Riva production plant in Patrica (FR) (87.5 km from Rome) and wooden elements are available from the Rubner Holzbau production plant in Calitri (AV) (249 km from Rome).

The LCA analysis of the Construction Phase leads to worse results for the wooden structure: indeed, such buildings affect the human health, the ecosystem quality, the climate, and resources, mainly due to the transportation phase.

3.4.3 Use Phase

Only ordinary maintenance is considered in the use-phase, being the study focused on the structural part of the building. With regard to the RC structure, steel reinforcement spalling on 5% of the surface of exposed beams and 5% of exposed columns is assumed. The steel elements of the steel structure are assumed to be subjected to zinc (ISO 14713-1:2009) and a powder coating during production to prevent environmental corrosion. Consequently, according to UNI EN ISO 14713 (ISO 14713-1:2009), given the typical consumption of this kind of coating (EPD, 2010), (ISO 12944, 1998) it is assumed that only limited maintenance would be required and, consequently, powder coating on just 20% of the total metallic surface is taken into account. For the glulam structure, a maintenance plan is necessary for steel connections and the glulam elements exposed to UV radiation. The presence of powder coatings on the surfaces of all the steel connections and the annual application of a wood-impregnating solvent with a long oil-high penetration alkyd resin on exposed surfaces of balconies, ridge beams, and columns are therefore assumed.

Such analysis leads to show that RC and steel structures require less maintenance than wood versions. Therefore, the ordinary maintenance of glulam structures has the greatest impact due to the need to frequently apply preservatives. Nevertheless, the impacts related to the use phase are negligible, respect the other phases.

3.4.4 End of Life (EoL) Phase

The EoL phase includes structure demolition and material disposal. Mechanical, conventional demolition is chosen for each structure, and is assumed to be carried out by one excavator with a hydraulic hammer (for foundation demolition) and jaw (for structure demolition and inert crashing), one wheel loader, and 28 m^3 lorries to remove demolition waste from the site. The RC structure demolition would need three 28 m^3 lorries carrying inert waste for 17 journeys and one such lorry carrying steel waste for one journey. The demolition of the steel structure would require two lorries carrying inert waste for 6 journeys and two carrying steel waste for 6 journeys. Finally, demolition of the wood structure would need one lorry carrying inert waste for 3 journeys, one carrying steel waste for 1 journey, and two carrying wood waste for 7 journeys.

After demolition, it is assumed that material separation would be carried out at the construction site and all the materials loaded to lorries and sent to other destinations 30 km away. The separated materials are sent for recycling, landfill, and/or to an incinerator according to national or European reference amounts. The recycled materials are computed in SimaPro as an avoided product. According to ANPAR (ANPAR 2012) and ISPRA data, 65% of inert waste is sent for recycling and 35% for landfill. Recycled inert materials are considered to be avoided gravel. According to ArcelorMittal (Constructalia, website), 65% of reinforcing steel is recycled and 35% is sent to landfill. Given that reinforcing steel is composed of 37% recycled steel, 28% (i.e. 65% minus 37%) of steel leaves the system boundary and is considered to be avoided new steel (from basic oxygen furnaces) (ISO/TR 14049:2000). Also according to ArcelorMittal (Constructalia, website), 98% of steel from steel beams and columns is recycled and 2% is sent for landfill. Given that low-alloyed steel is composed of 37% recycled steel, 61% (i.e. 98% minus 37%) of steel leaves the system boundary and is considered to be avoided new steel (from basic oxygen furnaces). According to TRADA reports (TRADA, 2008), 16% of glued laminated timber is recycled, 4% is sent to an incinerator, and 80% for landfill.

Results on Impact Assessment of the EoL phase show that demolition and transportation phases' contributions are negligible respect to the disposal phase. Steel structure has the lowest impact on all the four impact categories (human health, ecosystem quality, climate change, and resource use), being the recycling steel percentage, which is higher than the percentage of the other materials, the main parameter influencing the results.

3.4.5 Global Impact Assessment

In this section, results regarding the environmental impacts are shown for each life-cycle phase of the three buildings, in terms of midpoint and endpoint categories. For the following figures 7 and 9, for each category, impact values are divided by the maximum value achieved among the three buildings and are plotted in percentage, in order to effectively illustrate the building environmental performance comparison

Figure 7 reports the aggregated results of the LCA analysis over all the phases in terms of the midpoint categories related to the IMPACT 2002+ method. Results in terms of endpoint categories are shown in figure 8, when considering all the life-cycle stages of the buildings. As figures show, the greatest impact on human health is due to steel and wood structures, the highest impact on ecosystem quality and resource consumption is due to wood structure, and the greatest impact on climate change is caused by RC structure. Figure finally shows that RC structure has the least impact on each damage category, except climate change.

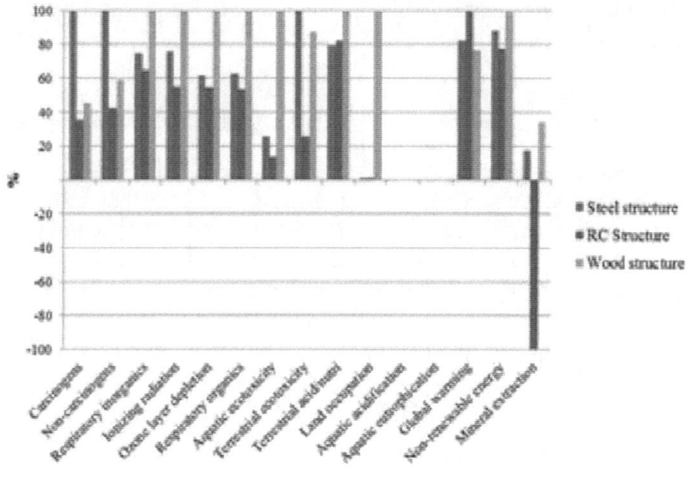

Figure 7: Comparative LCA results in terms of midpoint category according to the IMPACT2002+ method

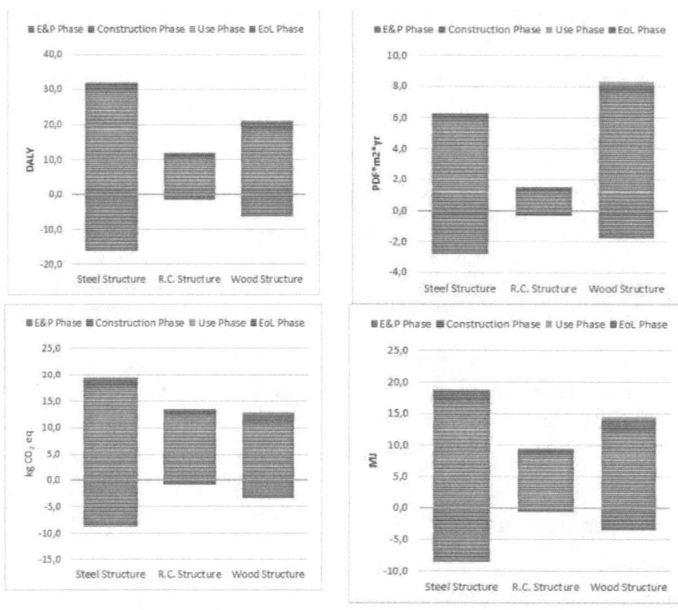

Figure 8: Comparative LCA results for each structural life-cycle phase in terms of endpoint category according to the IMPACT2002+ method: Human health (top-left); Ecosystem quality (top-right); Climate change (bottom-left); Resources (bottom-right)

A further elaboration of the environmental sustainability assessment has been provided for the case study. Indeed, data on the environmental impacts according to the EPD method (IEC, 2008) have been also computed and reported in Figure 9. In particular, the EPD method requires elaboration on some of the midpoint categories of the IMPACT 2002+ method. The EPD method provides impact assessment values regarding: global warming, ozone layer depletion, photochemical oxidation (equivalent to respiratory organics), acidification, eutrophication, non-renewables, and fossils (equivalent to non-renewable energy). Also according to this different Impact Assessment methodology, results show that the RC structure has the highest impact on global warming, followed by steel structure (80%) and then wood structure (75%). For all the other categories, a wood structure is responsible for the greatest impact, always followed by steel and RC structures.

Finally, Figure 10 shows the environmental impact according to the EPD method, considering each structural life-cycle phase. For the three structures, the greatest impact is generated by the extraction and production of building materials phase. Impacts due to transportation, construction, and maintenance amount to less than 10% of the impact due to extraction and the production of materials. The EoL phase's impacts equate to 20-50% of the extraction and production of materials phase. These results are similar to the ones shown in Figure 8.

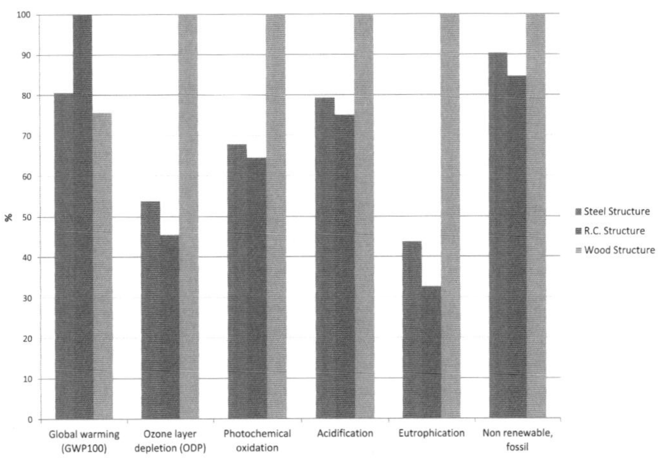

Figure 9: Environmental impact according to the EPD method

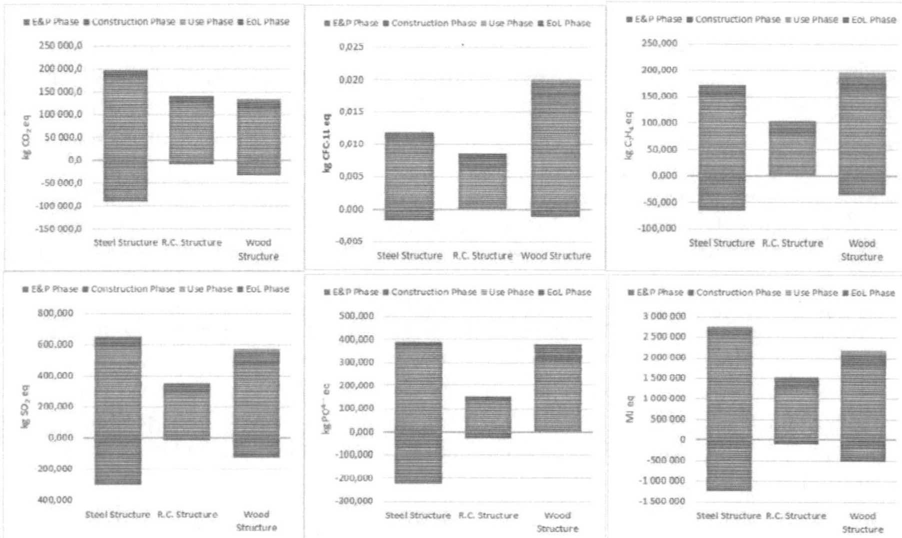

Figure 10: Environmental impact for each structural life-cycle phase according to the EPD method: Global warming (GWP) (top-left); Ozone layer depletion (ODP) (top-middle); Photochemical oxidation (POPC) (top-right); Acidification (AP) (bottom-left); Eutrophication (EP) (bottom-middle); Non-renewable, fossil (bottom-right)

4 Conclusions

In the present paper the authors have proposed a methodological approach to effectively perform a comparative environmental sustainability assessment of building structures, using an LCA-based analysis. The methodology has been applied to a case study, dealing with three alternative structural material options: RC, steel, and wood, and the related environmental impacts were quantified. Each option was designed to fulfil predefined structural, functional, and architectural requirements.

Authors can discuss some conclusions. Firstly, the environmental assessment of buildings has been analysed in different studies, but the presented one fixes external constraints on the minimum design performances, that should be considered for a comparative sustainable assessment. By comparing the different options, results revealed that RC structure is an environmentally worthy building solution. Indeed, RC structure has the highest impact only for one (climate change) out of four damage categories, according to IMPACT2002+, and one (climate change) out of six categories, according to EPD results, providing the lowest impacts for all the other categories. Steel structure is able to provide the widest benefit related to recycling. The study have revealed that there is no option that produces the best LCA-based environmental performance in all the impact categories. Indeed, the proposed methodology can support the decision-making process helping the designer to define the most sustainable alternative with respect to one of the environmental categories analyzed.

Moreover, authors want to point out that, besides the sustainability assessment methodology, the obtained environmental results depend on the case study considered and on the used databases. Future studies could include the enrichment of the proposed framework with energy performance indicators in the use-phase.

IOP Conf. Series: Materials Science and Engineering **442** (2018) 012010 doi:10.1088/1757-899X/442/1/012010

References

[1] Acca Software Tutorial and Edilus User Manual, 2013 (in Italian).
[2] Asdrubali F., Baldassarri C., Fthenaki V. 2013. Life cycle analysis in the construction sector: Guiding the optimization of conventional Italian buildings. Energy and Buildings 64 73–89.
[3] CEN (2004) Eurocode 8: design of structures for earthquake resistance - Part 1: general rules, seismic actions and rules for buildings. European Standard EN 1998-1:2004. Comité Européen de Normalisation, Brussels.
[4] CNR-DT 206/2007 Instruction for design, construction and control of wooden structures (in Italian).
[5] Constructalia - Steel - the most recycled material in the world – ArcelorMittal http://www.constructalia.com/
[6] Ecoinvent Centre.2007. The Life Cycle Inventory Data Version 2.2.
[7] EN 15978:2011 Sustainability of construction works - Assessment of environmental performance of buildings - Calculation method.
[8] EN 1993-1-1, 2005. Eurocode 3: Design of steel structures – Part 1-1: General rules and rules for buildings, May 2005.
[9] Environmental Product Declaration of zinc and powder coating systems in Italy by NordZinc S.p.A. company, 2010. Review II, 20th April 2010 (in Italian).
[10] Humbert S., De Schryver A., Bengoa X., Margni M., Jolliet O. 2012. IMPACT2002+ User Guide for vQ2 21 1 November 2012.
[11] ISO 12944 - Parts 1÷8: Paints and varnishes, 1998.
[12] ISO 14040: 2006 Environmental Management – Life cycle assessment - Principles and Framework.
[13] ISO 14044:2006 Environmental Management - Life cycle assessment - Requirements and guidelines.
[14] ISO 14713-1:2009 Zinc coatings -- Guidelines and recommendations for the protection against corrosion of iron and steel in structures -- Part 1: General principles of design and corrosion resistance.
[15] ISO/TR 14049:2000 Environmental management - Life cycle assessment - Examples of application of ISO 14041 to goal and scope definition and inventory analysis.
[16] Italy of Recycling 2012 – Annual report about waste recycling and recovery – Chapter 13: Inert waste from construction and demolition. ANPAR 2012 Report (in Italian).
[17] Menna, C., Asprone, D., Jalayer, F., Prota, A., & Manfredi, G. 2013. Assessment of ecological sustainability of a building subjected to potential seismic events during its lifetime. The International Journal of Life Cycle Assessment, 18(2), 504-515.
[18] Ministry of Infrastructure, D.M. 14.01.2008, Technical Codes for structures. Ordinary supplement to G.U. 04.02.2008 (in Italian).
[19] Peuportier B.L.P. 2001. Life cycle assessment applied to the comparative evaluation of single family houses in the French context. Energy and Buildings 33 443-450.
[20] Product Ecology Consultants SimaPro 7. 2010. Introduction to LCA.
[21] Superior Board of Public Works, Ministerial Circular n.617, 02.02.2009. Instruction for application of "Technical Codes for structures" according to Ministerial Decree of 14.01.2008. G.U. 26.02.2009 n.47, ordinary supplement n.27 (in Italian).
[22] The International EPD Cooperation (IEC). 2008. Introduction, intended uses and key program elements for environmental product declaration EPD. Version 1.0.
[23] TRADA. 2008. Recovering and minimizing wood waste, TRA-DA Technology wood information sheet 2/3 Sheet 59', re-vised June 2008.
[24] Vitale P., Torchia A., Menna C., De Gregorio F., Asprone D, Prota A., Arena U. 2014. Life cycle analysis of construction phase of a residential building. VIII Conference "Rete Italiana LCA", Florence 19-20 June 2014 (in Italian).
[25] Xing S., Xu Z., Jun G. 2008. Inventory analysis of LCA on steel- and concrete-construction office buildings. Energy and Buildings 40 1188–1193.

Eco-mechanical indexes for sustainability assessment of AAC blocks

D Ferretti[1], **B Gherri**[1] **and E Michelini**[1]

[1] Department of Engineering and Architecture, University of Parma, Italy

E-mail: daniele.ferretti@unipr.it

Abstract. This paper aims to provide a proper set of eco-mechanical indexes to evaluate both the mechanical performances and the environmental features of autoclaved aerated concrete blocks. To this purpose, a detailed review of existing sustainability indexes − originally developed for concrete − is first presented, and subsequently different possible eco-mechanical indexes are specifically developed for autoclaved aerated concrete masonry blocks, also in order to compare their performances with those of lightweight aggregate concrete blocks. The obtained results highlight that, based on currently available information, only few parameters appear to be effective in defining the overall sustainability performances of AAC blocks. While several researches were indeed carried out in these last years regarding material structural properties, there is still a lack of environmental data, which should be necessarily deepened in future research work to obtain more reliable results.

Keywords: Autoclaved Aerated Concrete, Eco-Mechanical Indexes, Ecological Performances, Mechanical Performances, Concrete Masonry.

1. Introduction

In the present work eco-mechanical indexes are proposed for the sustainability assessment of Autoclaved Aerated Concrete (AAC) blocks.

AAC material was introduced in the construction market since the mid-1920's, when the production process was patented. The material is obtained from a mix of sand, cement, water, gypsum, lime, and aluminum powder, which are poured in molds. Chemical reactions between aluminum, calcium hydroxide and water cause the production of hydrogen bubbles, which give the characteristic porosity of the material. When the material is sufficiently hardened, it is cut in blocks or panels. Their subsequent treatment in autoclave at 180 °C and 10 bar for 8-10 hours transforms sand and calcium hydroxide into calcium silicate hydrate (tobermorite), improving the mechanical properties of the material.

Porosity of AAC causes low compression strength (1-6 MPa) compared to normal-weight concretes, but permits to reach low density (about 350-800 kg/m^3) and interesting thermal properties (in terms of conductivity coefficient λ). As an example, AAC blocks with a density approximately equal to 400 kg/m^3 can reach thermal conductivity values that can be 10-20 times lower than normal-weight concrete [1]. Therefore, in some countries AAC blocks permit to build external walls (having or not a bearing function) without adding an extra-layer of insulation, thus reducing costs and environmental impact. Furthermore, contrary to multilayered walls, recycling AAC blocks is quite

IOP Conf. Series: Materials Science and Engineering **442** (2018) 012011 doi:10.1088/1757-899X/442/1/012011

simple, since site waste and rubble can be broken up, ground into powder and re-used in the manufacturing process [2]. For these reasons, AAC is gaining an increasing interest in the construction market as a valuable alternative in the manufacturing of products that are at the same time thermally attractive and characterized by a good construction efficiency.

In more details, AAC blocks are nowadays often used for the realization of non-structural walls (cladding and infill panels) within concrete frame structures, especially in seismic regions. In fact, the low specific weight of the material permits a reduction of seismic masses and, consequently, of seismic actions. Usually, this allows to save concrete and reinforcement of the structures with economic and environmental benefits. As a consequence, the advantages related to the use of AAC for external and partitioning walls should be evaluated considering the whole building, including structures. Some proper software tools have been proposed in the literature to the scope [3]. Since results depend on the features of the building, and on its geographical position, several representative case studies should be analyzed to obtain general conclusions. An analysis of this type could provide useful information on the importance of the involved variables, their relationships and weights. However, such a type of analysis is not so easy to be performed and is very time consuming.

AAC blocks are also used for bearing walls in low-to-medium rise masonry buildings in seismic and non-seismic areas [4-8]. Because of their low density, the size of the blocks can be large: this permits faster masonry work and a reduction of the number of mortar joints, which are usually thin.

In the present work, the attention is exactly focused on the use of AAC for masonry load bearing walls. The paper represents a first attempt in trying to provide a quick practical tool for the evaluation of material efficiency (in terms of both structural and environmental performances), also with respect to other alternative solutions, like blocks in lightweight aggregate concrete (LWAC), with or without integrated insulation. To the purpose, eco-mechanical indexes are introduced (see, e.g. [9] for a comprehensive review). This approach, first presented by Damineli et al [10] and subsequently modified by Fantilli and Chiaia [11] and Chiaia et al. [12], is commonly adopted for the sustainability assessment of concrete elements, through a concurrent combination of green aspects and mechanical properties.

As a first step of an ongoing research carried out by the Authors (focused on the definition of structural and environmental performances of AAC blocks), the concept of eco-mechanical index is herein extended to the case of AAC, starting from the work by Fantilli and Chiaia [11]. The attention is mainly concentrated to the individuation of the most significant variables to be considered, especially when indexes are adopted as a preliminary tool to choose the best construction strategy among different alternatives, like AAC vs. LWAC blocks.

2. Assessment of AAC performances through the definition of ecological and mechanical indexes

The assessment of sustainability performances of a given construction solution requires the introduction of a new holistic approach, which considers concurrently environmental and mechanical requirements. The building industry is currently touched by the ongoing sustainability debate, which is particularly vital for concrete structures, as the production of this material is responsible of a high energy consumption and causes large emissions of CO_2 [13]. An increasing sustainability of building structures therefore requires a reduction of the environmental impact associated with raw materials production, as well as with building erection, maintenance and operation processes. Moreover, an increase of the durability of the structure at its maximum technical performance is also desirable. For these reasons, environmental aspects must be considered along with other "classic" design aspects, such as strength and ductility requirements.

2.1 *Eco-mechanical indexes overview*

In case of traditional concrete structures, environmental issues have been recently addressed through the definition of proper sustainability indexes (among others, e.g., [10-12, 14-19]). As a generic statement, a set of environmental and mechanical indexes should be first evaluated and compared:

$$\textit{Material sustainability potential} = \frac{\textit{lifetime performance}}{\textit{environmental impact}} . \tag{1}$$

This definition [18] addresses clearly the three basic pillars of sustainability – considering the environmental aspects, expressed by the environmental impact, as well as social and economic aspects, embodied in the lifetime and performance parameters.

Broadening the basic pillars of sustainability, the objective of conciliating environmental with economic and social demands into the building sector have been chased since the early 2000s by the research group of Pons et al [19]. They have been working in this direction in order to define a general model for a complete sustainability analysis of concrete structures, named MIVES. MIVES is a Multi-Criteria Decision Making method able to consider all sustainability aspects, including economic, environmental and social ones, and incorporating a value function to homogenize all the indicators and consider the degree of satisfaction. Within MIVES framework, a Sustainability Index SI is defined according to Equation 2:

$$SI = V(P_x) = \sum_{i=1}^{N} \alpha_i \, \beta_i \, \gamma_i \, V_i(P_{i,x}) , \tag{2}$$

where $V(P_x)$ measures the degree of sustainability (value) of the alternative x weighed with respect to various criteria $P_x = (P_{1,x}; P_{2,x}; \ldots ; P_{N,x})$ considered; α_i are the weights of each requirement i, β_i are the weights of each criteria i and γ_i are the weights of the different indicators i. These weights are the preference, respectively, of these requirements, criteria and indicators. $V_i(P_{i,x})$ are the value functions used to measure the degree of sustainability of the alternative x with respect to a given criterion i.

Nevertheless, in the most of the real cases, social and economic aspects related to building construction and management are extremely arduous to be evaluated, especially during the preliminary design stage or even before, during the developing of the optimum mix-design of the raw materials. On the contrary, Equation 1 offers a simple way to assess the advantages and disadvantages of a certain material, taking into account its mechanical properties (lifetime performance), as well as its potential as a sustainable solution (environmental impact).

Based on this concept, simplified Eco-Mechanical Indexes (EMIs) were developed in recent years, as better discussed in [11]. These indexes were devoted to an easy and quick identification of the most virtuous construction solutions, characterized by the lower embodied energy and resource consumption, while ensuring adequate structural performances. EMIs were exclusively developed for structural concrete, as a tool for improving the mix-design of the material, so as to obtain a good balance between mechanical and ecological aspects. In more detail, a generic eco-mechanical index can be defined as:

$$EMI = \frac{MI}{EI} \tag{3}$$

where MI = Mechanical Index and EI =Ecological Index.

In the first works on this topic, MI comprised just the compressive strength of concrete, while other parameters were neglected. However, according to *fib* Bulletin 67 [14], concrete properties should be combined with structural performances, so including not only concrete strength, but also the results of other structural tests on materials. Similarly, EI was initially simply related to CO_2 emissions.

Based on these considerations, Chiaia et al [12, 17] proposed an improvement of Equation 3, by introducing also other significant mechanical and ecological parameters in the evaluation of MI and EI, and linking them through proper correlation functions $g()$ and $f()$:

$$EMI = \frac{f(MI)}{g(EI)}. \tag{4}$$

The correlation functions could be either a simple product of the different parameters, each one multiplied for a proper weighting coefficient, or logarithmic functions, so to take into account that the considered parameters could be of different order of magnitude [12].

2.2 Evaluation of ecological parameters for the definition of the ecological index EI

According to *fib* Bulletin 67 [14], the most crucial aspect in the determination of the ecological index *EI* is to define which environmental parameters should be taken into account to evaluate the sustainability of a given structure. So far, scientific investigations have been conducted only for structural concrete elements, while there is a lack of information specifically focused on AAC and LWAC masonry blocks.

According to several researchers and code rules [10, 15, 20], the most relevant ecological parameters are CO_2 footprint, embodied energy and water consumption used to produce concrete, but also other available parameters, such as biodiversity, toxic substances, resource depletion, as well as raw materials, electricity, safety, total cost and thermal conductivity can be use in the definition of *EI*. Consequently, the proposal of a general evaluation method for concrete blocks (AAC or LWAC) should first aim at determining the environmental impact in a standardized manner, using the so-called "eco-balance", as described in the European standards EN ISO:14040: 2006 [21], ISO 14044: 2006 [22], ISO 21930: 2007 [23]. The impact of every substance released to the environment can be so ascribed to one of the different impact categories, which have been internationally accepted.

Table 1 gives an overview on the most widely used environmental impacts for typical concrete blocks. For every country, exact data on a specific concrete block (included AAC and LWAC ones) can be derived from the Manufactures, using the Environmental Product Declarations EPD, that involves a Life-Cycle Assessment (LCA) for the evaluation of environmental aspects. CO_2 footprints (GWP), Embodied Energy (TNRE) and water consumption (W) of concrete components could be otherwise taken from environmental databases; however, in this work it has been preferred to directly derive these data from EPDs of Manufactures.

Among a very large variety of LCA parameters (see Table 1), it is first necessary to individuate the most significant factors to be included in the evaluation of the Environmental Index *EI* for the considered material. Comparative studies have shown that the environmental impacts of AAC manufacture are dominated both by the use of thermal energy and the connected emissions, and by bonding agent manufacture. Quicklime and cement manufacture are indeed based on energy-intensive combustion processes, that are responsible of relevant CO_2 emissions.

Table 1 Environmental data for assessing AAC ecological performances.

Parameter	Unit	
Global warming potential	(kg CO_2-Eq.)	GWP
Total non-renewable primary energy	(MJ)	TNRE
Cumulative energy demand (renewable)	(MJ)	TRE
Acidification potential of land and water	(kg SO_2-Eq.)	AP
Abiotic depletion potential for fossil resources	(MJ)	ADP
Depletion potential of the stratospheric ozone layer	(kg CFC_{11}-Eq.)	ODP
Use of secondary materials	(kg)	RM
Use of net fresh water resources	(m^3)	W
Non-hazardous waste disposed	(kg)	NHW
Hazardous waste disposed	(kg)	HW
Radioactive waste disposed	(kg)	RW
Components for re-use	(kg)	SM

The Global Warming Potential (GWP) associated to the production of 1 m^3 of AAC is indeed dominated more than 90% by carbon dioxide emissions, which are mainly originated from the manufacturing process and the production of thermal energy from natural gas. More than 2/3 of carbon dioxide emissions can be attributed to bonding agent manufacture and this is divided equally between quicklime and cement manufacture. Nonetheless, the manufacture of AAC requires less energy than other masonry products, thereby reducing the use of fossil fuels and associated emissions of carbon dioxide.

Acidification Potential (AP) from the manufacture of 1 m^3 of non-reinforced AAC is dominated up to 50% by sulphur dioxide emissions and to 40% by nitric oxides. Around 1/3 of the AP comes from the energy provision processes for manufacturing, while the production of quicklime and cement bonding agents causes around 40% of the AP.

Looking at the Embodied Energy, in case of AAC it is more appropriate to refer to Total Non-Renewable Primary Energy (TNRE). The dominance of the non-renewable energy produced in the works and used during the process of manufacturing of AAC blocks is indeed evident in this impact category: more than 1/3 of the Total Non-Renewable Primary Energy is directly caused by thermal energy requirements.

Secondary Raw Materials (RM) are not generally used during manufacture, neither secondary fuels. Moreover, all production offcuts are fed back into the production circuit; hence, RM parameter could not be taken into account for defining ecological index *EI* for AAC.

With regards to the use of fresh water (W), about 1/3 of total water amount is used in manufacture of AAC and it is mainly imputable to the electricity supply pre-chains. A further 1/3 is attributable to quicklime and cement production, and the proportion of direct water requirement for AAC manufacture directly in the plant is less than 1%.

Finally, the overall assessment of the accumulation of wastes, accounting for non-hazardous and hazardous ones (including mining waste materials, mineral processing residues, municipal waste and the domestic refuse, and commercial waste contained therein), needs to be evaluated. Non-Hazardous Waste NHW represents the greatest proportion in AAC production. Mining waste materials accumulate mainly in the electricity supply pre-chain from extraction of energy sources and also in the bonding agent manufacture pre-chains during raw material and energy source extraction. Radioactive waste accumulates exclusively through electricity production in nuclear power plants, therefore it is negligible for the considered case.

IOP Conf. Series: Materials Science and Engineering **442** (2018) 012011 doi:10.1088/1757-899X/442/1/012011

Based on this discussion, it is clear that several ecological LCA parameters reported in Table 1 can be neglected for the scope of this work. The most suitable parameters to be considered in the definition of eco-mechanical indexes for AAC blocks are only those summarized in Table 2.

Table 2. Environmental properties selected for the definition of *EI*.

Selected Parameter for EI	Unit	
Global warming potential	(kg CO_2-Eq.)	GWP
Total non-renewable primary energy	(MJ)	TNRE
Acidification potential of land and water	(kg SO_2-Eq.)	AP
Abiotic depletion potential for fossil resources	(MJ)	ADP
Use of net fresh water resources	(m^3)	W
Non-hazardous waste disposed	(kg)	NHW
Hazardous waste disposed	(kg)	HW

It should be however noted that the greatest limitation of the data reported in the Environmental Product Declarations for AAC and LWAC is that they are obtained considering as system boundary for LCA only the life-cycle phases of product manufacture (module A1-A3), while product installation (module A4-A5), the use stage (module B) and the disposal (module C) are not considered. This constraint is relevant since currently there is a global lack of a Cradle to Gate evaluation both for AAC and for LWAC blocks.

2.3 *Evaluation of mechanical parameters for the definition of the mechanical index MI*

Mechanical indexes can be conceived as synthetic parameters representative of the sustainability of the investigated material referring to social aspects [24]. Since AAC blocks are often used for the realization of load bearing masonry walls both in seismic and non-seismic areas, it is indeed crucial that their characteristics are able to secure adequate safety and serviceability (i.e. durability) margins to the resulting structure, so ensuring a satisfactory building performance. This means that the reduction of environmental impact in the design and production of the blocks should not compromise the mechanical performances of the material, and that the final product should be anyway characterized by well-defined mechanical properties.

Generally speaking, for masonry structures the most important mechanical property is the compressive strength f_c, which should be declared by Producers on the technical sheets of the block. To some extent, also the post-peak energy absorption capacity in compression (or, equivalently, the fracture energy in compression G_{fc}) can be a parameter of interest, since it is related to structural ductility. This is particularly true when it is necessary to evaluate the performances of new "recipes" with respect to more traditional solutions, like for example Fibre-Reinforced Aerated Concrete (FRAC) in spite of plain AAC. In this case, the effects related to the addition of short polymeric fibres in the admixture can be indeed hardly quantified by only measuring the material strength, since the latter is only slightly modified by the presence of fibres, and it also depends on the adopted curing procedure, which is different for FRAC and AAC (in the production of FRAC, the autoclaving process is substituted by curing at room temperature, to avoid fibre potential damage). The optimization of fibre amount in the admixture can be then evaluated by examining the post-peak response, since the presence of fibres increases the ductility, by reducing at the same time the crack opening.

However, the determination of the post-peak energy absorption capacity in compression is not so straightforward, since it requires the knowledge of the complete stress-strain relationship in compression. Traditional compression tests performed during material production and characterization are usually simply aimed at the determination of the pre-peak response of the material, until the reaching of the compressive strength f_c, and are performed under loading control. On the contrary, the determination of the post-peak branch can be only achieved working under displacement control, and also in this case the control of the softening branch can be difficult, especially in case of brittle materials like concretes. For this reason, fracture energy in compression G_{fc} is a parameter that can be certainly taken into account during the research and developing stages for tailoring a new mix-design of a given product, but that can be hardly used in practice for a quick comparison of alternative solutions (like for example AAC and LWAC blocks), since it is not available in the technical sheets of the products and cannot be approximately derived from other material properties.

Even if masonry structures are mainly subjected to compressive stresses, it could be useful to introduce also a parameter related to fracture toughness in the definition of the mechanical index *MI*. As highlighted in several research works and design codes (e.g. [25-29]), toughness exerts indeed a significant influence on the resistance of AAC against damage during transport and handling, and it affects the loadbearing behaviour under accidental or seismic loads. Fracture toughness represents an important material property also in presence of static loads, since it governs crack formation and propagation, and consequently it is strictly related to durability issues. Cracking of AAC walls is indeed a quite common problem, especially in case of building internal partitions, due to floor deformability.

A possible parameter representing fracture toughness, which derives from linear elastic fracture mechanics (LEFM), is the fracture energy in tension G_{ft}. This property is obtained experimentally, from the area under the complete force-crack mouth opening displacement (CMOD) curve. However, in absence of experimental data, G_{ft} can be roughly estimated through empirical expressions, on the basis of other material properties. For example, in case of lightweight concretes (like those used for the realization of LWAC blocks), G_{ft} can be calculated as a function of the mean value of material tensile strength, and depends on the type of sand adopted in the mix [30]. As far as AAC is concerned, only limited information is available, since G_{ft} is affected by autoclaving process conditions [31]; however, an almost linear relation with density can be adopted as a first approximation (as suggested, e.g., in [25, 27, 31, 32]).

3. Determination of mechanical and ecological parameters for AAC blocks

3.1. Assessment of AAC mechanical parameters

In this work, mechanical properties of AAC blocks (to be used for the definition of *MI*) are derived on the basis of the results of experimental tests carried out by the Authors on AAC samples with an average density $\rho \approx 550$ kg/m^3 produced by an Italian Manufacturer (see also [28] for more details about the experimental program).

3.1.1. Experimental characterization of the behaviour in compression. Material characterization in compression is performed on three prismatic AAC specimens with square basis, having an edge length of 40 mm and a height of 80 mm, cut from the central part of AAC blocks.

Compression tests are performed under displacement control, so to get the complete stress-strain curve, as reported in Figure 1a. Average deformations are measured over the whole specimen height, and over a central 25 mm gauge length through 4 LVDTS placed on specimen corners, as shown in Figure 2a. In all cases, failure occurs due to the spreading of diagonal frictional cracks (Figure 2b).

The most important parameters, which define specimen behaviour in the pre and post-peak stage, are the compressive strength f_c, the elastic modulus E_c, the plasticity number k (defined as the ratio between the elastic modulus E_c and the secant modulus from the origin to the peak stress, E_{c1}), and the fracture energy in compression G_{fc}. The corresponding values obtained from the tests are summarized in Figure 1b. It should be noted that the compressive strength obtained from tests on prisms is about 10% lower than that reported on technical sheets provided by Manufacturers, since during production f_c is usually derived from standard cubes cut from AAC blocks, according to UNI EN 772-1 and UNI EN 771-4 ([33-34], see also [28]).

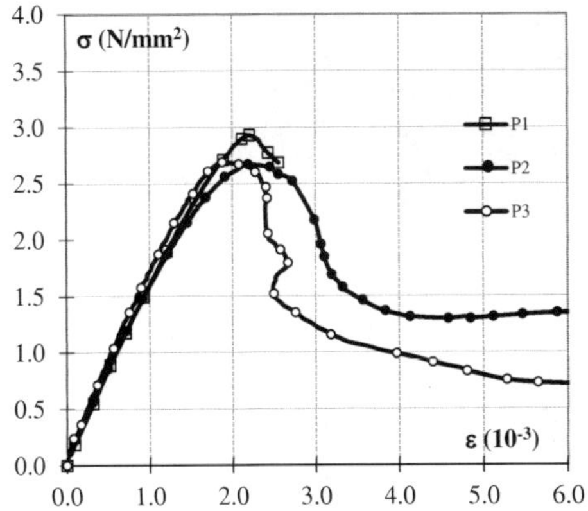

(a)

	f_c (MPa)	E_c (MPa)	k (-)	G_{fc} (N/m)
P1	2.93	1581	1.30	4.67
P2	2.67	1549	1.17	7.10
P3	2.69	1700	1.19	4.97
mean	**2.80**	**1610**	**1.22**	**5.58**

(b)

Figure 1. (a) Stress-strain curves resulting from compression tests on AAC prisms; (b) results of the uniaxial compression tests.

(a)

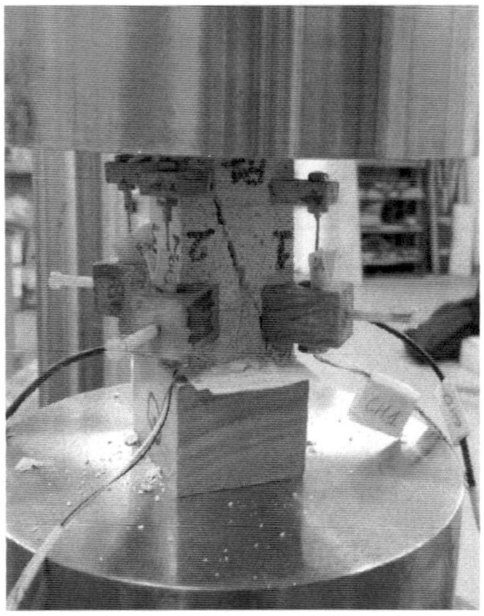

(b)

Figure 2. Compression tests on prismatic AAC samples: (a) test setup; (b) failure mode.

3.1.2 Experimental characterization of the behaviour in tension. The flexural tensile strength $f_{ct,fl}$ and the fracture energy in tension G_{ft} of the material are obtained from three-point bending tests on 3 AAC notched beams, having nominal dimensions L x H x b respectively equal to 620 x 251 x 100 mm. In all cases, the net span of the beams L_n is kept equal to 540 mm, while the average notch height is a = 12.8 mm. These tests are carried out under Crack-Mouth Opening Displacement (CMOD) control (with a speed of 1μm/min), by using an Instron 8862 universal testing machine. In order to measure the midspan deflection δ, a LVDT is applied on a specific device fixed onto supports (Figure 3a); moreover, an ESPI measurement system is used to observe cracking onset and propagation (Figure 3b; see also [28] for more details on the followed procedure).

(a) (b)

Figure 3. Three-point bending tests on AAC notched beams: (a) detail of the LVDT used for the monitoring of midspan deflection; (b) ESPI setup.

Figure 4a reports the load-CMOD curve for the three investigated specimens, while the main mechanical properties defining AAC behaviour in tension and obtained from the tests are summarized in Figure 4b. According to linear elastic analysis, the flexural strength $f_{ct,fl}$ can be deduced as:

$$f_{ct,fl} = \frac{3 P_u L_n}{2b(H-a)^2} \tag{5}$$

where P_u is the peak load, L_n is the net span of the beam, b is the specimen width, while $(H\text{-}a)$ is the distance between the tip of the notch and the top of the cross-section. The same beams without notch tested under three-point bending have provided an higher value of the flexural strength, which is in average equal to $f_{ct,fl} = 0.6$ MPa [28].

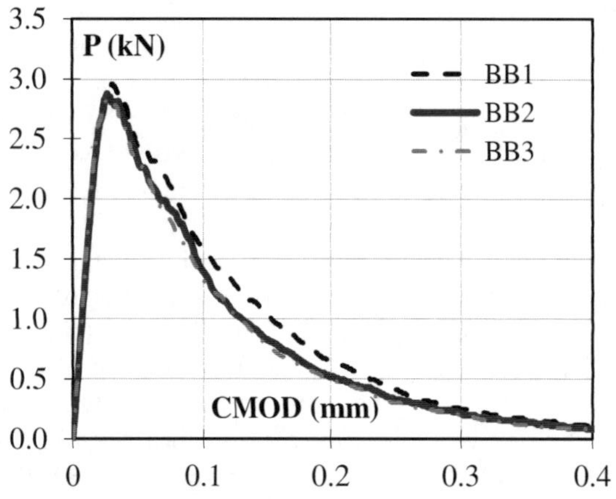

(a)

	$f_{ct,fl}$ (MPa)	G_{ft} (N/m)
BB1	0.42	4.8
BB2	0.40	4.6
BB3	0.39	9.3
mean	**0.41**	**6.2**

(b)

Figure 4. (a) Load-CMOD curves resulting from three-point bending tests on AAC notched beams; (b) summary of the main properties defining AAC behavior in tension, as obtained from the tests.

3.2. Assessment of AAC ecological parameters

As concerns the AAC samples with average density $\rho \approx 550$ kg/m^3 previously presented, no ecological data have been evaluated so far. Thus, the Authors do not have any EPD from the Manufacturer, neither environmental impact data to use for comparison with other AAC blocks. A complete Life

Cycle Assessment for the AAC blocks already characterized from a mechanical point of view is part of an ongoing research carried out by the Authors, and will be ready in the near future.

Meanwhile, the ecological parameters to be used in the definition of the environmental index *EI* presented in this work have been deduced by referring to data available in the literature and relative to comparable typologies of AAC masonry units. In more detail, three different commercial AAC blocks (named AAC1, AAC2 and AAC3 in the following) are chosen on the basis of three main criteria, that is same block thickness (equal to 300 mm), almost equal compressive strength ($f_c \approx 3.4$ MPa, as determined on cubes – being so comparable to that of blocks analysed in Section 3.1), and similar gross density (ranging between $445 \leq \rho \leq 600$ kg/m^3).

Finally, the selected blocks are characterized by thermal conductivity ranging between $0.1 \leq \lambda \leq 0.11$ W/(m·K). This last parameter is particularly interesting and makes AAC an eco-friendly material, since its low value allows to decrease the residential energy consumption and raise the energy efficiency in buildings constructed with AAC blocks. Therefore, external walls made of AAC masonry blocks with distributed thermal resistance can be a promising alternative for increasing the envelope level of thermal protection and decreasing the thickness of required thermal insulation.

4. Application of eco-mechanical indexes to different construction solutions: comparison between AAC and LWAC masonry block performances

As already mentioned, this work tries to summarize AAC performances through the definition of specific ecological and mechanical indexes, which can be combined with each other in order to obtain the most sustainable solution from both a social and an environmental point of view. Specifically, the main goal of the work is to understand which parameters should be considered in the definition of these eco-mechanical indexes, so as to identify a sort of "optimum" expression. Several possible alternative definitions are presented in Section 4.2.

Eco-mechanical indexes can serve two main scopes. First, they can be used in tailoring the mix adopted for the production of material, so as to reduce the environmental impact without compromising mechanical performances, which is crucial especially in case of blocks with a load bearing function. In addition, these indexes can represent a quick and easy tool to perform a preliminary evaluation of the most sustainable solution among different, but alternative construction techniques, such as for example AAC and LWAC masonry units.

Due to the abovementioned lack of original environmental data for the analyzed AAC blocks, the considered eco-mechanical indexes are not applied herein for the improvement of the material, while the attention is instead focused on the comparison among the performances of different concrete products available on the market for the realization of load-bearing masonry blocks in non-seismic areas. Once identified the "optimum" index, it can indeed represent a useful tool for the preliminary identification of the best building solution, orienting the construction workers in the choice of the "most virtuous" block among possible alternatives that seem formally equivalent to each other.

4.1. Eco-mechanical performances of AAC and comparison with other sustainable concrete masonry components

The attention is focused herein on three typologies of concrete masonry units (CMUs), that is to say AAC, LWAC with integrated thermal insulation made of mineral wool (indicated in the following as LWAC1), and LWAC hollow blocks without thermal insulation (named LWAC2).

As far as AAC blocks are concerned, mechanical indexes are defined on the basis of some of the experimental properties discussed in Section 3.1, while environmental parameters are taken from EPDs data of similar products, as discussed in Section 3.2. In this way, three possible solutions, named AAC1, AAC2 and AAC3 are obtained, characterized by the same mechanical parameters (the experimental ones), but with more or less efficient environmental performances (taken from the literature).

The data relative to the two considered LWAC blocks (LWAC1 and LWAC2) are directly taken from their EPDs and technical sheets provided by the Producers. On this point, it should be noted that there is a discrepancy between the amount of available environmental and mechanical parameters: while the first ones are more or less all provided in the technical documents, only the compressive strength of the blocks f_c is generally given. For this reason, in this work the other mechanical properties of LWAC1 and LWAC2 blocks are calculated starting from the declared mean value of f_c (equal to 2 MPa for both the types of units), according to the relationships provided for lightweight aggregate concretes by design codes. In more detail, the elastic modulus E_c and the tensile strength f_{ct} are calculated according to [35], while the fracture energy in tension G_{ft} is obtained according to [30], by hypothesizing that both the LWCAs contain lightweight sand.

Tables 3 and 4 briefly summarize the main data relative to both mechanical and environmental parameters for the 5 considered block typologies. For every block here presented, the selected system boundaries comprise manufacture of the concrete blocks, including the extraction of raw materials and auxiliaries and/or processing aggregates to form the product ready for shipping until it leaves the actual plant, according to the cradle to gate assessment. The results presented in Table 4 thus include the provision of all materials used – preliminary products – (phase A1 of LCA of the system boundary); transport processes to and in the plant (phase A2 of LCA of the system boundary); manufacturing processes including energy, waste, emissions (phase A3 of LCA of the system boundary). The usage and disposal stages of the block manufactured are not taken into consideration in this study, since no data are available for the time being, and therefore need to be supplemented for a further assessment within the context of the building use, when the blocks will be engaged.

Table 3. Mechanical properties of the 5 CMU typologies considered in this work.

	AAC1	*AAC2*	*AAC3*	*LWAC1*	*LWAC2*
ρ (kg/m^3)	445	600	480	450	600
f_c (MPa)	2.8	2.8	2.8	2.0	2.0
E_c (MPa)	1610	1610	1610	596	1060
G_{ft} (N/m)	6.2	6.2	6.2	2.51	2.71

Table 4. Environmental properties of the 5 CMU typologies considered in this work.

		AAC1	*AAC2*	*AAC3*	*LWAC1*	*LWAC2*
GWP	(kg CO_2-Eq.)	2.19E+02	1.68E+02	1.63E+02	8.40E+01	1.19E+02
AP	(kg SO_2-Eq.)	2.19E-01	2.08E-01	2.10E-01	1.59E-01	3.51E-01
ADP	(MJ)	1.70E+03	1.20E+03	1.13E+03	4.14E+02	7.74E+02
TNRE	(MJ)	1.80E+03	1.29E+03	1.21E+03	4.78E+02	8.91E+02
W	(m^3)	7.87E+01	5.28E-01	5.60E-01	n/a	1.39E+02
NHW	(kg)	4.69E+02	8.02E+00	1.56E+01	n/a	1.86E+01
HW	(kg)	n/a	1.10E-02	2.78E-04	n/a	2.58E-02

As concerns the Global Warming Potential (GWP), Table 4 shows that the most valuable block in terms of Low Carbon emission is the LWAC1 one, which registers the smaller amount of $KgCO_2$ - Eq. GWP is dominated by the processes associated with cement manufacturing, alongside with other essential factors that in case of LWAC1 also includes the impact of thermal insulation and aggregates.

Referring to Acidification Potential (AP), the block presenting the lower impact is once again LWAC1. In this case, AAC blocks register quite similar results, while LWAC2 block is characterized by the highest value of the examined parameter.

LWAC1 appears to be the better option also with regards to Abiotic Depletion Potential (ADP) for fossil resources. This parameter refers to the consumption of fossil resources, which embodies the energetic consumption during the manufacture of cement and aggregates.

As concerns the parameter of Non-Renewable Energy (TNRE) – which takes into account the processes of energy consumption during the manufacture of block, as well as during the manufacture of cement and aggregates – the most environmental responsible block is still LWAC1, having a TNRE just equal to 4.78E+02MJ.

Water consumption (W) registers the lower value in case of block AAC2 having a W just equal to 5.28E-01, but data for LWAC1 are not available.

Finally, Hazardous and Non-Hazardous Waste (HW and NHW) register the aggregated lower value for AAC2, even if some data are not available for AAC1, nor for LWAC1 blocks.

4.2. Application of different eco-mechanical indexes for the evaluation of the sustainability of AAC masonry blocks with respect to LWCA ones

In this Section, the Authors propose some Eco-Mechanical Indexes *EMIs* to assess the global environmental and mechanical performance of AAC blocks with respect to LWAC blocks. The proposed indexes are summarized in the left part of Table 5, which reports the corresponding values obtained for the five considered solutions, based on the mechanical end environmental data reported in Tables 3 and 4. Indexes have been chosen according to the criteria set out in the following.

As regards environmental parameters, GWP and TNRE are always considered in the definition of all *EMI*s, since they are recognized as the most significant Environmental Indexes (*EI*). Similarly, the most important Mechanical Index (*MI*), describing the behavior of blocks in compression, is recognized to be the compressive strength f_c.

The influence of mechanical parameters is investigated through the definition of the first 3 indexes EMI_1-EMI_3, which are characterized by the same denominator (i.e. by the same environmental parameters), while the terms at numerator are varied (that is, f_c for compressive behavior, G_{ft} for tensile behavior, f_c G_{ft} for both of them). On the contrary, indexes from EMI_4 to EMI_8 want to clarify the influence of different environmental parameters, for a given *MI* at numerator (i.e., f_c). Finally, EMI_9 represents a sort of "global index", which includes all the most significant mechanical and environmental parameters at the same time.

Table 5. Proposed eco-mechanical indexes and corresponding values for the considered solutions.

Eco-Mechanical Index (EMI)	Units	AAC1	AAC2	AAC3	LWAC1	LWAC2
$EMI_1 = f_c$ / (TNRE x GWP)	[MPa /(MJ $kgCO_2$-Eq)]	7.08E-06	1.29E-05	1.42E-05	4.99E-05	1.89E-05
$EMI_2 = G_{ft}$ / (TNRE x GWP)	[(N/m) / (MJ $kgCO_2$-Eq)]	1.58E-05	2.88E-05	3.17E-05	6.26E-05	2.55E-05
$EMI_3 = (f_c$ x $G_{ft})$ / (TNRE x GWP)	[MPa (N/m) / (MJ $kgCO_2$-Eq)]	4.41E-05	8.05E-05	8.86E-05	1.25E-04	5.10E-05
$EMI_4 = f_c$ / (TNRE x GWP x AP)	[MPa / (MJ $kgCO_2$-Eq $kgSO_2$-Eq)]	3.23E-05	6.21E-05	6.77E-05	3.14E-04	5.37E-05
$EMI_5 = f_c$ / (TNRE x GWP x W)	[MPa / (MJ $kgCO_2$-Eq m^3)]	8.99E-08	2.45E-05	2.54E-05	n/a	1.36E-07
$EMI_6 = f_c$ / (TNRE x GWP x NHW)	[MPa / (MJ $kgCO_2$-Eq kg)]	1.51E-08	1.61E-06	9.13E-07	n/a	1.01E-06
$EMI_7 = f_c$ / (TNRE x GWP x ADP)	[MPa / (MJ $kgCO_2$-Eq MJ)]	4.16E-09	1.08E-08	1.26E-08	1.20E-07	2.44E-08
$EMI_8 = f_c$ / (TNRE x GWP x HW)	[MPa / (MJ $kgCO_2$-Eq kg)]	n/a	1.17E-03	5.12E-02	n/a	7.31E-04
$EMI_9 = (f_c$ x $G_{ft})$ / (TNRE x GWP x AP x ADP)	[MPa (N/m) / (MJ $kgCO_2$-Eq $kgSO_2$-Eq MJ)]	1.19E-07	3.23E-07	3.73E-07	1.90E-06	1.88E-07

For reading convenience, the same data are plotted in terms of histograms in Figures 5-6. The graphs also report the maximum, minimum and average values of each considered *EMI*, with reference to the 3 different AAC solutions (AAC1-AAC3 in Table 5). As can be seen, even if these three solutions are characterized by the same mechanical parameters and by a comparable density, the environmental performances can differ significantly from each other, depending on the greater or lower "virtuosity" of the Producers. Moreover, the gap between maximum and minimum values is more pronounced when some specific environmental parameters are included in the definition of *EMI*. This is the case of water consumption, as well as of the production of hazardous and/or non-hazardous wastes, which can significantly vary from one Manufacturer to the other.

A more detailed examination of the results reported in Figure 5 leads to the following observations. With reference to index EMI_1 (Figure 5a), it can be stated that all the three AAC solutions, as well as LWAC2, are comparable to each other. LWAC1 clearly shows the best eco-mechanical performance, mainly because its GWP is negligible if compared to other GWP values. Similar results can be also obtained when considering indexes EMI_2 and EMI_3 (Figures 5b,c). In both cases, the average value obtained for AAC blocks is now equal or even higher than that corresponding to LWAC2, since the lower GWP and TNRE associated with LWAC2 are counterbalanced by the better behavior of AAC in tension (with higher values of G_{ft}).

The contribution of acidification potential is taken into account for the first time in the definition of an eco-mechanical index in EMI_4 (Figure 5d). Quite homogeneous values of this index are obtained for AAC1, AAC2, AAC3 and LWAC2 blocks. A higher value of EMI_4 is registered also in this case for LWAC1.

A comparison among the considered construction solutions is more difficult when referring to index EMI_5 (Figure 5e), since water consumption in different productions is extremely variable. LWAC2 production is certainly characterized by a larger water consumption (around 138 m^3) with respect to AAC manufacture (whose average values is around 27 m^3), and this represents an indubitable advantage in terms of sustainability for AAC blocks, also taking into account the limited available water resources. However, the same data relative to AAC production appear to be much dispersed, varying between 0.53 m^3 (AAC2) and 79 m^3 (AAC1), highlighting that this parameter should be better investigated in future research. For this reason, so far water consumption seems not to be a significant index to be considered in the comparisons.

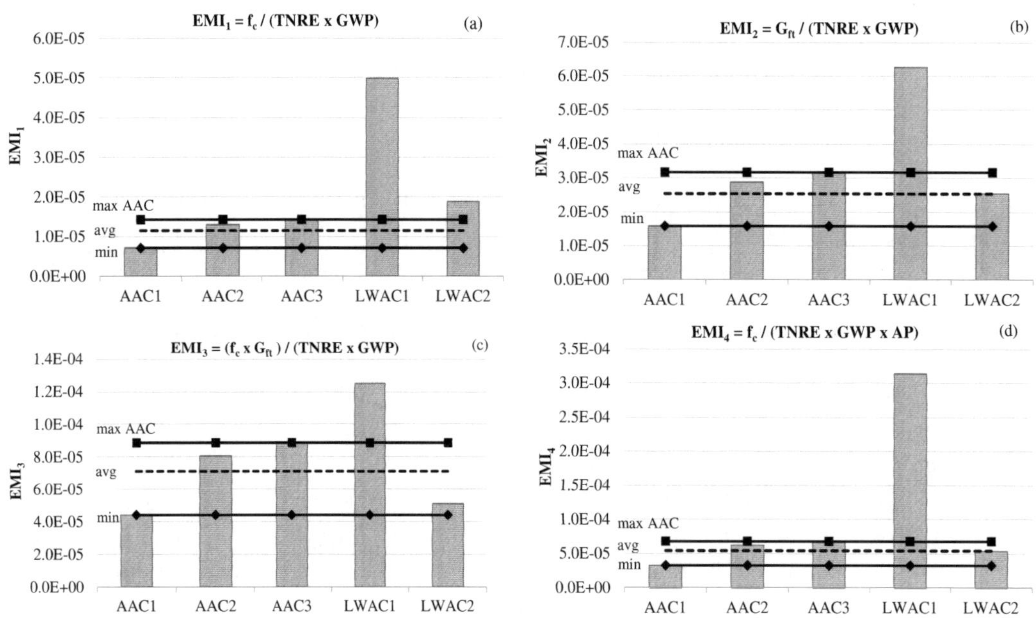

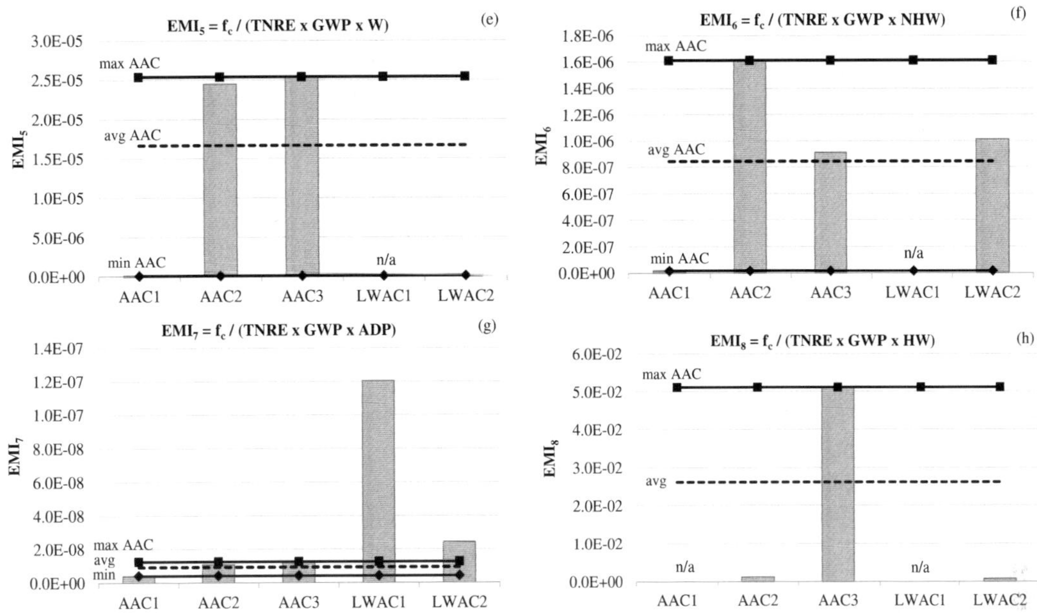

Figure 5. Comparisons of eco-mechanical performances of AAC and LWAC blocks, obtained by considering different EMIs.

EMI_6 and EMI_8 indexes (Figures 5f and 5h) includes the effect of wastes disposed in phase A1-A3 for each block considered. The assessment of the waste accumulation is usually shown separately in EPD charts for the three main categories of disposed non-hazardous waste (Figure 5f), hazardous waste for disposal (Figure 5h) and disposed radioactive waste (negligible for the considered cases). Non-hazardous waste represents the greatest proportion both in AAC and LWAC production, whilst the fraction of hazardous waste is generally negligible, as clearly shown in Table 4. Also in this case, the so obtained eco-mechanical indexes are hardly comparable to each other, since the range of variability of waste production is extremely large, even in case of similar productions (see, e.g., AAC1 and AAC2 for NHW, or AAC2 and AAC3 for HW). Moreover, no data are so far available for some block typologies (relatively to both NHW and HW for LWAC1, and relatively to HW for AAC1). Therefore, also the eco-mechanical indexes EMI_6 and EMI_8 appear to be not so significant and should be neglected.

Index EMI_7 takes into account the effect of ADP. As can be seen, the three AAC blocks are characterized by almost coincident values of this index, which are also comparable to that obtained for LWAC2. LWAC1 shows once again the best performance, related to its better ecological parameters (GWP, TNRE and ADP are indeed significantly lower with respect to the other examined products).

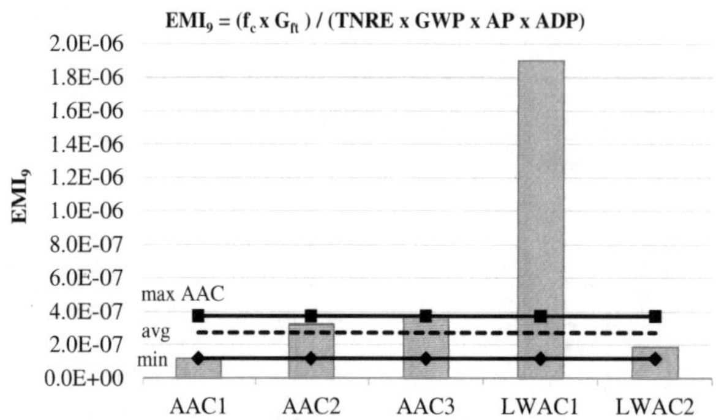

Figure 6. (a) Comparisons of eco-mechanical performances of AAC and LWAC blocks, obtained by considering all the significant mechanical and environmental parameters.

Finally, a sort of "global" index, taking into account all the most significant mechanical and environmental parameters, is reported in Figure 6. As can be seen, water consumption, as well as non-hazardous and hazardous waste are excluded from index definition for the abovementioned reasons. This global index shows that the average eco-mechanical performance of AAC is better than that of LWAC2, so evidencing that AAC block can actually represent a sustainable solution in the construction market. Further research work on the Life Cycle Assessment of AAC blocks already characterized by the Authors from a mechanical point of view could further confirm this assumption, providing more data to be analyzed.

However, based on the information so far available, LWAC1 blocks with integrated thermal insulation appear to be characterized by significantly lower values of all the considered environmental parameters with respect to both AAC and LWAC2 blocks, so seeming the most virtuous construction solution.

5. Conclusions

The present work represents a first attempt in the assessment of sustainability of masonry structures realized with AAC blocks, also in comparison with other competitive solutions, like those involving the use of LWAC blocks. To this aim, eco-mechanical indexes, already developed in the literature with reference to traditional concrete structures, are herein extended to the considered concrete masonry blocks.

The obtained results show that the parameters used to elaborate the eco-mechanical indexes should be carefully selected, by individuating the most significant mechanical and ecological aspects for the analyzed material. However, these results should be regarded as preliminary data, since they are affected by some main limitations, which are discussed in the following.

First of all, for both AAC and LWAC blocks, a general exertion in deriving complete mechanical data from the data sheets of the Manufacturers should be evidenced. For these reasons, in case of AAC, complete experimental data obtained by the Authors have been considered, while in case of LWAC, the main mechanical properties have been derived on the basis of the declared compressive strength, according to Code relations (but this has obviously limited the number of parameters to be included in the evaluation of *MI*).

Moreover, the lack of original environmental data provided by the Producer for the considered AAC blocks has just allowed the Authors to evaluate block ecological performances *EI* referring to literature data (i.e. based on the EPDs of similar AAC blocks available on the marked). In this sense, a further limitation is related to the reduced availability of EPDs, which often differ from each other for the analyzed life cycle stages, making comparative reading ineffective also for similar blocks. To obtain more significant results, it is therefore necessary to perform a detailed EPD evaluation for the

IOP Conf. Series: Materials Science and Engineering **442** (2018) 012011 doi:10.1088/1757-899X/442/1/012011

considered AAC blocks, already characterized by the Authors from a mechanical point of view. In this way, it will be possible to fully investigate the whole performances of a specific type of masonry block, by selecting among a wide range of both mechanical and environmental parameters.

Finally, it remains to be defined how much the environmental impact of the blocks is embodied in the remaining part of their life cycle, also considering the use phase. Some partial information, which are however inadequate, can be deduced by evaluating the insulation properties of the blocks themselves, but more detailed studies including other significant environmental aspects are certainly needed. Similarly, more information should be acquired on the disposal phase, for which no reliable data are so far available.

References

[1] Narayanan N and Ramamurthy K 2000 Structure and properties of aerated concrete: a review *Cem. Concr. Compos.* 20 321-9

[2] Hafez AH 2016 Integrating building functions into massive external walls *PhD Thesis* (TU Delft: The Netherlands) 294 pp.

[3] International Federation for Structural Concrete (fib) 2008 *Environmental design of concrete structures: general principles - fib bulletin 47* (Lausanne: fib), 48 pp.

[4] *Proc. 4th Int Conf on AAC — Autoclaved Aerated Concrete: Innovation and Development (Kingston)* ed MC Limbachiya and JJ Roberts (London: Taylor & Francis)

[5] Delmotte P, Rivillon P, Wesierski V and Hurez M 2003 Study on shear walls realized with autoclaved aerated concrete masonry (in French), *Cahier du CSTB* 3492, Livraison 445

[6] Klingner RE, Tanner JE, Varela JL and Barnett RE 2005 Autoclaved aerated concrete: innovative materials and civil infrastructure *Proc. Int. Work on Innovation in Materials and Design of Civil Infrastructure (Cairo)*, pp. 1–36.

[7] Tanner JE, Varela JL, Klingner RE, Brightman MJ and Cancino U 2005 Seismic testing of autoclaved aerated concrete shearwalls: a comprehensive review *ACI Struct. J.* 102(3) 374–82

[8] Ferretti D, Michelini E and Rosati G 2015 Mechanical characterization of autoclaved aerated concrete masonry subjected to in-plane loading: Experimental investigation and FE modeling *Constr. Build. Mater.* 98 353-65

[9] Miller SA, Monteiro PJM, Ostertag CP and Horvath A (2016) Comparison indices for design and proportioning of concrete mixtures taking environmental impacts into account, *Cem. Concr. Compos.* 68 131-43

[10] Damineli BL, Kemeid FM Aguiar PS and John VM 2010 Measuring the eco-efficiency of cement use *Cem. Concr. Compos.* 32 555-62

[11] Fantilli AP and Chiaia B 2013 Eco-mechanical performances of cement-based materials: an application to self-consolidating concrete *Constr. Build. Mater.* 40 189-196.

[12] Chiaia B, Fantilli AP, Guerini A, Volpatti G and Zampini D 2014 Eco-mechanical index for structural concrete *Constr. Build. Mater.* 67 386-92

[13] Ashby MF 2009 *Materials and the environment: eco-informed materials choice*, (Burlington: Butterworth Heinemann)

[14] International Federation for Structural Concrete (fib) 2013 *Guidelines for green concrete structures - fib bulletin 67* (Lausanne: fib), 60 pp.

[15] Habert G and Roussel N 2009 Study of two concrete mix-design strategies to reach carbon mitigation objectives *Cem. Concr. Compos.* 32 397–402

[16] Fantilli AP and Chiaia B 2013 The work of fracture in the eco-mechanical performances of structural concrete *J Adv. Concr. Tech.* 11 282–90

[17] Fantilli AP, Chiaia B and Gorino A 2016 Ecological and mechanical assessment of lightweight fiber-reinforced concrete made with rubber or expanded clay aggregates *Constr. Build. Mater.* 127 692–701

[18] Muller HS, Breiner R, Moffatt JS and Haist M 2014 Design and properties of sustainable concrete *Proc. Eng.* 95 290–30

[19] Pons O, de la Fuente A and Aguado A 2016 The use of MIVES as a sustainability assessment MCDM Method for architecture and civil engineering applications *Sustainability* 8 460-75

[20] Flower DJM and Sanjayan JG 2007 Green house gas emissions due to concrete manufacture *Int. J. Life Cycle Assessment* 12(5) 282-8

[21] EN ISO 14040: 2006 *Environmental management - Life cycle assessment - Principles and framework*

[22] ISO 14044: 2006 *Environmental management -- Life cycle assessment -- Requirements and guidelines*

[23] ISO 21930: 2007 *Sustainability in building construction -- Environmental declaration of building products*

[24] Sakai K, Shibata T, Kasuga A and Nakamura H 2016 Sustainability design of concrete structures *Struct Concr* 17(6) 1114-24

[25] RILEM Recommended Practice 1993 *Autoclaved Aerated Concrete: Properties, testing and design*, ed S Aroni et al. (London: E&FN Spon) pp 15-17

[26] Wittmann FH and Gheorghita I 1984 Fracture toughness of autoclaved aerated concrete *Cem. Concr. Res.* 14 369−74

[27] Trunk B, Schober G, Helbing AK and Wittmann FH 1999 Fracture mechanics of autoclaved aerated concrete *Cem. Concr. Res.* 29 855−9

[28] Ferretti D, Michelini E and Rosati G 2015 Cracking in autoclaved aerated concrete: experimental investigation and XFEM modeling *Cem. Concr. Res.* 67 156−67

[29] Bonakdar A, Babbitt F and Mobasher B 2013 Physical and mechanical characterization of fiber-reinforced aerated concrete (FRAC) *Cem. Concr. Compos.* 38 82−91

[30] International Federation for Structural Concrete (fib) 2013 *fib Model Code for Concrete Structures 2010* (Weinheim: Wiley-VCH Verlag GmbH & Co. KGaA), pp 78-9

[31] Isu N, Teramura S, Ishida H and Mitsuda T 1994 Mechanical property evolution during autoclaving of aerated concrete using slag: II fracture toughness and microstructure *J Am. Ceram. Soc.* 2 2093-6

[32] Eden NB, Manthorpe AR, Miell SA, Szymanek P H and Watson K L 1980 Autoclaved aerated concrete form slate waste – part 1: some property/density relationship *Int. J Cem. Compos. Lightweight Concr.* 2 95−100

[33] UNI EN 772-1: 2002 Methods of test for masonry units — part 1, Determination of Compressive Strength.

[34] UNI EN 771-4: 2005 Specification for masonry units — part 4, Autoclaved Aerated Concrete Masonry Units.

[35] UNI EN 1992-1-1: 2015 Eurocode 2: Design of concrete structures – Part 1-1: General rules and rules for buildings.

Case Study – The CARES Sustainable Construction Steel Certification Scheme for Steel Reinforcement in Concrete

Lee Brankley[1], Ayhan Tugrul[1], Ladin Camci[1] and Dave Knight[2]

[1] CARES, U.K.
[2] One Planet Limited, U.K.

E-mail: leebrankley@ukcares.com

Abstract. The expectations of stakeholders across the construction industry value chain have evolved significantly in recent years. Driven by scientific trends, a greater understanding of sustainability impacts and new legislation, construction clients are demanding transparent, reliable data and comparable sustainability information about competing construction materials, management of their production and performance trends. Standard setting organisations and building rating systems are maturing in their requirements and third-party certification bodies have responded with improved certification options that enable the provision of data collection, auditing and reporting to meet construction client demands. The CARES Sustainable Constructional Steels (SCS) scheme is one option that applies to reinforcing carbon steel, reinforcing stainless steel, structural steel and hot rolled flat steel. Frequently specified as part of structural concrete building systems, the accredited scheme has been developed with the inputs of a wide range of stakeholders. It is based on foundations of technical specifications, traceability, product quality and the sustainability principles of inclusivity, integrity, stewardship and transparency. SCS Certification provides confidence to specifiers, designers and wider stakeholders that the sustainability impacts of constructional steel products throughout their supply chain are being actively managed and continually improved.

Keywords Steel, Certification; Environmental Product Declarations (EPDs); Life-Cycle Analysis (LCA); Concrete Reinforcement; Socio-Economic Impacts

1. Introduction

Constructional steels are a key part of any major construction project and are often used to reinforce concrete construction systems. Its supply chain covers; material sourcing, steel manufacture, distribution, fabrication and delivery to a construction site. It is international and complex. Global purchasing to meet local project specifications requires effective management of the supply chain to ensure projects are delivered on time, on budget and with the required performance.

The last ten years has seen a significant shift in the understanding and application of sustainability across this supply chain. Initially the preserve of greener design specifications and the management of environmental impacts on site, stakeholder feedback has resulted in a broader understanding of sustainable construction. It now encompasses; labour rights, community considerations, business ethics as well as low carbon, environmentally improved buildings and infrastructure. Additionally, these expectations run across the entire construction value chain from raw material sourcing, through product manufacture and transport to the construction site, use and end of life of construction materials.

This shift has fuelled an increasing demand for accurate, accessible and timely information on the environmental, social and ethical impacts of different materials and construction products. Legislation, a strong business case for efficiency and cost reduction are all supporting this drive for change in construction. BREEAM, LEED, Estidama, Greenstar and other building rating frameworks provide the structure and incentives for the specification of sustainable materials and periodically they introduce new criteria and raise their expected performance standards.

Expanding scope and improving standards has not been without challenges. To achieve the objectives of lower carbon usage, efficient use of natural resources and confidence that labour and welfare standards are maintained in the supply chain, there is a need for relevant, reliable and trusted data. Third-Party Certification Bodies play a key role in providing independent data validation, auditing and reporting. This information not only provides confidence to the designers, specifiers and owners of buildings and infrastructure but also enables them to compare competing suppliers' performance and management practices.

This paper describes how CARES, a not for profit certification body and competent authority set up by the UK government and industry, has worked with its stakeholders to develop and enhance its Sustainable Constructional Steel (SCS) certification scheme to help meet this demand and drive improvements to sustainability performance.

Based on full traceability, through the supply chain, the scheme provides an objective and workable approach to the identification, collection, auditing and reporting of sustainability data and management approaches at a supplier level. It also enables sector performance to be communicated. Companies under the scheme can be confident that their data is being produced in accordance with industry and international standards. Through the CARES Rosette Rating Scheme, they can now demonstrate higher levels of performance. Buyers have the confidence that they are selecting products which are produced under a system seeking to continuously improve sustainability performance.

2. Drivers of sustainability – Why act now?

Sustainability is increasingly well understood, yet still considered in many different ways. For clarity, we refer to the commonly accepted definition of sustainable development from the World Commission on Environment and Development, 1987, which states that *"Sustainable development meets the needs of the present without compromising the ability of future generations to meet their own needs."*

Whilst legislation is now common in many markets, the UK, the EU and certain American states, including California, have typically been at the forefront of the introduction of sustainability legislation and standards. Recent legislative and standardisation trends indicate a broadening of the range of issues which require systematic management and increased performance expectations.

The UK Modern Slavery Act[1], for example, introduced in March 2016, requires any business with a worldwide annual turnover above £36m to publicly report, in a statement, its efforts to eradicate slavery and human trafficking across its entire value chain. The statement, signed by a Director, must be easy to find from the home page of companies' website. This has similar requirements to the California Transparency in Supply Chains Act[2], which took effect in 2012.

The EU Non-Financial Reporting Directive (2014/95/EU)[3], which should have been transposed into national legislation by 31st December 2016, obliges large public-sector organisations, listed companies and certain other entities with more than 500 employees to disclose in their annual management report relevant and useful information on their policies, main risks and outcomes relating to at least:

- environmental impacts,
- social and employee aspects,
- respect for human rights,
- anticorruption and bribery issues, and
- diversity in their board of directors.

There is some flexibility in how to report, with allowances for following existing international guidelines and standards in this area, such as ISO 26000[4], OECD Guidelines for Multinational Enterprises[5] and the Global Reporting Initiative (GRI)[6].

The Paris Accord[7] regarding climate change is another significant example of increased demands for evidence of sustainability. The agreement of 195 nation states (ratified at time of writing by 174) is to keep a global temperature rise this century well below 2 degrees Celsius *(3.6 degrees Fahrenheit)* and to drive efforts to limit the temperature increase even further to 1.5 degrees Celsius *(2.7 degrees Fahrenheit)* above pre-industrial levels. Additionally, the agreement aims to strengthen our ability to deal with and recover from the impacts of climate change.

Whilst allowing nation states to present their own 'Nationally Determined Contributions' (their action plans), it essentially lays the foundations for an accelerated transition towards a low carbon economy and ultimately a fossil free future. Construction companies will have to respond and demonstrate their contribution to national reductions.

Furthermore, major infrastructure and construction projects are increasingly incorporating sustainability into procurement frameworks. For example, High Speed 2, the first new railway to be built north of London, UK, in 120 years, will evaluate all its contractors and their subcontractors against a balanced scorecard[8]. This scorecard incorporates social, environmental and economic management and performance, which is used when selecting suppliers as well as when monitoring performance during project delivery.

Being able to demonstrate sustainability management and performance is becoming a prerequisite of doing business on leading projects.

3. The CARES Sustainable Constructional Steel (SCS) Scheme

In response to these drivers, CARES initially set up a Sustainable Reinforcing Steel Certification scheme in 2012. The scheme initially covered reinforcing carbon steel and reinforcing stainless steel. Its scope was expanded in 2013 to include other structural steels and in 2015, hot rolled flat steel. It applies to primary producers of steel and fabricators who process that steel into building products that are used in a construction project.

The SCS scheme aims to deliver improved sustainability and resilience in the construction sector by providing assurance to construction product buyers that sound sustainability management practices are in place throughout their constructional steel supply chain. It provides full traceability, following the chain right back to the steel scrap or other raw material suppliers. It also provides a way for suppliers to clearly demonstrate sustainability management practice and outcomes.

The SCS scheme is subject to periodic review and further to stakeholder consultation, has recently been enhanced. The scheme criteria have been strengthened and a new rating system has been included - the CARES Rosette Rating System - which enables producers to differentiate their performance against their peers and buyers to benchmark their suppliers.

The SCS scheme is accredited to BS 8902:2009 (Responsible Sourcing Sector Certification Schemes for Construction Products)[9] by UKAS, the official UK accreditation body and is recognised by BREEAM UK New Construction 2014[10]. Company-specific verified Environmental Product Declarations (EPD's) to EN 15804[11], which are a requirement of the scheme, are accepted in the USGBC's LEED certification[12].

3.1 The concept of an extended product and how it applies to CARES SCS certified reinforcing bar

The concept of the extended product can be used to illustrate the demands placed upon a modern construction material supply chain, as shown in Figure 1.

At the centre of the diagram is the product's core benefit: the reinforcement of concrete within a construction project. Shown around the core are other attributes relating to that product's performance, e.g. strength and ductility. These attributes are stated in a product standard, e.g. BS 4449[13] and hence constitute the 'specified product'. All companies in the CARES-approved supply chain must have an environmental management system certified to ISO 14001: 2015[14], a quality management system certified to ISO 9001[15], a Health and Safety Management System certified to OHSAS 18001 or ISO 45001[16], a traceability mechanism and an independently verified environmental product declaration (EPD to EN 15804).

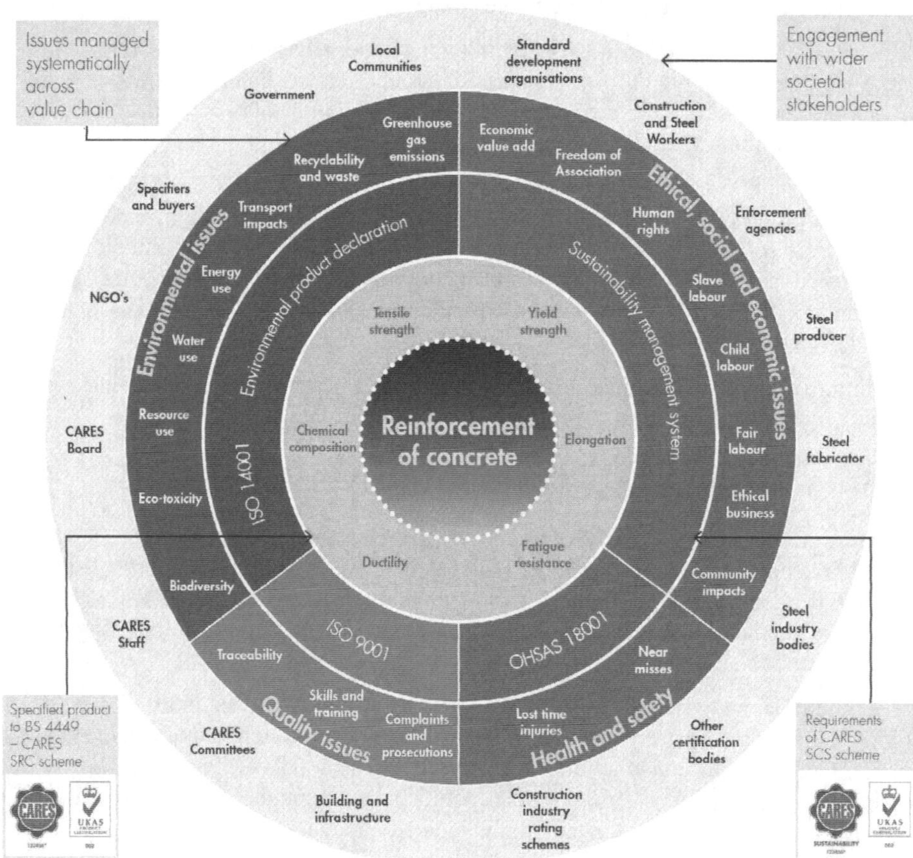

Figure 1. Concept of the 'extended' product applied to reinforcing steel

Beyond the physical product, buyers must be able to assess the broader impacts of its supply chain. The concept of 'Materiality' – the issues and impacts that are most significant and relevant to the organisation and its stakeholders along the entire value chain - is used to inform the selection of issues shown in the outer rings of Figure 1. Companies must demonstrate effective management of these (and other issues) as well as improving performance in these issue areas, as buyers are increasingly making purchase decisions based on these factors.

Finally, the whole construction network, and specifically in this case, the steel supply chain, is informed by the inclusion and consideration of a wide range of stakeholder opinion. This should be actively sought, to promote continual improvement towards sustainability performance. This is represented as the background outermost circle in Figure 1.

3.2 Basis of the scheme

The principles of; inclusivity, integrity, stewardship and transparency provide the basis for the CARES SCS Scheme. Companies must be able to demonstrate how these are incorporated into their management practices. They are also measured against a series of criteria covering the issues illustrated in Table 1 below and required to report and demonstrate continual improvement of sustainability performance.

The scheme now also includes a maturity matrix, which provides characteristics that demonstrate an evolution of management practices, enabling organisations to track their progress over time.

3.3 Certification process and operation

To achieve SCS certification, a company's product, quality, environmental, health and safety and sustainability management systems are all examined. A two-stage audit process is completed based on checking evidence on site against a self-completed Sustainability Key Performance Indicator (KPI) workbook and Environmental Product Declaration. CARES auditors audit over 100 criteria and performance indicators for each approved supplier under the headings shown below in Table 1. If all requirements are met, SCS certification is then awarded.

While the supplier remains responsible for compliance with legal requirements and standards, the scheme enables credits to be obtained from going beyond compliance. Credits can be gained for performance that meets specified criteria above the mandatory certification level as part of the 'Rosette Rating System'. Beyond basic certification, there are 3 additional performance levels reflecting good, better and best practices – 1 Rosette, 2 Rosettes and 3 Rosettes.

CARES conduct annual audits to ensure the scheme requirements are being maintained and performance continually improved. It collects, audits, collates and reports the sector sustainability performance data in line with requirements set out in BS 8902 - Responsible Sourcing of Construction Products and the Scheme Principles. The scheme itself is subject to accreditation audits by UKAS, which confirm its independence and that it meets all requirements of BS 8902. The Environmental Product Declarations are valid for three years and subject to an additional layer of independent verification by The UK Building Research Establishment (BRE).

3.4 Sustainability criteria alignment

A set of sustainability indicators are used within the audit and assessment framework of the SCS scheme. These are a result of extensive engagement with stakeholders, including with the construction industry, the steel industry, academics and government and have been approved by the CARES Sustainability Committee. They are aligned to those used in leading private and public-sector initiatives and standards as shown in Table 1, below.

Table 2 breaks out two of the CARES headline indicators to show how they work. In this case the indicator relates to Global Warming Potential (GWP) covering greenhouse gas emissions in production and transport.

Table 1. Comparison of relevant sustainability indicators across selected initiatives and standards (summarised)

ISO 26000 Social Responsibility Standard	BS 8902:2009 Responsible sourcing of construction products	BES 6001:2015[17] Framework Standard for Responsible Sourcing	CARES SCS scheme Indicator areas
Human Rights	Workers' conditions	Fundamental rights at work	Human Rights
Human Rights	Workers' conditions	Fundamental rights at work	Human Rights
Labour practices	Freedom to join trade unions	Fundamental rights at work	Freedom of Association
Labour practices	Slave labour	Fundamental rights at work	Slave Labour
Labour practices	Child labour	Fundamental rights at work	Child Labour
Labour practices	Safe and healthy working conditions	Health and safety	Safe and Healthy Working Conditions
Fair operating practices	Equality re. gender, ethnicity, religion, politics	Fundamental rights at work	Fair Labour Conditions
Fair operating practices	Fair wages Working hours and holidays	-	Fair Labour Conditions
Fair operating practices	Ethical business practice	Business Ethics	Ethical Business Practices
The environment	Recyclability, recycled content, renewability. Eco-t	Life-cycle assessment Ecotoxicity	Biodiversity and Eco-toxicity
The environment	Waste management	Waste prevention and management	Waste, Recyclability and Recycled Content
The environment	Harvesting or extraction impacts	Resource Use	Primary Material Use and Materials Efficiency
The environment	Greenhouse gas emissions	Greenhouse gas emissions	Greenhouse Gas Emissions/GWP
The environment	Energy usage	Energy Use	Energy and Water Use
The environment	Transport impacts	Transport impact	Transport Impacts
The environment	Water usage	Water abstraction	Energy and Water Use
The environment	Biodiversity	Site stewardship	Biodiversity and Eco-toxicity
The environment	Land remediation	-	Land remediation
-	Skills and training	Employment and skills	Skills and Training
-	Complaints and prosecutions	Complaints and prosecutions	Stakeholder Complaints and Prosecutions
-	Contribution to the built environment	Contribution to the built environment	
Community involvement	Community relations	Local Communities	Community Relations and Initiatives
-	Diversity and stability of the local economy	-	Diversity and Stability of the Local Economy
-	Long-term financial viability	Financial stability	Long-term Financial Viability

Table 2. Example indicators: Greenhouse gas emissions including from transport

Disclosure	Description	Indicator and Evidence required
Does the organization have a system for monitoring and reduction of the Global Warming Potential (GWP) and the Greenhouse Gas Emissions (GHG)?	Description: Global warming potential (GWP 100 years): an impact assessment level with global effect and a relative measure of how much heat a GHG traps in the atmosphere; GWP is mainly caused from CO_2 and CH_4 emissions which account for over 98% of GHG emissions from the steel industry. In exploratory impact assessments, other GHGs account for less than 2% of the steel industry GHG emissions, on a CO_2 equivalent basis[18]. CO_2 is an unavoidable in the steel industry due to energy usage. The process needs to be continually improved to increase energy efficiency and reduce the GWP.	• kg of CO_{2e} per tonne of steel billet • kg of CO_{2e} per tonne of delivered product Environmental Product Declaration (EPD) Report.
Has the organization completed "Transport Impact" worksheet within the workbook?	Description: The environmental impacts of transport are significant due to major use of energy and natural resources (e.g. petroleum). This creates air pollution, including nitrous oxides and particulates and is a significant contributor to global warming through emission of CO_2. The 'Transport Impact' sheet records the number of miles travelled per mode (road, rail or ship) of transport for; raw materials; for billets for further processing and for delivered product.	• kg of CO_{2e} per tonne of delivered product Completed "Transport Impact" worksheet.

4. Future Developments

A future vision for economic transition to a more sustainable state is gaining momentum. This includes adopting circular economy principles of preserving and enhancing nature capital (finite and renewable resources). This involves the circulation of products, components and materials keeping them at the highest utility at all times and designing out externalities. Industry is realising the benefits of this approach as it increases efficiencies, reduces costs and reduces negative environmental impacts.

Public procurement is becoming more sophisticated with expectations of reductions in embodied carbon from capital assets and operations. The UK's Infrastructure Carbon Review[19] highlighted the importance of the infrastructure sector in carbon reductions. Approximately 16% of UK emissions are associated with construction, operation and maintenance of infrastructure assets and the review says that by 2050, 24mt *(26.46 USmt)* of carbon can be reduced, saving the economy £1.4 billion. Part of enabling this process is the creation of common standards. This is being codified in the UK through the Publicly Available Specification (PAS) 2080 Carbon Management in Infrastructure[20].

Social impacts associated with procurement for any large infrastructure project, are not being neglected by the UK Government. It's 'Procuring Growth'[21] white paper encourages all of its departments to adopt a balanced scorecard approach, such as that adopted by the London 2012 Olympics. Many construction firms are seeking to calculate 'Social Value' associated with their projects. Ethical labour issues are also increasingly being codified, for example in BES 6002: Ethical Labour Standard[22], published in February 2017, which deals primarily with issues of fair employment practices and modern-day slavery.

These new developments will impact upon steel producers in the future. CARES' accreditation scheme will ensure that new expectations such as these will be met.

5. Conclusion

The sustainability agenda is maturing, not only in demanding that a broader range of issues are being systematically managed but also to see that performance in these areas is being improved. Independent, third party certification bodies play a key role in enabling industry to clearly demonstrate it is meeting these demands. They help to deliver confidence to construction project designers and specifiers through the provision of more transparent and reliable data and comparable information about competing construction materials.

Whilst there has been a convergence in the scope of international and sector standards relating to sustainable development, there remain significant differences in the structure and approaches taken. Therefore, clearly defined boundaries for construction sector value chain elements and clear methodologies for the collection and collation of material sustainability indicators are crucially important.

CARES, the constructional steel industry and other stakeholders have developed an objective and workable approach to the identification, collection, auditing and reporting of sustainability data and management system processes to deliver continual performance improvement. The Scheme uses industry best practice and International Standards throughout the whole constructional steel supply chain.

Fundamentally, CARES Approved Sustainable Constructional Steel producers and processors are helping to deliver more sustainable construction and infrastructure.

References

[1] Her Majesty's Stationery Office (http://www.legislation.gov.uk/ukpga/2015/30/contents/enacted), accessed 22nd Feb 2018
[2] California Attorney General (https://oag.ca.gov/sites/all/files/agweb/pdfs/cybersafety/sb_657_bill_ch556.pdf) accessed 22nd Feb 2018
[3] European Commission (https://ec.europa.eu/info/business-economy-euro/company-reporting-and-auditing/company-reporting/non-financial-reporting_en) accessed 22nd Feb 2018
[4] International Standards Organisation (https://www.iso.org/iso-26000-social-responsibility.html) accessed 22nd Feb 2018
[5] Organisation for Economic Cooperation and Development (www.oecd.org/corporate/mne/1922428.pdf) accessed 22nd Feb 2018
[6] The Global Reporting Initiative (https://www.globalreporting.org/standards/) accessed 22nd Feb 2018
[7] The United Nations Framework Convention on Climate Change (http://unfccc.int/paris_agreement/items/9485.php) accessed 22nd Feb 2018
[8] High Speed 2, Supplier Guide (http://assets.hs2.org.uk/sites/default/files/inserts/CS086%20Supplier%20Guide%20V11_Web%20-%20Interactive%20-%20final_0.pdf) accessed 22nd Feb 2018
[9] British Standards Institution (https://shop.bsigroup.com/ProductDetail/?pid=000000000030191223) accessed 22nd Feb 2018
[10] The Building Research Establishment (https://www.breeam.com/BREEAMUK2014SchemeDocument/) accessed 22nd Feb 2018
[11] British Standards Institution (https://shop.bsigroup.com/ProductDetail/?pid=000000000030279721) accessed 22nd Feb 2018
[12] United States Green Building Council (https://new.usgbc.org/leed) accessed 22nd Feb 2018
[13] British Standards Institution (www.ukcares.com/downloads/guides/PART3.pdf) accessed 22nd Feb 2018
[14] International Standards Organisation (https://www.iso.org/standard/60857.html) accessed 22nd Feb 2018
[15] International Standards Organisation (https://www.iso.org/standard/62085.html) accessed 22nd Feb 2018
[16] International Standards Organisation (https://www.iso.org/standard/63787.html) accessed 22nd Feb 2018
[17] Building Research Establishment (http://www.greenbooklive.com/filelibrary/responsible_sourcing/BES-

IOP Conf. Series: Materials Science and Engineering **442** (2018) 012012 doi:10.1088/1757-899X/442/1/012012

6001--Issue-3.1.pdf) accessed 22nd Feb 2018
[18] World Steel Association, Methodology report Life cycle inventory study for steel products, 2011 (http://www.hare.com/wp-content/uploads/2015/11/WHG-Worldsteel-LCA-Report.pdf) accessed 22nd Feb 2018
[19] Infrastructure Cost Review and Green Construction Board, 2013, (https://www.gov.uk/government/publications/infrastructure-carbon-review) accessed 22nd Feb 2018
[20] Construction Leadership Council's Green Construction Board (GCB) (https://shop.bsigroup.com/PAS2080) accessed 22nd Feb 2018
[21] Crown Commercial Service and Cabinet Office (https://www.gov.uk/government/publications/procurement-policy-note-0916-procuring-for-growth-balanced-scorecard) accessed 22nd Feb 2018
[22] Building Research Establishment (http://www.greenbooklive.com/filelibrary/ELS/BES-6002-Ethical-Labour-Sourcing-Standard.pdf) accessed 22nd Feb 2018

IOP Conf. Series: Materials Science and Engineering **442** (2018) 012013 doi:10.1088/1757-899X/442/1/012013

Contribution of concrete structures to sustainability – challenge for the future

Petr Hajek

Czech Technical University in Prague, Czech Republic

E-mail: petr.hajek@fsv.cvut.cz

Abstract. New situation in natural and socio-economic environment requires new technical solutions for construction of new and reconstruction and modernization of existing structures. Structures and all built environment should be better prepared for new conditions – they should be sustainable and resilient. Concrete gradually becomes building material with high potential for new technical solutions resulting in needed environmental impact reduction and consequent social and economic improvement. The paper presents potential contribution of concrete and concrete structures to solution of Sustainability Development Goals specified in UN 2030 Agenda for Sustainable Development and brings basic principles of implementation of sustainability approach into fib Model Code 2020.

Keywords: sustainability, concrete structures, environmental impact, resilience, SDG

1. Background and context

Conditions for life on the Earth are changing. Biodiversity including human life is affected by climate changes, decrease of quality of environment and depletion of natural resources. Changes of these conditions is innate and everlasting process. Although Earth goes through natural regular cycles of warming and cooling, human activities evidently accelerated these climate changes. Global climate changes are now happening faster than anticipated. The impacts of climate changes include warming temperatures, changes in precipitation, rising sea levels and increases in the frequency or intensity of some extreme weather events (Hajek 2016a). Floods, tropical storms, hurricanes, tornados, wildfires, heat and cold waves, longer periods of drought etc. are more and more frequent and with higher intensity. These impacts threaten human health by affecting the quality of food, water, air, and the entire quality of environment in which we live. The frequency of climate-related natural disaster is now nearly five times higher as it was in the 1970s (WMO 2015) – see Figure 1.

As a consequence, socio-economic problems graduated in recent years, including human migration. These aspects are in mutual interaction. Depletion of natural sources influences climate and causes economic and social stresses. Quality of the environment directly influences biodiversity, social life of humans and existence of animals on the Earth.

The natural and socio-economic effects and changes are different (in qualities and quantities) in different regions. It is thus essential to consider regional specifics.

New situation in natural and socio-economic environment requires new technical concepts and solutions for construction of new and reconstruction and modernization of existing structures – buildings and infrastructure, like roads, bridges, dams etc. Structures and all built environment should be better prepared for new conditions – they should be sustainable and resilient. Advanced codes for structural design based on a holistic approach and considering a wide range of sustainability issues are thus essential, and urgently needed.

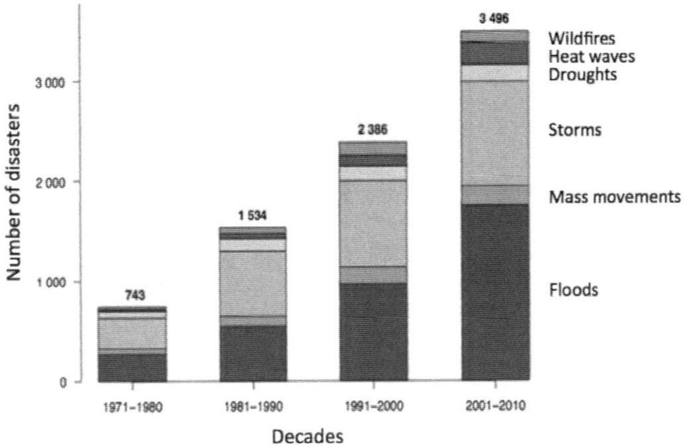

Figure 1. Increasing number of disasters by hazard type since 1970 (based on WMO data)

2. Sustainability perspective

Sustainability represents in general savings of biodiversity including survival of humans on the Earth. In International Standard ISO 15392 (ISO 2008) are presented general principles of sustainability related to buildings and other construction works. There are three primary aspects (pillars) of sustainability: social, environmental and economic. These pillars are formed by groups of aspects with associated performance levels and goals:

- Social aspects: social performance – social responsibility
- Environmental aspects: environmental performance – environmental impact
- Economic aspects: economic performance – economic efficiency

The performance level of structures could be optimized and assessed by wide range of criteria (described by indicators and benchmarks) in the framework of these three pillars of sustainability. Figure 2 presents the framework of sustainability perspective covering three pillars of sustainability. Basic criteria of structural performance – safety, serviceability, durability – are social aspects, they are part of social responsibility. Environmental impact could/should be assessed using LCA standard methodology. Economic efficiency could/should be assessed using LCC standard methodology. Resilience is going across all three pillars of sustainability.

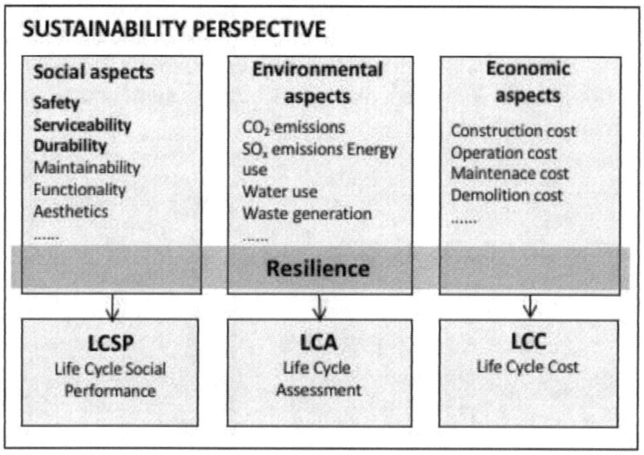

Figure 2. Framework of Sustainability perspective and assessment

It is essential to consider changes and development of performance within the entire life of a structure. Life Cycle approach is thus important part of any criteria assessment. This could be performed using standard assessment methods:

- Social criteria: LCSP – Life Cycle Social Performance – standard technical requirements and assessment (e.g. safety, serviceability, durability, … etc.). These requirements and methodologies for their assurance are specified in wide range of technical standards.
- Environmental criteria: LCA – Life Cycle Assessment – defined in standards ISO, CEN, ACI ... (e.g. ISO 14040 and associated standards).
- Economic criteria: LCC – Life Cycle Cost – standard LCC procedure defined in standards ISO, CEN, ACI, … (e.g. ISO 15686-5 and associated standards).

3. Role of concrete and concrete structures

3.1 Environmental impact of concrete and concrete structures

During the last century, concrete evolved into the most important building material in the world. At present concrete is the second most used material on the planet, after water. Annual production of concrete in the industrialized world amounts to 1.5–3.0 tonnes per capita (fib 2013). World cement production increased 12-fold in the second half of the last century (Hajek et al. 2011). Nevertheless, compared with other building materials, the specific magnitude of harmful impact embodied in a concrete unit is relatively small, see Figure 3.

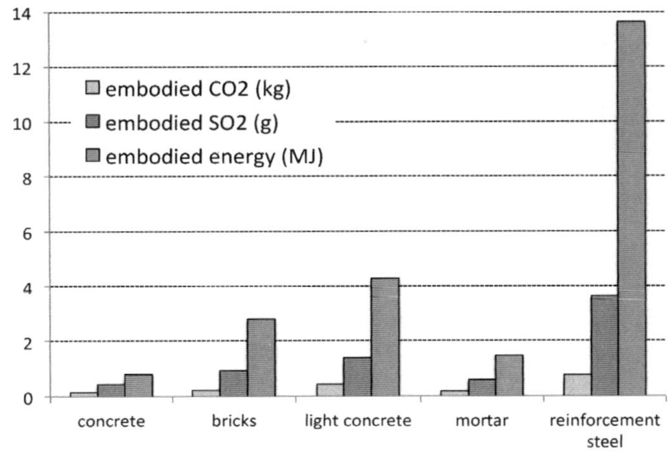

Figure 3. Environmental impacts of basic building materials based of generic data (data source: GEMIS 2010)

However, due to the high production of cement and concrete structures (the cement industry now accounts for approximately 8% of global man-made CO_2 emissions), the final negative cradle-to-gate global warming potential of all concrete structures is high (fib 2004). Due to recent development of concrete technology, concrete gradually becomes building material with high potential for new technical solutions, resulting in needed environmental impact reduction and consequent social and economic improvement.

With respect to specifics of concrete presented as a strong and durable material it is possible to design and construct robust structures with high level of resilience when faced to the exceptional natural or man-made disasters. High structural safety and reliability, and higher fire resistance of concrete (compared to other materials) results in a better resistance to extreme conditions during natural disasters (floods, storms, winds, hurricanes, tornados, fires, earthquakes, etc.) and terrorist attacks.

To gain from these advantages of concrete, it requires a better knowledge about technological processes and their impacts from wide variety of sustainability aspects within entire life cycle – from acquisition of materials, through production of concrete and concrete components, construction, use, up to demolition of concrete structure and recycling.

3.2 Advantages of concrete structures

Concrete is from many aspects structural material convenient for design of sustainable and resilient buildings and other engineering structures. Concrete structures are, in comparison to other materials, usually robust and strong to resist to exceptional loads during storms, hurricanes, tornados, floods, earthquakes and man-made explosions. Concrete is incombustible material with high fire resistance. High thermal mass of concrete could help to keep internal environment during heat and cold waves and can save energy needed for cooling of buildings. Concrete floors or walls can also significantly contribute to acoustic solutions in buildings.

Principal advantages of concrete structures from sustainability and resiliency viewpoints could be summarized into following points (Hajek 2016b):

- structural safety, including high resistance to exceptional effects like natural disasters (floods, storms, winds, hurricanes, tornados, fires, earthquakes, etc.) or man-made effects;
- thermal mass - concrete can contribute to energy savings associated with cooling and heating;
- acoustic properties – due to high specific weight of concrete, air-born sound protection of concrete floors or walls is good;
- high fire resistance – important especially for division of building into different fire compartments;
- long term durability enabling longer service life of buildings;
- flexibility – concrete technology enables high level of design flexibility;

- maintainability reducing operational cost and increasing overall building quality;
- use of recycled materials substituting natural materials in concrete mix.

These advantages could be significant in designing new constructions as well as in renovation of existing concrete structures. With respect to specifics of concrete presented as a strong and durable material it is possible to design and construct on this material bases robust structures with high level of resilience when faced to the exceptional natural or man-made disasters.

4. Contribution of concrete to sustainable Development Goals

United Nations (UN) summit Rio+20 held in 2012 in Brazil initiated a process of defining of a unified set of sustainable development goals. In October 2015 UN adopted Resolution 70/1: *Transforming our world: the 2030 Agenda for Sustainable Development* (UN 2015). This Agenda is a plan of action for people, planet and prosperity and should stimulate action up to 2030 in areas of critical importance for humanity and the planet. The principal goal is to end poverty, protect the planet and ensure prosperity for all.

There are specified 17 Sustainable Development Goals (SDGs) with 169 associated targets which are integrated and indivisible and balance three dimensions of sustainable development: economic, social and environmental. These new goals and targets came into effect on 1 January 2016 and guide the actions up to 2030. It is expected, that all stakeholders, all of us, will work to implement the SDGs Agenda within own countries and at the regional and global levels, taking into account national specifics and priorities.

One of the priority is sustainable urban development as a crucial aspect influencing the quality of life of people. Construction industry as a main stakeholder responsible for the use of material and energy resources has a key role in the implementation of SDG actions in the process of design, development and operation of buildings, roads, bridges and other infrastructure creating urban built environment. Concrete and concrete structures play in this process significant role.

In the following subchapters of this paper there are presented selected targets, where concrete industry and concrete structures can contribute to solution of SDG goals. Nevertheless, this selection is not enclosed and can serve as an inspiration and motivation for other stakeholders involved in concrete industry and in development of built environment. For further rethinking you can use whole SDG system presented at UN web pages (www.un.org).

4.1 Contribution of concrete industry to SDG 1: End poverty in all its forms everywhere

One of the target of SDG 1 is to build the resilience of the poor and those in vulnerable situations and reduce their exposure and vulnerability to climate-related extreme events and other economic, social and environmental shocks and disasters.

In case of extreme events and disasters concrete structures provide:
- high level of resistance;
- high level of resilience;
- concrete structures can serve as shelters for people in case of extreme events.

4.2 Contribution of concrete industry to SDG 3: Ensure healthy lives and promote well-being for all at all ages

Concrete structures can contribute to following targets of SDG 3: (a) substantially reduce the number of deaths and illnesses from hazardous chemicals and air, water and soil pollution and contamination and (b) halve the number of global deaths and injuries from road traffic accidents by following aspects:
- standard concrete is non-toxic, there are not used any hazardous chemicals;
- concrete can positively contribute to acoustic quality in buildings;
- high performance quality of buildings and infrastructure can improve people‘s health and wellbeing.

4.3 Contribution of concrete industry to SDG 7: Ensure access to affordable, reliable, sustainable and modern energy for all

One of the principal target of SDG 7 is to double the global rate of improvement in energy efficiency by 2030. Concrete structures can contribute to this target by:

- optimization of concrete mix, shape optimization and effective structural and technological concepts of concrete structure; this can result in a reduction of embodied energy in the concrete structure;
- concrete structures in buildings can due to high thermal mass reduce heating/cooling energy demands – especially reduce cooling energy demand in hot areas.

4.4 Contribution of concrete industry to SDG 8: Promote sustained, inclusive and sustainable economic growth, full and productive employment and decent work for all

Important target of SDG 8 is to improve progressively, through 2030, global resource efficiency in consumption and production.

- Recent development of concrete and new types of silicate composites including use of recycled components represents a great potential for resource efficiency in consumption and production.

4.5 Contribution of concrete industry to SDG 9: Build resilient infrastructure, promote sustainable industrialization and foster innovation

Concrete industry can contribute to several SDG 9 targets: (a) develop quality, reliable, sustainable and resilient infrastructure, to support economic development and human well-being, (b) promote inclusive and sustainable industrialization, significantly raise industry's share of employment and gross domestic product, (c) upgrade infrastructure and retrofit industries to make them sustainable, with increased resource-use efficiency and greater adoption of clean and environmentally sound technologies and industrial processes and (d) enhance scientific research, upgrade the technological capabilities of industrial sector, encouraging innovation and substantially increasing number of research and development workers.

Contribution to solution of these targets by concrete industry is essential and could be supported by

- the use of concrete structures for development of high quality, reliable, resource efficient and resilient built environment – housings and infrastructure;
- development and adoption of clean and environmentally sound technologies and industrial processes in production of concrete, concrete elements and structures and operation of concrete structures;
- continuing scientific research of concrete structures and associated technologies for design, construction and operation;
- support of dissemination of advanced technologies, research and innovation into developing countries.

4.6 Contribution of concrete industry to SDG 11: Make cities inclusive, safe, resilient and sustainable

SDG 11 covers, besides others, following targets (a) to ensure access to adequate, safe and affordable housing and basic services for all, (b) to provide access to safe, affordable, accessible and sustainable transport systems for all, improving road safety, (c) to enhance inclusive and sustainable urbanization and capacity for participatory, integrated and sustainable human settlement planning.

- Concrete structures play/and will play in the development of sustainable and resilient cities and their infrastructure crucial role.
- Concrete structures provide high level of safety, reliability and resilience of buildings and infrastructure in cities.

4.7 Contribution of concrete industry to SDG 12: Ensure sustainable consumption and production patterns

SDG 12 is focused to implementation of the 10-Year Framework of Programmes on Sustainable Consumption and Production Patterns. Targets relevant to concrete industry are (a) to achieve by 2030 the sustainable management and efficient use of natural resources, (b) substantially reduce waste generation through prevention, reduction, recycling and reuse and (c) encourage companies to adopt sustainable practices and to integrate sustainability information into their reporting cycle. This could be achieved through:

- optimization of concrete mix, resulting in a more efficient use of natural resources;
- increasing use of different kinds of recycled materials incl. recycled concrete aggregate in a concrete mix – and thus saving natural resources and reducing negative environmental impact (reduction of GHG emissions etc.).

4.8 Contribution of concrete industry to SDG 13: Take urgent action to combat climate change and its impacts

Main goal of SDG 12 is to support actions to reduce impacts of climate change by strengthening of resilience and adaptive capacity to climate-related hazards and natural disasters.

- Concrete structures have in case of climate-related hazards and natural disasters high level of resilience and adaptive capacity.

4.9 Contribution of concrete industry to SDG 17: Revitalize the global partnership for sustainable development

Targets of the last SDG 17 are focused to different actions in five areas: Finance, Technology, Capacity-building, Trade and Systemic issues. The leading principle is to promote the development, transfer, dissemination and diffusion of environmentally sound technologies to developing countries.

Implementation of advanced concrete technologies with reduced environmental impacts and increased performance quality could significantly contribute to solution of global problems. This requires international cooperation, access to science, technology and innovation, dissemination of knowledge, support in education and cooperation in science.

It is evident from the above overview that optimized concrete structures using new types of concrete in advanced technologies can significantly contribute to fulfilment of some specific targets of Sustainable Development Goals, and especially can contribute to reduction of global environmental impacts and can support increase of sustainability and resilience of built environment and entire society. The role of concrete industry in the process of sustainable development is thus essential.

5. Implementation of sustainability approach into Model code

5.1 General – fib attitude and process

The need for consideration of sustainability in the design and assessment of concrete structures has been already specified in previous *fib* Model Code 2010 (fib 2012). This was initiated by *fib* SAG8 Sustainability initiative and *fib* Commission 3 both chaired by Koji Sakai.

Preparation of the new *fib* Model Code 2020 (MC2020) started in 2016 From the very beginning sustainability was considered as one of the leading principles. Convener of Task Group 10.1 "Model Code 2020" Stuart Matthews expressed in Editorial for journal Structural Concrete (Matthews 2017) following:

- MC2020 will take sustainability as a fundamental requirement, based upon a holistic treatment of societal needs and impacts, life-cycle cost and environmental impacts.
- MC2020 provides an important opportunity to further advance our approach to the sustainable design, construction and through-life care of concrete structures.

This clearly shows fib attitude to implementation of sustainability approach into design, construction and operation of any concrete structure.

fib Commission 7 Sustainability is responsible for taking care about specific chapters implementing sustainability approach into Model Code 2020. The goal of *fib* is to publish new Model Code MC2020 in the year 2020.

5.2 Concept of sustainability approach in MC2020

Sustainability should become a basic concept of any design, construction process, operation, maintenance and repair of concrete structure. Sustainability should be presented in MC2020 as a conceptual approach – as an "umbrella" for efficient/high quality design and operation of concrete structures through entire life cycle – considering all three pillars – social, environmental and economic.

Sustainability approach should be implemented into MC2020 in accordance with international standardization – ISO 15392 General principles. Environmental assessment should be solved using existing standard methodologies of LCA and economic pillar using standard methodologies of LCC (see Chapter 2). There are already standardized methodologies in ISO, CEN or other standard systems. The MC2020 should to specify the need for these assessments (LCA, LCC) and provide references to existing standards.

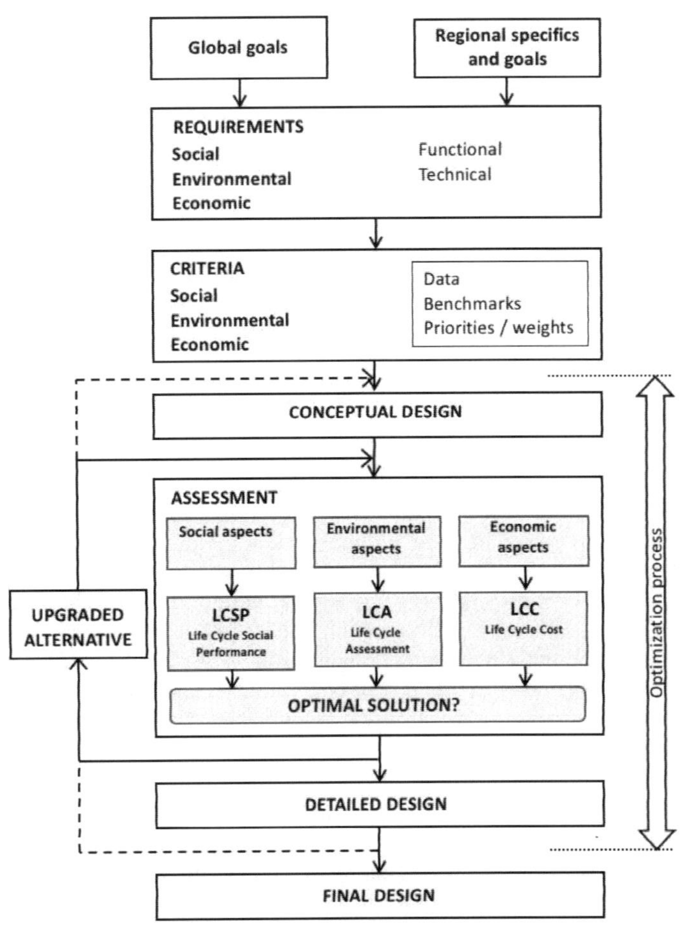

Figure 4. General Framework of sustainable design and assessment of concrete structures

Basic framework of sustainable design and assessment of concrete structures is shown in Figure 4. Based on global goals (given in SDGs) and regional specifics and goals are specified requirements and corresponding criteria associated with three sets of sustainability aspects. Optimization process is cyclic iterative prosecco starting with conceptual design, searching optimal solution based on assessment and resulting with detailed design and final design.

Safety, serviceability and durability of concrete structure is a part of social pillar of sustainability (see Figure 2). Specific needs for increase of reliability in specific regional conditions (climatic, geomorphological etc.) should be respected and solved within the framework of reliability concept of design of concrete structures. In this sense reliability concept should be extended also for the environmental consequences and needs for resilience.

Mixing of three pillars of sustainability into "simple" decision: what is the "most sustainable" solution could result in wrong decision and could be misused by different parties in order to support some preferred solution. E.g. increase of structural reliability could be in general in contradiction with optimum solution in environmental and economic pillar of sustainability. Mathematically expressed: solution of sustainable optimal result is typical multicriterial problem and could be searched in the form of Pareto optimum set of solutions – however, this is not practical and manageable for design code to be used by professional engineers. The code should be independent and should leave decision about weights between economy, environment and social pillars to decision makers or politics.

5.3 Regional specifics

There is a different ability of various societal and environmental systems in individual regions to mitigate or adapt to specific changes, including the extent of global change effects and potential natural or man-made disasters.

Considering resiliency, especially of buildings, the climate conditions and potential impacts and loses are essential for the design of sustainable structures/buildings. In different regions there are different major risks of natural disasters.

In general, the process of design and construction of structures varies in different countries and regions. They are more or less determined by regional specifics due to different material basis, energy mix, different climate conditions, different technology (based on the local labor cost, tradition in organization of work and different climate conditions) and differences in cultural traditions. Regional specifics should be considered when collecting the embodied environmental data of different materials. The type of material sources, mining technologies, transport means, transport distance and technology of production have a significant influence on the final unit embodied values.

As noted in chapter 3.2, concrete is from many aspects convenient structural material from viewpoints of sustainability and resiliency. The basic components of concrete are available in the most regions of the world. However, the level of technology and quality of construction is based on long-term tradition in construction and on economy aspects in the specific region. The implementation of standards for design and construction of concrete structures is especially in developing countries essential. Only thus could be ensured high performance quality and long-term functionality of concrete structures – from the perspective of sustainability and resiliency of buildings and other infrastructure in the built environment.

6. Conclusion

Sustainability should become a basic concept - an "umbrella" for efficient/high quality design and operation of concrete structures through entire life cycle – considering all three pillars – social, environmental and economic. Environmental assessment should be solved using existing standard methodologies of LCA and economic pillar using standard methodologies of LCC. There are already standardized methodologies in ISO, CEN or other standard systems. The goal of *fib* is to publish new Model Code MC2020 in the year 2020.

Acknowledgements

This work has been supported by the Ministry of Education, Youth and Sports within National Sustainability Programme I (NPU I), project No. LO1605 - University Centre for Energy Efficient Buildings – Sustainability Phase.

References

[1] fib 2004. *fib* Bulletin 28. Environmental design – State-of-the-art, ISBN 2-88394-068-1
[2] fib 2012. *fib* Bulletins 65 and 66. Model Code 2010 - Final draft, Volume 1, Volume 2, ISBN 978-2-88394-105-2, ISBN 978-2-88394-106-9
[3] fib 2013. *fib* Bulletin 71. Integrated Life Cycle Assessment of Concrete Structures, Hajek, P. et al. State-of-the art report, ISBN 978-2-88394-111-3
[4] GEMIS 2010. *Global Emission Model for Integrated Systems* - version 4.6, database, www.oeko.de/service/gemis/, Öko Institut.
[5] Hajek, P., Fiala, C. & Kynclova, M. 2011. Life Cycle Assessment of Concrete Structures - Step towards Environmental Savings, *Structural Concrete*, Journal of the fib, Volume 12, Number 1, ISSN 1464-4177.
[6] Hajek, P. 2016a. Sustainability of concrete structures in changing World, In proc. *International Conference on Concrete Sustainability ICCS16*, Madrid
[7] Hajek, P. 2016b. Sustainable Concrete and Concrete Structures – Challenge for Sustainable Future, In proc. *fib Symposium 2016, Performance-based approaches for concrete structures*, Cape Town
[8] ISO 15392 - Sustainability in building construction — General principles
[9] Matthews, S. 2017. fib Model Code 2020 - A new development in structural codes. *Structural Concrete*, Journal of the fib, Volume 18, Issue 5, Pages 651–652
[10] UN 2015. Resolution adopted by the General Assembly on 25 September 2015. RES 70/1 Transforming our world: the 2030 Agenda for Sustainable Development
[11] WMO 2015. World Meteorological Organisation 2015. *www.wmo.int* and *www.theguardian.com*

The Utilization of Recycled Materials for Concrete and Cement Production- A Review

Tereza Pavlů

University Centre for Energy Efficient Buildings, Technical University in Prague, Czech Republic

E-mail: tereza.pavlu@cvut.cz

Abstract. The utilization of recycled materials is one of the seven principles of sustainable construction. These principles based on the efficient use of resources were defined by the International Council of Building in 1994. Recycled materials such as recycled concrete aggregate or recycled brick aggregate from construction and demolition waste, waste glass from municipal waste or recycled waste gypsum from gypsum boards are able to use as replacement of primary raw materials in cement and concrete production. Concrete and cement production totally depends on the natural resources. The world production of concrete has been twelve times increased during last six decades. Nowadays, nearly one ton of concrete is produced each year for every human being in the world on average. On one hand, the use of recycled materials in cement and concrete production helps to reduce raw materials and urban land occupation. However, on the other hand, the recycled materials used as partial replacement of raw materials influence properties of final product. This paper reviews the different uses of recycled materials in cement and concrete production and the effect of the properties of these materials to the cement and concrete quality.

Keywords: Construction and demolition waste, Recycled concrete aggregate, Recycled brick aggregate, Waste glass, Recycled cement powder, Fine recycled aggregate

1. Introduction

Concrete is the most widely used construction material that is caused by its excellent mechanical and durability properties. Nearly one ton of concrete is produced for every human being in the world each year [1], which means almost ten billion tons of concrete annually. Concrete and concrete structures, which are used for buildings, bridges, dams etc. have a large impact on the environment, due to their enormous production and utilization. The most important environmental impacts of concrete production are progressive depletion of natural resources, high consumption of energy, large amount of emissions of greenhouse gasses and large amount of construction and demolition waste production [2]. The consumption of energy is mainly connected with cement production which accounts for 83% of total energy use in the production of non-metallic minerals [3] and represents around 90% of total embodied energy of concrete [4]. Furthermore, the emissions of greenhouse gasses which are responsible for climate change originate mainly from cement production and emissions of SO_X which is responsible for acidification mainly originates from the concrete production [4]. Aggregate represents only 13% to 20% of total CO_2 emissions of concrete in contrast with cement production which is responsible for 74 % to 81 % of total CO_2 emissions of concrete [5] and the cement industry accounts for approximately 5-7 % of total world CO_2 emissions [2], [6]. Finally, the large amount of construction and demolition waste (CDW) is produced. As the result of increasing construction industry, widespread urbanization and the economic condition, the old constructions are being demolished to build new structures. Due to these activities, the huge amount of the CDW is generated all over the world. The amount of CDW represents 25% -30% [7] of the overall waste generation and 850 million tons were generated in EU per year [8]. The CDW originates from the construction, renovation and demolition of buildings, bridges, roads and other structures. The CDW typically contains inert materials such as concrete, bricks, plasters etc. and also hazardous particulars such as asbestos, particulate matters etc. [9]. Inert waste is considered to have a priority to be recycled according to the EU Waste Strategy [10]. The waste concrete accounts for about 45% of construction waste, which is the highest percentage representation [11].

The concrete production is totally dependent on primary raw materials such as natural sands, natural gravel, cement and water. The reduction of environmental impacts of cement and concrete production could be occurred by utilization of waste materials in building industry. One of the main assumption for the use of recycled CDW is selective demolition, which means very thorough sorting of materials during demolition process [5]. This could lead to the good quality input materials for recycling process and subsequently to the secondary raw materials, which is possible to use as replacement of primary raw materials for new applications.

2. The possibilities of natural aggregate replacement in concrete production

High and constantly and rapidly increasing concrete production causes still growing natural resources depletion and amount of inert CDW. It was estimated that three billion tons of aggregate were consumed in European Union [12]. The use of recycled aggregate (RA) as partial or full replacement for natural aggregate in concrete structures belongs to one of the most effective approaches for recycled CDW utilization.

The research of the replacement of natural gravel by coarse recycled aggregate has become very complex over the last four decades. Recycled aggregate characteristics and their influence to concrete properties have been investigated during these decades. Mechanical, physical and durability properties of recycled aggregate concrete were examined and describes in many studies. The use of fine RA replacement as partial or full replacement of natural sand is mainly complicated by the higher mortar and impurity contents of the fine RA as compared to coarse RA [13]. This is one of the main reason that the replacement of sand in concrete mixture by secondary raw materials has not been defined in Standards yet. Although, the sand extraction from seaside has high environmental impact such as increasing seaside erosion, changing wave behavior, local fauna and flora ecosystems and other environmental aspects. The sand extraction from riverbed has negative impact to local ecosystems [14].

Recycled concrete aggregate (RCA), which contains more than 70% of concrete particles, is according to standards [15], [16] possible to use as partial replacement of coarse natural aggregate in certain applications (see Figure 1.). The use of recycled brick aggregate (RBA), which contain more than 50% of brick particles, fine RCA and RBA, originated from concrete or bricks CDW, and glass aggregate has not been defined as partial replacement of aggregate in concrete in Czech standards yet. As it was written before, the main assumption of high quality recycled materials, which are possible to use as partial replacement in new structural applications is sophisticated demolition and recycling process.

Figure 1. RCA from Czech recycling centre of fraction 63/120, 8/16, 4/8 and 0/4 mm

2.1. The utilization of recycled concrete aggregate (RCA) in concrete mixture
The possibilities of utilization of RCA are mainly connected with quality and properties of recycled aggregate. The main barriers in utilization of RCA are their high water absorption which negatively influence workability of fresh concrete and unwanted impurities which could degrade mechanical properties of concrete [13]. The water absorption of coarse RCA is between 0.5% and 14.75% which is up to twenty times higher in comparison with natural aggregate. The dry density of coarse RCA is generally lower than that natural gravel and ranges from 1900 to 2700 kg/m^3 [13]. The water absorption of fine RCA is between 4.3% and 13.1% and the dry density of fine RCA ranges from 1900 to 2360 kg/m^3 [17]. Higher water absorption and lower density of RCA is caused by old mortar attached on the surface of origin aggregate, due to that mortar is more porous and is less dense than the particles of aggregate [13].

2.1.1. The utilization of fine RCA in concrete mixture. The studies of utilization of fine RCA as partial or full replacement of sand show the main problems with its use such as higher water absorption due to high content of cement mortar and unwanted and undefined impurities. The water absorption of recycled aggregate concrete with partial replacement of sand and gravel of RCA is higher in comparison with conventional concrete. The main reason is high porosity of recycled concrete aggregates, which leads to mixes with higher w/c ratios, in order to achieve a constant workability [18]. On one hand, the fines content (particles finer than 75 μm) of aggregate has larger surface area, which leads to higher water consumption. On the other hand, the fines content would fill pores between larger particles for better aggregate skeleton of concrete mixture [19]. The particles size between 125 – 500 μm shows high content of cement mortar [14]. This could lead to the better mechanical and permeability properties of concrete. The maximal replacement rate of sand by fine RCA in concrete mixtures without significant effect to the mechanical properties is up to 30% [20]. However, in this study the laboratory prepared fine RCA was examined, which could show different results than fine RCA from CDW. The results of study where natural sand was replaced by fine RCA show decrease of compressive strength between 25% -30%, However, the mixtures with fine RCA contained less cement in this study. The results of compressive strength for replacement rate of sand by fine RCA was 25%, 50% and 75% were similar. The lowest compressive strength were examined for full replacement of fine fraction [21].

2.1.2. The utilization of coarse fraction of RCA in concrete mixture. Higher water absorption, the rougher surface and more irregular shape of RCA negatively influence the workability of fresh concrete [22]. Moreover, higher water absorption influence effective w/c ratio of concrete mixture [23]. Due to these findings, it is necessary to add more water to concrete former to attain a required workability. There are few ways how to reach it. One way is to pre-saturate the aggregate [23]–[25], further way is immersion of RCA 30 days in water [26], the third way is two-stage mixing approach [27]. The saturation level of the recycled aggregate could affect the mechanical properties of concrete, since at higher saturation levels the mechanical bonding between the cement paste and the recycled aggregate becomes weaker [28], [29].

The compressive strength of recycled aggregate concrete (RAC) is influenced by replacement rate of RCA and its quality and composition [30], [31]. Further, the decline of compressive strength depends on the presence of two interfacial transition zones, which is normally between aggregate and new cement mortar, but in RAC is also between attached old mortar and new cement mortar [27], [32], [33].

There were presented different maximal replacement rates of coarse aggregate in concrete by RCA without compressive strength degradation [13], [34], [35]. As a maximal replacement rate of natural gravel was stated 30% [13], [35] and 50% [13], [34]. The range of compressive strength decrease of concrete with full replacement of coarse natural aggregate by RCA is from 10% to 35% [2], [20]. The decline of compressive strength could be compensated by additional cement while maintaining the same amount of water. It is needed 8.3% of additional cement in concrete mixture with full replacement of natural aggregate [23]. Higher strength development of compressive strength of RAC after age 28 days is observed [20], [29]

The durability properties of concrete are an essential for its usage in structural applications. Concrete structures are very often exposed to the effect of thermal changes. The durability properties of concrete, especially freeze-thaw resistance and drying shrinkage could be negatively influenced by attached cement mortar and thereby higher porosity and water absorption of RCA. Drying shrinkage is higher for concrete containing RCA [21], [36], [37]. The water absorption by capillarity influence the freeze-thaw resistance of fine aggregate concrete. Higher capillary water absorption leads to worse freeze-thaw resistance [38]. The frost resistance of concrete is affected by its porosity, the water content in porous system and the environmental conditions. The frost resistance of RA can be assessed according to the method based on the change in the Los Angeles coefficient after 25 direct freezing–thawing cycles on saturated sample. Generally, the frost resistance of saturated RAC is not satisfying, and their use in structures exposed to severe climate is not recommended [39].

2.2. The utilization of recycled brick aggregate (RBA) in concrete mixture

The possibilities of utilization of RBA are mainly connected with properties and composition of recycled aggregate. The main barriers in utilization of RBA are their high water absorption which negatively influence workability of fresh concrete and unwanted impurities which could degrade mechanical properties of concrete [13]. The range of water absorption of coarse RBA is from 10.1% to 18.9%, which is up to twenty-five times higher in comparison with natural aggregate. The dry density of coarse RBA is generally lower than that natural gravel and ranges from 1800 to 2700 kg/m^3 [40]–[43]. One of the main difference between sand and fine RBA is density and water absorption. The density of fine RBA mainly originated from crushed bricks was between 2000 and 2500 kg/m^3 that is lower value in comparison with natural sand. The values of water absorption of fine RBA were measured between 12 and 15%, which is more than 10 times higher in comparison with natural sand [21], [40], [44].

Figure 2. The utilization of fine RBA in fine aggregate concrete mixture (50% replacement rate)

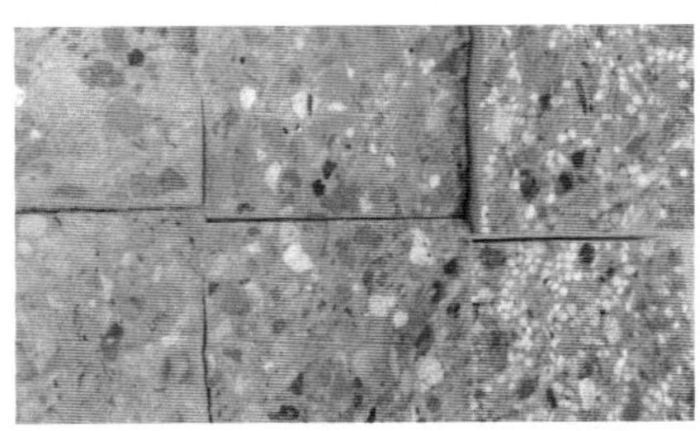

Figure 3. The utilization of coarse RBA in concrete mixture (up to 98% replacement rate)

2.2.1. The utilization of fine RBA in concrete mixture. The main characteristics of fine FBA responsible to the concrete properties were presented in many studies of utilization of fine RBA as partial or full replacement of sand [21], [40], [44]. High water absorption of fine RA causes worse workability of fresh concrete [21]. This finding have to be considered during mixture design and mixing, due to higher water absorption the additional water have to be calculated and add during mixing to concrete mixture to achieve the similar workability with conventional concrete mixture [44] or the aggregate were soaked into water for 24 hours before mixing [40]. The replacement ratios of fine natural aggregates by fine RBA were 25%, 50%, 75% and 100% [21], [40] and 20%, 50% and 100% [44]. As a coarse fraction was used natural aggregate. The compressive strength at age 28 days decreases with increasing replacement rate of sand [21], [40], [44]. The compressive strength between the age of 28 and 90 days increase for all mixtures containing fine RBA and the increase was higher in comparison with conventional concrete and with concrete containing fine RCA. Due to author's conclusions, the reason could be the silica and alumina contents in crushed bricks which could leads to the pozzolanic reactions [21]. The other reason could be lower w/c ratio of concrete mixtures with fine RA, due to non-compensation of high water absorption of RA and thereby lower effective w/c ratio. This assumption was proven by other studies where the w/c ratio was calculated according water absorption of RA. The results of these studies show no appreciable changes in the rate of strength development between the age of 28 and 90 days [40], [44]. The fine aggregate concrete mixture with 50 % replacement rate of sand by fine RBA is shown on Figure 2.

2.2.2. The utilization of coarse RBA in concrete mixture. The workability of fresh concrete with partial replacement of natural gravel by RBA is influence by its higher water absorption. The workability could be improved by pre-soaked RBA or adding additional water to concrete mixture during mixing process. Workability and effective w/c ratio influenced the compressive strength of concrete [45]. It was found that no significant changes in compressive strength of concrete containing coarse RBA with replacement rate 15%. The decline of compressive strength of concrete with containing coarse RBA replacement rate 30% was found [41]. The compressive strength of concrete with full replacement of coarse natural aggregate by RBA decreases up to 35% [40]. The concrete mixture with 0%, 66%, 75% and 98% replacement rate of gravel by coarse RBA is shown on Figure 3. Two shown concrete mixtures contain 46% and 58% of coarse RBA and recycled expanded polystyrene for better thermal properties.

2.3. The utilization of waste glass in concrete mixture

The EU countries produce approximately 30% of the total world's production of glass. However, the recycling rate of glass is around 60%. The utilization of clear glass without impurities is mostly into new glass. The clear waste glass is possible to recycled many times without significant degradation of quality and physical and chemical properties [46]. However, the large amount of unrecycled glass is still disposed in landfills [47]. The potential use of mixed glass with impurities is for concrete production. The utilization of waste glass as partial replacement of aggregate in concrete mixture is related with its chemical composition. The utilization of crushed or milled glass as partial replacement of sand or fines content in concrete mixtures is limited by potential of alkali-silica reaction (ASR) [48]–[50]. With regard to the risk of ASR, the maximum replacement rate of the aggregate is connected with the particle size of waste glass.

There have been published many studies which examine the possibility of the utilization of waste glass as aggregate in concrete. In these studies, the basic properties of aggregate and waste glass are usually compared. Further, the possibility of alkali-silica reaction (ASR) is verified. Furthermore, the influence on the mechanical properties of concrete with glass aggregate is examined. It was presented that the density of the glass is 2190 kg/m^3 that is about 15% lower than the density of the sand. The water absorption of the waste glass is 0.39% which is 14% lower than the water absorption of sand [51].

2.3.1. The utilization of fine crushed glass in concrete mixture. The results of many studies showed that with the increasing particle size of the waste glass, the risk of ASR increases. It was verified that the maximum grain size that can be used is 0.9 to 1 mm and the maximal replacement rate is 20%. The replacement rate of recycled glass with the particle size up to 0.15 mm is 40% and for the size between 0.036 and 0.05 can be used up to 70%. The utilization of fine particles of waste glass as a replacement of sand positively influence properties of concrete. Some studies indicated that glass powder (the finest part of waste glass) is used to increase pozzolanic properties. Further, the reaction of glass powder and cement leads to the production of cementitious materials. These aspects positively affects the properties of concrete [52]. The very important for utilization of glass in concrete mixture is chemical composition of waste glass. Due to its high content of SiO_2, the grinding waste glass is possible to use as partial replacement of silica fume or silica flour in HPC [53].

3. Possibilities of cement replacement in concrete production

Many more or less researched technologies how to reduce cement in concrete or how to reduce environmental impact of cement production or both were described. Among of emerging technologies are included high activation grinding, fluidized bed kiln, calcareous oil shale as an alternative raw material, use of steel slag as raw material for kiln, use of carbide slag etc [54]. Another method is the utilization of industrial by-products and waste materials, which have been used as supplementary cementing materials such as fly ash [55], silica fume [56], granulated blast-furnace slag [57], ceramic powder [58]–[60], recycled cement powder from waste concrete [61]–[63] and recycled gypsum (from recycled waste gypsum boards) instead of natural gypsum for cement production [64]. The main aim of the substitution is to reduce negative impact of cement production, due to its consumption of natural resources. The chemical composition of the concrete components and glass are cited in Table 1.

Table 1: The chemical composition of cement, glass, sand, silica fume, concret powder and brick powder [50], [60], [65].

Chemical composition	Cement [53]	Sand [53]	Silica fume [63]	Clear glass [53]	Crushed glass [53]	Glass powder [53]	Concrete powder [63]	Brick powder [68]
SiO_2 [%]	20.2	78.6	94.5	72.4	72.6	72.2	53.8	88.5
Al_2O_3 [%]	4.7	2.6	0.9	1.4	1.4	1.5	13.2	0.1
CaO [%]	61.9	7.1	0.4	11.5	11.7	11.4	13.6	1.0
Fe_2O_3 [%]	3.0	2.5	0.1	0.1	0.5	0.5	2.2	5.7
MgO [%]	2.6	0.5	1.6	0.3	0.6	0.8	2.6	0.8
Na_2O [%]	0.2	0.4	0.1	13.6	13.1	12.9	0.7	-
SO_3 [%]	3.9	-	-	0.2	0.1	0.1	-	-

The examples of materials which is possible to use as partial replacement of cementitious materials are shown on Figure 4 and 5.

Figure 4. The milled and grinding glass powder

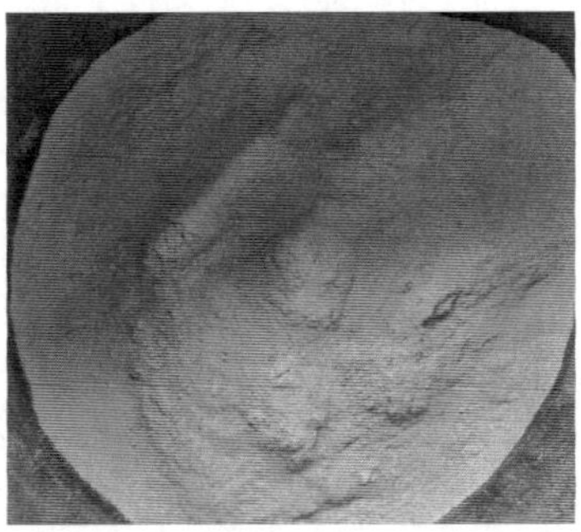

Figure 5. The recycled cement powder from waste concrete

3.1. The utilization of recycled cement powder from waste concrete as partial replacement of cement

The use of recycled cement powder from waste concrete is one way to utilization of the finest fraction from recycled concrete. The major constituent of the fines content of recycled concrete is cement paste, which is possible to use as clinker raw material, because contains high concentration of CaO [11]. There were published some possibilities how to separate aggregate and attached mortar. One of them is the heat treatment and abrasion method. This method decreases aggregate content in recycled cement powder to 33%. Nevertheless, this procedure is not so efficient for reduction of CO_2 emissions because the saving is only up to 2%. If the new technique will be developed and aggregate content decreases to 2.4% the reduction rate of CO_2 emissions would be reduce to 10% [11]. Further, the complete separation of the cement paste and aggregate is a heat treatment between 300 and 500°C. The attached cement mortar is separated from aggregate during this process and then separated cement mortar is milled to cement powder and use as partial replacement of cement. It was verified that recycled cement powder is possible to use instead of 10% of cement in concrete mixture in this study [66]. The other way of attached cement mortar separation from aggregate surface is by microwave heating [67].

3.2. The utilization of brick and ceramic powder as partial replacement of cement

There were found that the use of fine aggregate from waste bricks and sanitary ware contribute to pozzolanic activity [68]. The replacement rate of cement in concrete or mortars by ceramic waste up to 20% had no effect to mechanical properties and modulus of elasticity [69], [70]. The other study showed that the maximal replacement rate of cement in concrete mixture is 15% when natural aggregate is used and 5% when 30% of natural aggregate is replaced by recycled aggregate from waste bricks [59].

3.3. The utilization of grinding waste glass as supplementary cementing material in concrete

The glass production is provided by melting silica, soda ash and calcium carbonate at high temperature [71]. Furthermore, the glass powder (fraction 0,038 – 0,045 mm) has high silica content ($SiO_2 > 70\%$), high surface area and amorphous character [72]. These all factors suggested the potential of its utilization as an alternative supplementary cementing material in concrete [47].

4. Conclusion

The utilization of recycled materials for concrete and cement production is reviewed in this contribution. The results of cited references show many possibilities of utilization of waste materials from construction and demolition waste in cement and concrete production. On one hand, the use of these secondary materials as a full replacement of natural materials mostly leads to the mechanical properties decrease. On the other hand, when only the part of the natural materials is replaced, the quality of cement or concrete could be maintain. The clear benefit is the natural resources savings and often energy and CO_2 emissions savings that are depleted during cement and concrete production process.

Acknowledgements

This work has been supported by the Ministry of Education, Youth and Sports within National Sustainability Programme I, project No. LO1605 and by MPO TRIO FV 10397 – RENCO Recycled Environmental Concrete for Building Construction.

References

[1] B. C. Lippiatt and S. Ahmad, "Measuring the life-cycle environmental and economic performance of concrete: the BEES approach," in *Proceedings of the International Workshop on Sustainable Development and Concrete Technology*, 2004, pp. 213–230.

[2] S. B. Marinković, "Life cycle assessment (LCA) aspects of concrete," in *Eco-Efficient Concrete*, Elsevier, 2013, pp. 45–80.

[3] R. K. Srivastava, S. Vijay, and E. Torres, “Reduction of Multi-pollutant Emissions from Industrial Sectors: The U.S. Cement Industry – A Case Study,” in *Global Climate Change - The Technology Challenge*, Springer, Dordrecht, 2011, pp. 241–272.

[4] S. B. Marinković, M. Malešev, and I. Ignjatović, “11 - Life cycle assessment (LCA) of concrete made using recycled concrete or natural aggregates,” in *Eco-efficient Construction and Building Materials*, Woodhead Publishing, 2014, pp. 239–266.

[5] F. Pacheco Torgal and S. Jalali, *Eco-efficient Construction and Building Materials*. London: Springer London, 2011.

[6] “Cement Technology Roadmap 2009: Carbon emissions reductions up to 2050 | WBCSD publications library.” .

[7] “Construction and demolition waste - Environment - European Commission.” [Online]. Available: http://ec.europa.eu/environment/waste/construction_demolition.htm. [Accessed: 27-Feb-2018].

[8] C. Fischer, M. Werge, and A. Reichel, “EU as a Recycling Society,” *Eur. Top. Cent. Resour. Waste Manag. Work. Pap. 22009*, 2009.

[9] Z. Wu, A. T. W. Yu, L. Shen, and G. Liu, “Quantifying construction and demolition waste: An analytical review,” *Waste Manag.*, vol. 34, no. 9, pp. 1683–1692, Sep. 2014.

[10] “Policies and strategies - Waste - Environment - European Commission.” [Online]. Available: http://ec.europa.eu/environment/waste/strategy.htm. [Accessed: 27-Feb-2018].

[11] E. Gartner and H. Hirao, “A review of alternative approaches to the reduction of CO2 emissions associated with the manufacture of the binder phase in concrete,” *Cem. Concr. Res.*, vol. 78, pp. 126–142, Dec. 2015.

[12] “Effectiveness of environmental taxes and charges for managing sand, gravel and rock extraction in selected EU countries,” *European Environment Agency*. [Online]. Available: https://www.eea.europa.eu/publications/eea_report_2008_2. [Accessed: 27-Feb-2018].

[13] K. P. Verian, W. Ashraf, and Y. Cao, “Properties of recycled concrete aggregate and their influence in new concrete production,” *Resour. Conserv. Recycl.*, vol. 133, pp. 30–49, Jun. 2018.

[14] L. Evangelista, M. Guedes, J. de Brito, A. C. Ferro, and M. F. Pereira, “Physical, chemical and mineralogical properties of fine recycled aggregates made from concrete waste,” *Constr. Build. Mater.*, vol. 86, pp. 178–188, Jul. 2015.

[15] “CSN EN 206-1 Concrete: Specification, performance, production and conformity, (in czech), Prague.” CEN, 2014.

[16] “CSN EN 12620+A1 Aggregate for concrete, (in czech), Prague.” CEN, 2008.

[17] L. Evangelista and J. de Brito, “Concrete with fine recycled aggregates: a review,” *Eur. J. Environ. Civ. Eng.*, vol. 18, no. 2, pp. 129–172, Feb. 2014.

[18] D. Pedro, J. de Brito, and L. Evangelista, “Structural concrete with simultaneous incorporation of fine and coarse recycled concrete aggregates: Mechanical, durability and long-term properties,” *Constr. Build. Mater.*, vol. 154, pp. 294–309, Nov. 2017.

[19] A. K. H. Kwan, P. L. Ng, and K. Y. Huen, “Effects of fines content on packing density of fine aggregate in concrete,” *Constr. Build. Mater.*, vol. 61, pp. 270–277, Jun. 2014.

[20] L. Evangelista and J. de Brito, “Mechanical behaviour of concrete made with fine recycled concrete aggregates,” *Cem. Concr. Compos.*, vol. 29, no. 5, pp. 397–401, May 2007.

[21] J. M. Khatib, “Properties of concrete incorporating fine recycled aggregate,” *Cem. Concr. Res.*, vol. 35, no. 4, pp. 763–769, Apr. 2005.

[22] R. Kurda, J. de Brito, and J. D. Silvestre, “Influence of recycled aggregates and high contents of fly ash on concrete fresh properties,” *Cem. Concr. Compos.*, vol. 84, pp. 198–213, Nov. 2017.

[23] M. Etxeberria, E. Vázquez, A. Marí, and M. Barra, “Influence of amount of recycled coarse aggregates and production process on properties of recycled aggregate concrete,” *Cem. Concr. Res.*, vol. 37, no. 5, pp. 735–742, May 2007.

[24] I. Martínez-Lage, F. Martínez-Abella, C. Vázquez-Herrero, and J. L. Pérez-Ordóñez., “Properties of plain concrete made with mixed recycled coarse aggregate,” *Constr. Build. Mater.*, vol. 37, pp. 171–176, Dec. 2012.

[25] D. Pickel, S. Tighe, and J. S. West, “Assessing benefits of pre-soaked recycled concrete aggregate on variably cured concrete,” *Constr. Build. Mater.*, vol. 141, pp. 245–252, Jun. 2017.

[26] M. Gesoglu, E. Güneyisi, H. Ö. Öz, I. Taha, and M. T. Yasemin, “Failure characteristics of self-compacting concretes made with recycled aggregates,” *Constr. Build. Mater.*, vol. 98, pp. 334–344, Nov. 2015.

[27] V. W. Y. Tam and C. M. Tam, "Diversifying two-stage mixing approach (TSMA) for recycled aggregate concrete: TSMAs and TSMAsc," *Constr. Build. Mater.*, vol. 22, no. 10, pp. 2068–2077, Oct. 2008.
[28] M. B. de Oliveira and E. Vazquez, "The influence of retained moisture in aggregates from recycling on the properties of new hardened concrete," *Waste Manag.*, vol. 16, no. 1, pp. 113–117, Jan. 1996.
[29] C. S. Poon, Z. H. Shui, L. Lam, H. Fok, and S. C. Kou, "Influence of moisture states of natural and recycled aggregates on the slump and compressive strength of concrete," *Cem. Concr. Res.*, vol. 34, no. 1, pp. 31–36, Jan. 2004.
[30] I. B. Topçu, "Physical and mechanical properties of concretes produced with waste concrete," *Cem. Concr. Res.*, vol. 27, no. 12, pp. 1817–1823, Dec. 1997.
[31] İ. B. Topçu and S. Şengel, "Properties of concretes produced with waste concrete aggregate," *Cem. Concr. Res.*, vol. 34, no. 8, pp. 1307–1312, Aug. 2004.
[32] V. W. Y. Tam, "Economic comparison of concrete recycling: A case study approach," *Resour. Conserv. Recycl.*, vol. 52, no. 5, pp. 821–828, Mar. 2008.
[33] D. Kong, T. Lei, J. Zheng, C. Ma, J. Jiang, and J. Jiang, "Effect and mechanism of surface-coating pozzalanics materials around aggregate on properties and ITZ microstructure of recycled aggregate concrete," *Constr. Build. Mater.*, vol. 24, no. 5, pp. 701–708, May 2010.
[34] S. Kou, C. Poon, and F. Agrela, "Comparisons of natural and recycled aggregate concretes prepared with the addition of different mineral admixtures," *Cem. Concr. Compos.*, vol. 33, no. 8, pp. 788–795, Sep. 2011.
[35] V. W. Y. Tam and C. M. Tam, "Assessment of durability of recycled aggregate concrete produced by two-stage mixing approach," *J. Mater. Sci.*, vol. 42, no. 10, pp. 3592–3602, May 2007.
[36] K. K. Sagoe-Crentsil, T. Brown, and A. H. Taylor, "Performance of concrete made with commercially produced coarse recycled concrete aggregate," *Cem. Concr. Res.*, vol. 31, no. 5, pp. 707–712, May 2001.
[37] K. P. Verian, N. Whiting, J. Olek, J. Jain, and M. Snyder, "Using Recycled Concrete as Aggregate in Concrete Pavements to Reduce Materials Cost," Purdue University, Dec. 2013.
[38] T. Pavlů and M. Šefflová, "Study of the Freeze-Thaw resistance of the fine-aggregate concrete," presented at the EAN 2016 - 54th International Conference on Experimental Stress Analysis, 2016.
[39] R. Zaharieva, F. Buyle-Bodin, and E. Wirquin, "Frost resistance of recycled aggregate concrete," *Cem. Concr. Res.*, vol. 34, no. 10, pp. 1927–1932, Oct. 2004.
[40] F. Debieb and S. Kenai, "The use of coarse and fine crushed bricks as aggregate in concrete," *Constr. Build. Mater.*, vol. 22, no. 5, pp. 886–893, May 2008.
[41] P. B. Cachim, "Mechanical properties of brick aggregate concrete," *Constr. Build. Mater.*, vol. 23, no. 3, pp. 1292–1297, Mar. 2009.
[42] J. Yang, Q. Du, and Y. Bao, "Concrete with recycled concrete aggregate and crushed clay bricks," *Constr. Build. Mater.*, vol. 25, no. 4, pp. 1935–1945, Apr. 2011.
[43] M. T. Uddin, A. H. Mahmood, M. R. I. Kamal, S. M. Yashin, and Z. U. A. Zihan, "Effects of maximum size of brick aggregate on properties of concrete," *Constr. Build. Mater.*, vol. 134, pp. 713–726, Mar. 2017.
[44] A. V. Alves, T. F. Vieira, J. de Brito, and J. R. Correia, "Mechanical properties of structural concrete with fine recycled ceramic aggregates," *Constr. Build. Mater.*, vol. 64, no. Supplement C, pp. 103–113, Aug. 2014.
[45] H.-J. Chen, T. Yen, and K.-H. Chen, "Use of building rubbles as recycled aggregates," *Cem. Concr. Res.*, vol. 33, no. 1, pp. 125–132, Jan. 2003.
[46] A. Shayan and A. Xu, "Value-added utilisation of waste glass in concrete," *Cem. Concr. Res.*, vol. 34, no. 1, pp. 81–89, Jan. 2004.
[47] A. Omran and A. Tagnit-Hamou, "Performance of glass-powder concrete in field applications," *Constr. Build. Mater.*, vol. 109, no. Supplement C, pp. 84–95, Apr. 2016.
[48] H. Du and K. H. Tan, "Use of waste glass as sand in mortar: Part II – Alkali–silica reaction and mitigation methods," *Cem. Concr. Compos.*, vol. 35, no. 1, pp. 118–126, Jan. 2013.
[49] C. D. Johnston, "WASTE GLASS AS COARSE AGGREGATE FOR CONCRETE.," *J. Test. Eval.*, vol. 2, no. 5, pp. 344–350, 1974.
[50] Y. Jani and W. Hogland, "Waste glass in the production of cement and concrete – A review," *J. Environ. Chem. Eng.*, vol. 2, no. 3, pp. 1767–1775, Sep. 2014.
[51] Z. Z. Ismail and E. A. AL-Hashmi, "Recycling of waste glass as a partial replacement for fine aggregate in concrete," *Waste Manag.*, vol. 29, no. 2, pp. 655–659, Feb. 2009.

[52] R. Idir, M. Cyr, and A. Tagnit-Hamou, "Use of fine glass as ASR inhibitor in glass aggregate mortars," *Constr. Build. Mater.*, vol. 24, no. 7, pp. 1309–1312, Jul. 2010.
[53] D. Harbec, A. Zidol, A. Tagnit-Hamou, and F. Gitzhofer, "Mechanical and durability properties of high performance glass fume concrete and mortars," *Constr. Build. Mater.*, vol. 134, no. Supplement C, pp. 142–156, Mar. 2017.
[54] A. Hasanbeigi, L. Price, and E. Lin, "Emerging energy-efficiency and CO2 emission-reduction technologies for cement and concrete production: A technical review," *Renew. Sustain. Energy Rev.*, vol. 16, no. 8, pp. 6220–6238, Oct. 2012.
[55] K. Šeps and I. Broukalová, "Mechanical Properties of Cement Composites with Alternative Binders," *Adv. Mater. Res.*, vol. 1106, pp. 37–40, 2015.
[56] M. Ženíšek, T. Vlach, and L. Laiblová, "Dosage of Silica Fume in High Performance Concrete," *Key Eng. Mater.*, vol. 677, pp. 98–102, 2016.
[57] M. Ženíšek, T. Vlach, and L. Laiblová, "Options for Improving the Workability of High Performance Concrete," *Adv. Mater. Res.*, vol. 1106, pp. 53–56, 2015.
[58] O. Holčapek, "Resistance of refractory cement composite to cyclic temperature loading," *Key Eng. Mater.*, vol. 677, pp. 23–28, 2016.
[59] V. Letelier, E. Tarela, and G. Moriconi, "Mechanical Properties of Concretes with Recycled Aggregates and Waste Brick Powder as Cement Replacement," *Procedia Eng.*, vol. 171, pp. 627–632, 2017.
[60] P. Zhu, X. Mao, W. Qu, Z. Li, and Z. J. Ma, "Investigation of using recycled powder from waste of clay bricks and cement solids in reactive powder concrete," *Constr. Build. Mater.*, vol. 113, pp. 246–254, Jun. 2016.
[61] T. Pavlů, V. Kočí, and M. Šefflová, "Study Replacement of Cement with Recycled Cement Powder and the Environmental Assessment," *Solid State Phenom.*, vol. 249, pp. 136–141, 2016.
[62] H. A. Abdel-Gawwad, E. Heikal, H. El-Didamony, F. S. Hashim, and A. H. Mohammed, "Recycling of concrete waste to produce ready-mix alkali activated cement," *Ceram. Int.*, vol. 44, no. 6, pp. 7300–7304, Apr. 2018.
[63] E. Kwon, J. Ahn, B. Cho, and D. Park, "A study on development of recycled cement made from waste cementitious powder," *Constr. Build. Mater.*, vol. 83, pp. 174–180, May 2015.
[64] C. Chandara, K. A. M. Azizli, Z. A. Ahmad, and E. Sakai, "Use of waste gypsum to replace natural gypsum as set retarders in portland cement," *Waste Manag.*, vol. 29, no. 5, pp. 1675–1679, May 2009.
[65] B. M. V. Kumar, H. Ananthan, and K. V. A. Balaji, "Experimental studies on cement stabilized masonry blocks prepared from brick powder, fine recycled concrete aggregate and pozzolanic materials," *J. Build. Eng.*, vol. 10, no. Supplement C, pp. 80–88, Mar. 2017.
[66] H. Shima, H. Tateyashiki, R. Matsuhashi, and Y. Yoshida, "An advanced concrete recycling technology and its applicability assessment through input-output analysis," *J. Adv. Concr. Technol.*, vol. 3, no. 1, pp. 53–67, 2005.
[67] H. Choi, M. Lim, H. Choi, R. Kitagaki, and T. Noguchi, "Using Microwave Heating to Completely Recycle Concrete," *J. Environ. Prot.*, vol. 05, no. 07, pp. 583–596, 2014.
[68] T. Vieira, A. Alves, J. de Brito, J. R. Correia, and R. V. Silva, "Durability-related performance of concrete containing fine recycled aggregates from crushed bricks and sanitary ware," *Mater. Des.*, vol. 90, no. Supplement C, pp. 767–776, Jan. 2016.
[69] F. Pacheco-Torgal and S. Jalali, "Reusing ceramic wastes in concrete," *Constr. Build. Mater.*, vol. 24, no. 5, pp. 832–838, May 2010.
[70] F. Pacheco-Torgal and S. Jalali, "Compressive strength and durability properties of ceramic wastes based concrete," *Mater. Struct.*, vol. 44, no. 1, pp. 155–167, Jan. 2011.
[71] S. B. Park, B. C. Lee, and J. H. Kim, "Studies on mechanical properties of concrete containing waste glass aggregate," *Cem. Concr. Res.*, vol. 34, no. 12, pp. 2181–2189, Dec. 2004.
[72] Y. Shao, T. Lefort, S. Moras, and D. Rodriguez, "Studies on concrete containing ground waste glass," *Cem. Concr. Res.*, vol. 30, no. 1, pp. 91–100, Jan. 2000.

The influence of sulphur slime on the properties of alkali binding material from biomass bottom ashes

D Vaičiukynienė[1], R P Borg[2], A Kantautas[3], V Bocullo[1] and V Vaičiukynas[4]

[1] Faculty of Civil Engineering and Architecture, Kaunas University of Technology, Lithuania
[2] Faculty for the Built Environment, University of Malta, Malta
[3] Faculty of Chemical Technology, Kaunas University of Technology, Lithuania
[4] Institute of Water Resources Engineering Faculty of Water and Land Management, Lithuania

E-mail: danute.palubinskaite@ktu.lt, ruben.p.borg@um.edu.mt

Abstract. Alkali-activated materials are potential alternatives for Portland cement. Their use leads to reductions in CO_2 emissions and recycling of various industry by-products. These new alternative binders have a wide range of uses and high technological properties. In the research presented in this paper, the influence of additional sulphur slime powder (as accelerator) was investigated. The biomass bottom ash (BBA) was used as raw material and sodium hydroxide was used as the alkaline activator. The sulphur slime is a by-product in a fertilizer production plant. The samples were prepared with different amounts of sulphur slime; the levels of sulphur slime additive were 0%, 0.5%, 1.0%, 3.0% and 5.0% by weight in the raw material mixtures. Alkali-activated binders were mixed and conditioned at a temperature of 60 °C for 48 h, followed by curing at room temperature for 26 days. The microstructure of the material was analyzed through scanning electron microscopy (SEM) and the composition of the materials was analyzed using X-ray diffraction (XRD) and X-ray fluorescence (XRF) spectroscopy. The compressive strength of hardened alkali activated paste was measured after 28 days. In all the cases investigated, the compressive strength of hardened cement paste samples increases with the presence of Sulphur slime powder. It was found that during the hardening process sulphur reacts with sodium hydroxide forming Na_2SO_4, which acts as an accelerator. According to some researchers, Na_2SO_4 shows a significant accelerating influence in alkali-activated binder systems. The optimal amount of sulphur slime in raw materials mixtures was also evaluated.

Keywords: alkali-activated materials, Sulphur slime waste, biomass bottom ash

1. Introduction

The biomass bottom ash is a waste material generated by combustion as residue, and is collected in the bottom hopper of the burning furnace. There are various cases reported where bottom ash was used as a construction material. Carrasco et al. [1] investigated the addition of BBA in cement-based material. The results demonstrated that the addition of BBA increased the material's porosity, thereby decreasing its thermal conductivity and compressive strength. The best mechanical characteristics (compressive strength of 61.11 MPa) with an acceptable thermal conductivity value (0.773 W/(m·K)) were determined. Eliche-Quesada et al. [2] studied the feasibility of using fly or bottom biomass ash as a partial replacement in the production of eco-friendly fired bricks. The bricks do not present environmental problems according to the leaching study. Sklivaniti et al [3], reported on the characterization of a bottom ash derived from the combustion of olive plants trimmings. The authors examined its suitability as a substitute in the production of composite cements. According to the results reported, the woody bottom ash could be used as a substitute for Portland cement, in the production of composite cements of the strength classes 42.5 and 32.5 of EN 197-1. Carrasco et al. [4] studied the effect of the addition of bottom ash from the combustion process of biomass plants to calcium-silicate samples, and its potential as a construction material. The calcium-silicate samples were manufactured by mixing bottom ash with different sources of lime and Portland cement. The amount of bottom ash added ranged from 10% to 90% of the dry weight of the mixtures. The compressive strength reported covered a range from 25.21 MPa (bottom ash / lime) to 61.11 MPa (bottom ash / Portland cement).

The application of BBA from thermal power plants as precursors for the production of alkali-activated binders has been reported in different studies. Matalkah et al. [5] suggested the use of non-wood biomass combustion ash in the production of alkali-activated concrete and concluded that non-wood biomass ash could be an effective raw material in the production of alkali-activated concrete. Qiao et al. [6] investigated the fine fraction (<14 mm) of incinerator bottom ash. It was milled and thermally treated at 600, 700, 800 and 880 °C. The treated materials were activated with $Ca(OH)_2$ (10 wt%). Significant volumes of gas were generated during curing and this produced a macro-porous microstructure that limited the compressive strength to 2.8 MPa. The new materials may have potential for use as controlled low-strength materials.

In the geopolymer synthesis, a large variety of materials may be used as a source of aluminosilicate in amorphous or semi-crystalline state. Chen et al. [7] presented a study on geopolymer bricks manufactured using bottom ash from circulating fluidized bed combustion. The highest compressive strength of 21.9MPa was reached when the modulus was 1.5. Kalaw et al [8], reported that geopolymer samples were formed from the mixture of coal ash, using coal fly ash and coal bottom ash, and rice hull ash. Geopolymers were produced with a compressive strength of 18.5 MPa at 28-day of curing, when pre-cured at 80 °C for 24 h. Santa et al. [9] used pure bottom ash obtained after the mineral coal burning, as a source of aluminosilicate for geopolymer synthesis. The curing was performed at room temperature. The use of bottom ash for the geopolymer production exhibited good results, and the technology may be reproducible and environmentally viable for some samples analysed. Zhu et al. [10] used incineration bottom ash as the gas-forming agent, to replace costly aluminum powder, in the production of lightweight aerated metakaolin geopolymer. Results show that the gas generation capacity of incineration bottom ash is about 1/250 that of aluminum powder. Addition of incineration bottom ash remarkably increases the yield stress and viscosity of fresh pastes. Ul Haq et al [11] carried out research to develop geopolymer mortars and concrete from fly ash and bottom ash and compared the characteristics of the products obtained from either of these materials. Mechanical characterization indicated that geopolymer processed from fly ash reached a compressive strength of 61.4 MPa, whereas bottom ash geopolymer reached a compressive strength of 55.2 MPa. The mechanical characterization indicated that bottom ash geopolymers are almost equally viable as fly ash geopolymer.

In the present paper the application of sulphur slime as activation accelerator was studied. The Sulphur slime is a by-product of sulphuric acid (H_2SO_4) production waste from a fertilizer production plant. The slime occurs as a result of the cleaning operation of sulphur filters in the smelters, collectors, and sulphur repository. The amount of slime formed is calculated according to the norm (1t H_2SO_4 – 0.45 t slime). The aim of this work is to determine the main properties of alkali activated biomass bottom ash binder containing dissolved sulphur slime.

2. Experimental techniques

It is known from previous work that the addition of crushed raw sulphur slime into alkali - activated material can cause formation of undesirable expansive materials, leading to rapid expansions and the destruction of samples within 24 hours [12]. To produce a strong and durable material the crushed sulphur slime was crushed further using a ceramic ball mill to the fineness of cement. The specific surface area was determined using Blaine´s method according to the EN 196-6: 2005 standard.

Five series of samples were formed with each material. The composition of alkali-activated biomass bottom ashes samples varied with varying content of sulphur slime powder as additional material: 0 %, 0.5 %, 1 %, 2 % and 5 % (Table 1).

Table 1. Mixture design and main oxide molar ratios of alkali-activated biomass bottom ashes

Sulphur slime powder, wt%	Mass parts, %				SiO_2/Na_2O, mol	SiO_2/Al_2O_3, mol
	$Al(OH)_3$	Fly Ash	NaOH	H_2O		
0	16.7	51.4	13.8	18.2	2.0	1.0
0.5	16.7	51.4	13.8	18.2	2.0	1.0
1	16.7	51.4	13.8	18.2	2.0	1.0
2	16.7	51.4	13.8	18.2	2.0	1.0
5	16.7	51.4	13.8	18.2	2.0	1.0

Fine sulphur powder was dissolved in the alkali solution prior to the mixing of the alkali-activated material. In the alkali solution, sulphur reacts with some NaOH and sodium sulphate is formed.

$$NaOH + S + H_2O \rightarrow Na_2SO_4 + H_2O. \quad (1)$$

Then the alkali solution was added to the dry components - BBA and aluminum hydroxide. The components were mixed together until a homogenous paste was obtained. Then 20 × 20 × 20 mm cubes were formed in order to investigate the mechanical characteristics of the hardened samples. The moulds were filled and sealed to retain humidity for 24 hours. The samples were then placed in an oven at 60 °C, where they were kept for a period of 24 hours. After curing and drying, the bulk density and compressive strength of samples were evaluated. Elevated temperature was chosen because, according to various studies reported in the literature, this is more effective for alkali activation. The samples were kept sealed at ambient temperature for an additional 26 days. The cubes were tested for compression using an ELE AutoTest machine. The crushed samples were used for the XRD (powder samples) and SEM investigation.

SEM investigation was performed with a high-resolution scanning electron microscope FEI Quanta 200 FEG with a Schottky field emission gun (FEG).

The X-ray diffraction analysis of the materials was performed using the X-ray diffractometer DRON-6, with CuKα radiation and Ni filter. The powder X-ray diffraction patterns were identified with references available in PDF-2 database (PDF – 2 International Centre for Diffraction Data, 12 Campus Boulevard Newtown Square, PA 19073-3273 USA). XRFA elemental analysis was used to determine the chemical composition of biomass bottom ash and sulphur slime. The elemental composition was recalculated to oxides (Fig. 1, B and Fig. 3, B).

3. Materials

BBA from a Lithuanian combustion plant was used. The mineral composition of BBA was evaluated using XRD analysis (Fig. 1, A). It was concluded that the BBA contain significant amounts of: quartz, mullite and hematite. The bottom ashes consisted mainly of silicon, aluminum, potassium, calcium and iron (Fig. 1, B).

Sulphur slime is a grey solid material (Fig. 2, A). The SEM imagines indicate that sulphur slime has the compact and continuous structure, of a polymeric material. Furthermore, a small number of prisms, gypsum crystals are visible (Fig. 2, B).

The sulphur slime used in this research was obtained from the Lithuanian chemical industry plant SC "Lifosa". On average the plant produces 45.3 t of sulphur slime every month and 544 t annually. Sulphur slime consists mainly of sulphur (50%), with a moisture content of 3% and a density of 1 500 kg/m^3. According to the XRD analysis (Fig. 3) elemental sulphur and small amounts of $CaSO_4 \cdot 0.5H_2O$ and $CaSO_4$ are present in the mineral composition of this material. XRF analysis of chemical composition indicated that the major component of the material is sulphur (55.85%). The rest of the sulphur is contained in the $CaSO_4 \cdot 0.5H_2O$ and $CaSO_4$ composition. After recalculation to SO_3, it makes up 26.97% of the composition.

A

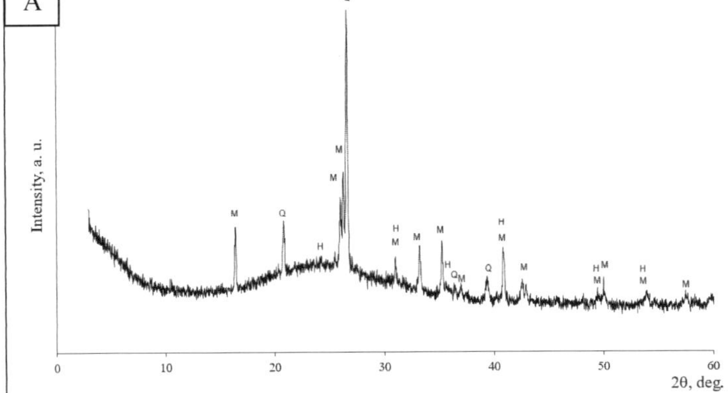

B

Oxide	Amount, %	Oxide	Amount, %
CaO	5.68	P_2O_5	1.31
SiO_2	49.47	SrO	0.11
Na_2O	0.95	ZrO_2	0.15
Al_2O_3	27.45	SO_3	0.92
MnO	0.06	ZnO	0.05
MgO	1.70	TiO_2	1.66
K_2O	4.54	CuO	0.03
Fe_2O_3	5.38	NiO	0.03

Figure 1. XRD and XRF analysis of fly ashes (FA). Notes Q – quartz SiO_2 (83-539), M – mullite $Al_{4.59}Si_{1.41}O_{9.7}$ (83-1881), H – hematite Fe_2O_3 (24-72).

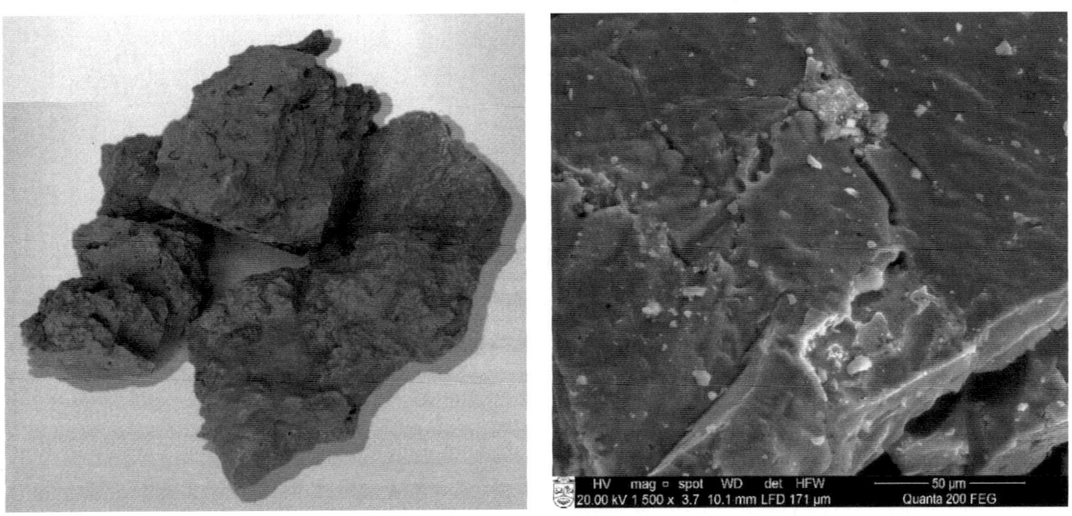

A B

Figure 2. Image of sulphur slime material (A) and SEM image of sulphur slime (B).

IOP Conf. Series: Materials Science and Engineering **442** (2018) 012015 doi:10.1088/1757-899X/442/1/012015

The main chemical element in sulphur slime is sulphur (55.85%). In addition, calcium oxide is identified at about 10.79% with small amounts of iron and silicon oxides (Fig. 3, B).

In alkali activation binding materials the most widely available and economical chemicals are the following: NaOH, Na_2CO_3, $Na_2O \cdot nSiO_2$ and Na_2SO_4. In this research a strong acid salt, namely Na_2SO_4 is used in addition to the NaOH.

4. Results and discussion

The compressive strength was measured after 28 days of curing in ambient temperature. The sample bulk density was around 1250 kg/m^3. In general, it does not appear that the sulphur slime had a significant influence on the sample density.

The compressive strength test results are presented in Fig. 4. As seen in figure 4 sulphur slime powder additive had positive effects on the compressive strength of samples within the additive range 0.5-2.0 %. The increasing amount of additive up to 1,0 % led to an increase in compressive strength reaching a peak of 16.94 MPa which is more than twice the compressive strength of the control samples (7.33 MPa). With 2.0 % additive, the compressive strength reached 16.82 Mpa, which is close to the compressive strength of samples with 1.0 % additive. Increasing the additive content up to 5.0 % has negative effects on the compressive strength (13.02 MPa). In general, it can be concluded that a small amount of additive can significantly increase the compressive strength of alkali activated biomass bottom ash. Therefore, the use of Na_2SO_4 improves the strength development of biomass bottom ash materials.

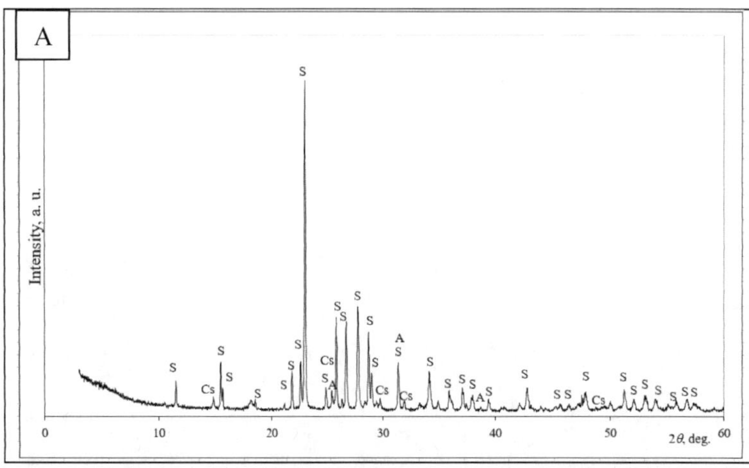

B

Oxide, %	Amount, %
SiO_2	3.05
Fe_2O_3	3.34
CaO	10.79
SO_3	26.97
S	55.85

Figure 3. XRD and XRF analysis of sulphur slime. Notes Cs is semi-hydrate gypsum $CaSO_4 \cdot 0.5H_2O$ (1-999), S is alfa elemental sulphur (24-733), A is anhydrite $CaSO_4$ (80-787).

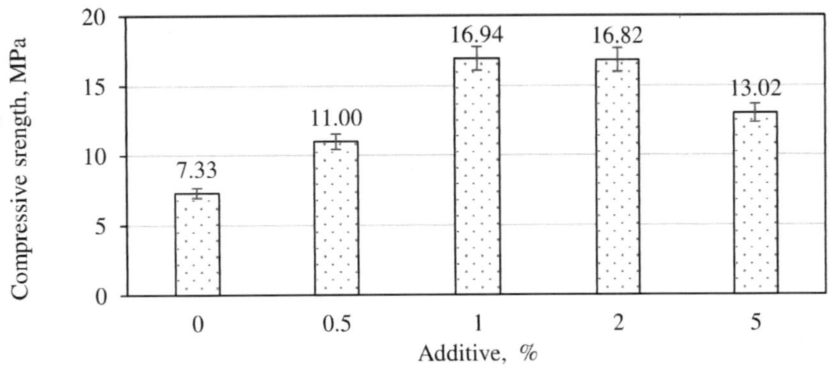

Figure 4. Compressive strength of alkali-activated biomass bottom ashes: without (reference sample) and with 0.5%, 1%, 2% and 5% sulphur slime additive.

The main products of the alkali activation reaction were amorphous alkali aluminosilicate gel and a crystalline phase (Fig. 5 and Fig. 6). Figure 5 shows the XRD patterns of the products prepared after alkali activation of BBA. Hydrosodalite $(Na_8SiO_{24})(OH)_2(H_2O)$ and calcium aluminium silicate hydroxide hydrate $CaAl_2(Si_2O_7)(OH)_2(H_2O)$ were produced. The height of the diffraction peaks for $CaAl_2(Si_2O_7)(OH)_2(H_2O)$ increased with the addition of Na_2SO_4. These correlate with the compressive strengths results. Qiao et al. [13] obtained similar results by using Na_2SO_4. Qian et al. [14] investigated the activation of fly ash in different blended cements. The addition of Na_2SO_4 can significantly increase the strength of all these blended cements. Donatello et al. [15] stated that the Na_2SO_4 salt acts as a safe and convenient in situ source of alkali to activate fly ash glassy phases. Na_2SO_4 reduced setting times, shortened the induction period, and compressive strength development, but also restricted ettringite formation. In a recent study, Lee et al [16] compared the strength and microstructure of Portland fly ash cements, which contained 40% fly ash and which was activated with Na_2SO_4. The research results support the hypothesis that the addition of small amounts of activators is a viable solution for increasing the early-age compressive strength of fly ash concrete.

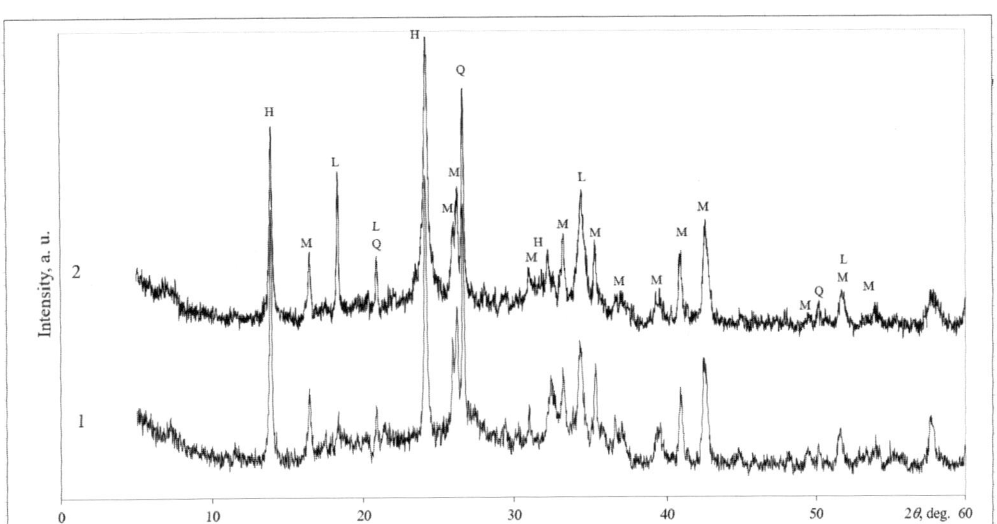

Figure 5. X-Ray diffraction pattern of alkali-activated biomass bottom ashes: without (reference sample) and with 5% sulphur slime. Notes: Q – quartz SiO_2 (83-539), M – mullite $Al_{4.59}$ $Si_{1.41}$ $O_{9.7}$ (83-1881), H – hydrosodalite $Na_8SiO_{24})(OH)_2(H_2O)$ (72-2329), L – calcium aluminium silicate hydroxide hydrate $CaAl_2(Si_2O_7)(OH)_2(H_2O)$.

IOP Conf. Series: Materials Science and Engineering **442** (2018) 012015 doi:10.1088/1757-899X/442/1/012015

Activation with a combination of Na_2SO_4 and NaOH yields the double amount of hydration products. In this case hydrosodalite and calcium aluminium silicate hydroxide hydrate formed (Fig 5, curve 2).

The activation of ash has been studied widely and the addition of Na_2SO_4 has been proved to be a successful method. In the research reported by Naik et al [17] the optimized blended cements consisted of 2 to 4% sodium sulfate anhydrite as a chemical activator. X-ray diffraction (XRD) and scanning electronic microscopy (SEM) analyses indicate that the addition of sodium sulfate accelerates the formation of CSH gel. Many studies have been conducted on accelerating the pozzolanic properties of fly ash in the blended cement systems.

The microscopic analysis, shown in Fig. 6, d, also confirmed the results as samples with 2% Sulphur slime were found to have greater transformation of the amorphous compound to the crystalline compound making its microstructure even denser than the other mixture without Sulphur slime powder, presented in Fig. 6a. Hydrosodalite prevailed in all investigated samples without and with sulphur slime additive [18].

Therefore, the higher strength of the alkali sulphate-activated pastes is attributed to acceleration of the pozzolanic reactions and the additional hydration products.

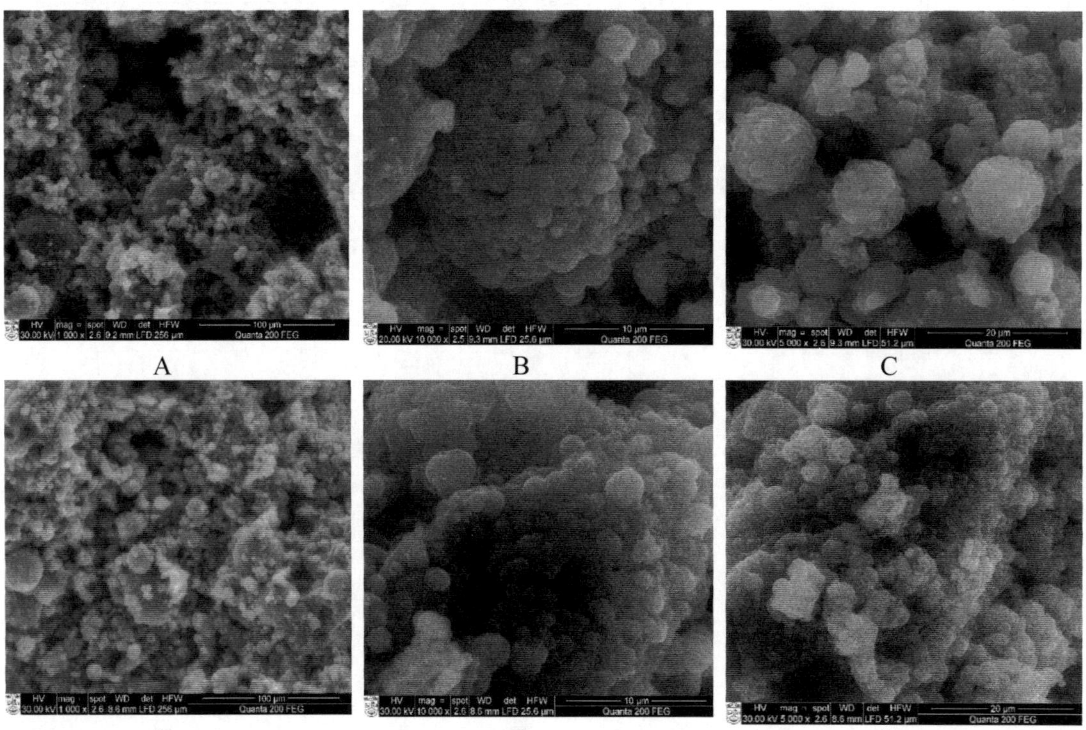

Figure 6. SEM pictures of alkali-activated biomass bottom ashes: without (A, B, C) and with 2% sulphur slime (D, E, F).

5. Conclusions

Based on the results obtained in this research, the following conclusions were made:

- The additive of sulphur slime is effective in increasing the compressive strength of alkali activated biomass bottom ash. Small amounts of the additive (up to 2%) can double the value of compressive strength. The addition of sulphur slime (Na_2SO_4) has beneficial effects on the compressive strength development of the samples.
- The application of Na_2SO_4 changes the chemical composition of alkali-activated material - calcium aluminium silicate hydroxide hydrate occurs in the binder's system, which can be attributed to the increase in the compressive strength.

The resulting Alkali Activated Material (AAM) microstructure was denser, consisting of amorphous gel-like structures.

References

[1] Carrasco B, Cruz N, Terrados J, Corpas FA and Pérez L 2014 *Fuel* 118 272
[2] Eliche-Quesada D., Felipe-Sesé MA, Moreno-Molina AJ, Franco F and Infantes-Molina A 2017 *Appl. Clay Scienc.* 135 333
[3] Sklivaniti V, Tsakiridis PE, Katsiotis NS, Velissariou D, Pistofidis N, Papageorgiou D and Beazi M 2017 *J. Environm. Chem. Engineer.* 5 205
[4] Carrasco-Hurtado B, Corpas-Iglesias FA, Cruz-Pérez N, Terrados-Cepeda J and Pérez-Villarejo L 2014 *Constr. Build. Mater.* 52 155
[5] Matalkah F, Soroushian P, Abideen SU and Peyvandi A 2016 *Constr. Build. Mater.* 121 491
[6] Qiao XC, Tyrer M, Poon CS and Cheeseman CR 2008 *Wast. Managem.* 28 1955
[7] Chen C, Li Q, Shen L and Zhai J 2012 *Environm. Technol.* 33 1313
[8] Kalaw ME, Culaba A, Hinode H, Kurniawan W, Gallardo S and Promentilla MA 2016 *Mater.* 9 580
[9] Santa RAAB, Soares C and Riella HG 2017 *Constr. Build. Mater.* 157 459
[10] Zhu W, Rao XH, Liu Y and Yang EH 2018 *J. Clean. Product.*
[11] ul Haq E, Padmanabhan SK and Licciulli A 2014 *Ceram. Internat.* 40 2965
[12] Vaičiukynienė D, Bocullo V, Kantautas A and Bistrickaitė R 2016 *Proc. Int. Conf. on Industrial & Hazardous Waste Management* (Chania: Technical University of Crete) p 114
[13] Qiao XC, Zhou B and Yu JG 2008 *Proc. Conf. on held at the Kingston University* (London: Taylor & Francis) p 501
[14] Qian J, Shi C and Wang Z 2001 *Cem. Concr. Res.* 31 1121
[15] Donatello S, Fernández-Jimenez A and Palomo A 2013 *J. Amer. Ceram. Soc.* 96 900
[16] Lee CY, Lee HK and Lee KM 2003 *Cem. Concr. Res.* 33 425
[17] Naik S, Vremec D, Wu L, O'Keeffe M and Shortman K 2003 *Blood* 102 601
[18] Part WK, Ramli M and Cheah CB 2015 *Constr. Build. Mater.* 77 370

IOP Conf. Series: Materials Science and Engineering **442** (2018) 012016 doi:10.1088/1757-899X/442/1/012016

The application of Natural Organic Additives in Concrete: Opuntia ficus-indica.

Anne Aquilina[1], Ruben Paul Borg[1], Joseph Buhagiar[2]

[1] Faculty for the Built Environment, University of Malta, Malta.
[2] Department of Biology, University of Malta, Malta

E-mail: ruben.p.borg@um.edu.mt, joseph.buhagiar@um.edu.mt

Abstract. The use of synthetic additives as setting retarders and plasticisers has over the past years become a standard practice in mortar mixtures, despite their potential adverse environmental effects. Since antiquity, man has used organic additives ranging from plant and animal-derived extracts, to improve the setting properties of mortars. These practices are slowly being revived and investigated scientifically to promote the use of materials with lower impacts on the environment. Amongst the many different types of plant-derived extracts that have been used as mortar and concrete additives, the prickly pear (*Opuntia ficus-indica*) mucilage extract, is popular in Meso- and South-America. The scientific basis of these additives lies in the hydrating properties of the mucilage polysaccharide complex. The purpose of this research was to prepare *Opuntia ficus-indica* (OFI) extracts in different forms and incorporate them in cement pastes and mortar mixtures by either replacing the water in the mixture with OFI mucilage or by replacing the cement in the mixture with OFI lyophilised powder. The inclusion of *Opuntia ficus-indica* additives in cement-based mortars increased their performance in terms of strength for both water and powder replacements. Conversely results in cement pastes showed that *Opuntia ficus-indica* additives weakened the cement paste samples in terms of strength in both water and power replacements. Results for both cement-based mortars and cement pastes showed an increase in setting time for both water and powder replacements, indicating that *Opuntia ficus-indica* additives could be potentially used as retarding agents.

Keywords: Natural organic additives, Opuntia ficsu-indica, setting time, retarder

1. Introduction

Concrete structures are expected to resist weathering action, chemical attacks, and physical action while still maintaining their desired engineering properties over their expected lifetime. The development of micro-cracks and subsequent penetration of fluids into these structures often leads to degradation and failure before the expected life-time. Durability performance is in part related to the setting and curing properties of the concrete, cement paste or mortar at the time of manufacture. Rapid setting of the cement affects the micro-structure of these materials.

The setting properties of mortars and concrete can be controlled using admixtures including different types of organic extracts. Retarding admixtures as the name implies help to retard the setting of the cement especially in the case of high ambient working temperatures associated with hot weather concreting. By using retarders, the heat generated from the exothermic hydration reactions is allowed to dissipate and the desired crystalline structure can develop.

This retarding action is usually provided by the addition of carbohydrates including cane sugar, as well as other organics and soluble zinc salts amongst many others. The dosage of any retarding admixture is determined through trial mixes. The effect of these admixtures depends on the chemical composition of the cement and the admixture performance. For example, a very low dosage of cane sugar (0.05 %) can delay setting time by as much as 4 hours. However, a larger quantity can completely prevent the cement from setting. It is also interesting to note that research has shown that some admixtures can act as retarding agents at low percentages but can also act as accelerators if higher percentages are used (Neville and Brooks 1987). Though the mechanism of retardation is not fully understood, it is known that admixtures modify crystal growth. The inclusion of these retarding admixtures influences greatly the rate of strength gain of concrete, and there is some evidence that the final strength may be improved. Several researchers have observed a higher ultimate strength of concrete when the rate of strength gain at the early stages is retarded (Chandra et al 1998). Furthermore, if the retarding admixture is also a water-reducing admixture, a higher strength can be achieved by decreasing the water to cement (W/C) ratio (Neville and Brooks 1987).

Despite their potential adverse environmental effects, the use of synthetic additives as setting retarders and plasticisers has over the past years become a standard practice in mortar mixtures. Organic additives have been used since ancient times to improve the working properties and durability of lime-based mortars and concrete. These practices are slowly being revived and investigated scientifically to promote the use of materials with lower impacts on the environment. Amongst the many different types of plant-derived extracts that have been used as mortar and concrete additives, the prickly pear (*Opuntia ficus-indica*) mucilage extract, is popular in Meso- and South-America. Previous studies have identified *Opuntia ficus-indica* (OFI) extract as a potential concrete and mortar additive. It is an ideal organic additive as the plant is widely available in warm climates and it can be easily harvested and processed to yield the additive. The scientific basis of these additives lies in the hydrating properties of the mucilage polysaccharide complex.

Chandra et al., (1998) studied the potential benefits of cactus extracts in mortars. They used X-ray diffraction and the scanning electron microscopy to understand the hydration process of the mortar with mucilage addition. Moreover, they also studied flexural strength, compressive strength and water absorption with regards to OFI addition. From the diffractograms, it was noted that the calcium hydroxide peaks decrease with the addition of the OFI extract (Chandra et al 1998). The conclusion was that the reduction in peaks was an indication that the calcium hydroxide could be reacting with components of the OFI extract forming calcium-based organic complexes. Chandra et al (1998) have shown that the natural OFI polymer influences the hydration process of tri-calcium silicate and hinders the formation of big crystals of calcium hydroxide. This research thus confirmed that the influence of the mucilage on the hydration process is not an artifact.

Martínez-Molina, Torres-Acosta, Celis-Mendoza, & Alonso-Guzman (2015) investigated the physical properties of cement-based mortars containing OFI extracts. In this case, a dehydrated powder was used as a replacement for the cement. This research was conducted over a period of 900 days focusing on four types of tests, total void content, compressive strength, ultrasound wave propagation and wet electrical resistivity. The authors concluded that the mortar fluidity was observed to decrease with the increase of the organic extracts and therefore more water was needed to obtain mortars with the same workability. Nevertheless, the hardened properties of the mortars with dehydrated OFI showed a marginal improvement in hardened properties even at low percentage replacements.

Several techniques are recorded in the literature to obtain the mucilage from the *Opuntia* pads (cladodes) which are modified stems that serve as a water storage and the main photosynthetic organ in the absence of true leaves. Each method of OFI extract production has its own set of constraints varying from low to the high energy input and this is in turn related to the amount of residual water in the final product.

The purpose of this research was to produce *Opuntia ficus-indica* (OFI) extract by two different techniques and incorporate them in cement pastes and mortar mixtures by either replacing the water in the mixture with fresh OFI mucilage, or by replacing the cement in the mixture with OFI lyophilised powder.

2. Materials and Methods

2.1. Opuntia ficus-indica Additions

Water and powder replacement methods were investigated. For both methods, the inner white section of the *Opuntia ficus-indica* pads or cladodes was used to produce the mucilage. These cladodes have an estimated water content of 90%. For the water replacement, the inner part of the cladode was finely diced, weighed and boiled. The resulting mucilage was then sieved to remove the large fibers, and was refrigerated until used in the cement pastes and mortar mixes. For the powder replacements, a freeze-drying process was used on the inner white section of the pads, followed by crushing of the material using a mortar and pestle. The weight of powdered mucilage was recorded before and after sieving through a 150 μm mesh. The process of lyophilsation is illustrated in Figure 1.

Figure 1. Opuntia ficus-indica lyophilising process: a) Opuntia ficus-indica plant; b) inner white section of the cladode; c)inner part of the cladode subject to the freeze-drying process; d) Opuntia

IOP Conf. Series: Materials Science and Engineering **442** (2018) 012016 doi:10.1088/1757-899X/442/1/012016

2.2. Paste and Mortar Specimen Preparation

2.2.1. Paste Sample Preparation

A total of sixty 20 x 20 x 20 mm paste cubes, with a water to powder (W/P) ratio of 0.28 and 0.35 were prepared in accordance with EN 196-1:2016. The consistency of the control sample was obtained through a number of trial mixtures until a consistency of (6 ± 2) mm was obtained. The same mixture was used for the different percentage replacements. In the case of the cement pastes containing powder replacements, the mixture was not workable for the same W/P Ratio. Therefore, a higher W/P Ratio was used for these samples. Once cast, the samples were demolded after 24 hours and testing carried out after 3, 7, 28 and 90 days from casting. In the case of the power replacements, the samples were demolded after 48 hours, due to a prolonged retardation of setting time. Once demolded, samples were placed in water at a temperature of ±20°C until testing. Tables 1 and 2 give details of the composition of the cement pastes.

Table 1. Mixture proportions for cement pastes containing OFI Water Replacements at 0.28 W/P Ratio

Type	OFI Percentage (%)	OFI Content(g)	Mucilage(g)	Water (g)	Cement (g)	W/P Ratio
Liquid [a]	0		0	140.00	500	0.28
Liquid [a]	2.5	3.34	3.45	136.51	500	0.28
Liquid [a]	5	3.34	7.00	133.06	900	0.28
Liquid [a]	7.5	3.34	10.50	129.56	900	0.28
Liquid [a]	10	3.34	14.00	126.82	900	0.28
Liquid [a]	20	3.34	28.00	112.07	900	0.28
Liquid [a]	40	3.34	56.00	84.02	900	0.28
Liquid [a]	60	3.34	84.00	54.04	900	0.28

[a] Boiled + Liquidised

Table 2. Mixture proportions for cement pastes containing OFI Powder Replacements at 0.35 W/P Ratio

Type	OFI Percentage (%)	Powder(g)	Water(g)	Cement(g)	W/P Ratio	W/C Ratio
Powder [a]	0	0.0	175	500	0.35	0.350
Powder [a]	0.5	2.5	175	497.5.	0.35	0.352
Powder [a]	1	5.0	175	495	0.35	0.354
Powder [a]	1.5	7.5	175	492.5	0.35	0.355
Powder [a]	2	10	175	490	0.35	0.357

[a] Lyophilised Powder

2.2.2. Mortar Prism Sample Preparation

Mortar prisms were prepared, in accordance to EN 196-1:2016. Each mixture was used to cast 72, 40 x 40 x 160 mm prisms to be used for the determination of the mechanical properties including the ultrasonic pulse velocity, flexural strength and compressive strength of the mortars. Tables 3 and 4 give details of the composition of the mortar mixtures. The mortar prisms were produced using Portland cement (Type CEM 1, 42.5 R) and a Coralline Limestone fine aggregate (4 mm maximum nominal size), with a W/P ratio of 0.7.

Table 3. Mixture proportions for mortar samples containing OFI Water Replacements at 0.7 W/P Ratio

Type	OFI Percentage (%)	OFI Content(g)	Mucilage(g)	Water (g)	Cement (g)	Sand(g)	W/P Ratio
Liquid [a]	0	0	0	630	900	2700	0.7
Liquid [a]	2.5	3.34	15.75	614.25	900	2700	0.7
Liquid [a]	5	3.34	31.5	598.5	900	2700	0.7
Liquid [a]	10	3.34	63	567	900	2700	0.7
Liquid [a]	20	3.34	126	504	900	2700	0.7
Liquid [a]	40	3.34	252	378	900	2700	0.7
Liquid [a]	60	3.34	378	252	900	2700	0.7

[a] Boiled + Liquidised

Table 4. Mixture proportions for mortar samples containing OFI Powder Replacements at 0.7 W/P Ratio

Type	OFI Percentage (%)	Powder(g)	Water(g)	Cement(g)	Sand(g)	W/C Ratio	W/P Ratio
Powder [a]	0	0.0	630	900	2700	0.7	0.0
Powder [a]	0.5	4.5	630	895.5.	2700	0.7	4.5
Powder [a]	1	9.0	630	891	2700	0.7	9.0
Powder [a]	1.5	13.5	630	886.5	2700	0.7	13.5
Powder [a]	2	18.0	630	882	2700	0.7	18.0

[a] Lyophilised Powder

3. Results and Discussion

3.1. Fresh and Hardened Properties of Cement Pastes

Results indicate that all cement pastes samples containing both liquid and powder OFI replacements, exhibited a general decrease in flow, which effect was exacerbated with increased percentage replacement. With a decrease in flow, an increase in consistency was also noted for all the percentage replacements. Results are illustrated in Figures 2. This increase in cement paste consistency may be attributed to the presence of polysaccharides in OFI (Ribeiro et al. 2010). This follows because, the type of branching polysaccharides present in OFI gives rise to a matrix structure with high water-retaining characteristics which indeed is referred to as mucilage. Therefore, and it is safe to assume that some of the water present in the mixture finds itself in the polysaccharide matrix thus reducing flow and increasing consistency.

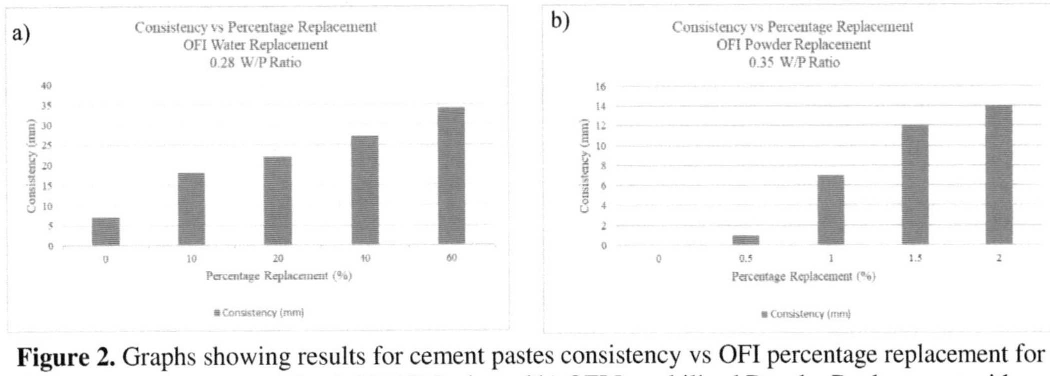

Figure 2. Graphs showing results for cement pastes consistency vs OFI percentage replacement for a) OFI water replacement with a 0.28W/P Ratio and b) OFI Lyophilised Powder Replacement with a

Graphs in figure 3 refer to the setting times for pastes containing OFI additions. As can be observed, all samples containing OFI additions exhibited an increase in setting time. In the case of OFI water replacements, the low percentages led to values close to the control sample. The largest increase was observed for 40% and 60% water replacement with an increase of 31% and 43% respectively. In the case of the cement pastes containing OFI powder replacement, the increase in setting time was considerably higher. The greatest increase was observed for 1% percentage replacement with a 75% increase for both the initial and final setting time. However, even though the results show that the highest increase in setting time was observed at the 1% percentage replacement, visual observation showed that the higher percentages (1.5%- 2%) took longer to consolidate. Indeed, these samples were still fragile when removed from the mould after 2 days of curing, when compared to the other percentage replacements.

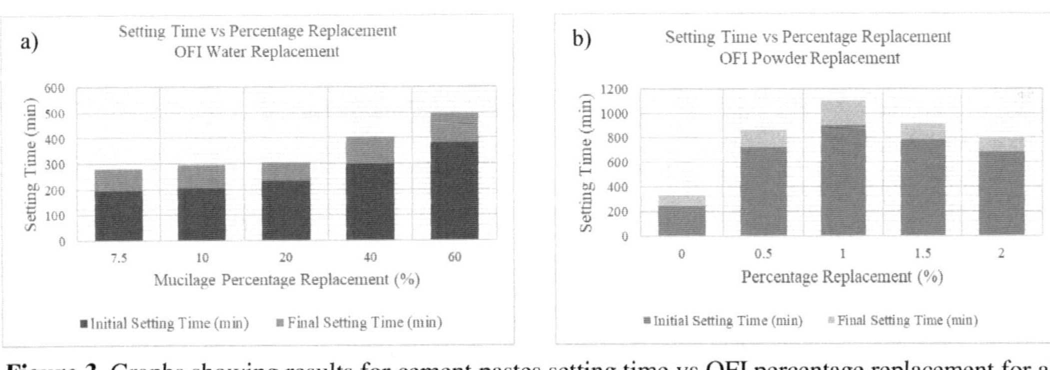

Figure 3. Graphs showing results for cement pastes setting time vs OFI percentage replacement for a) OFI water replacement with a 0.28W/P Ratio and b) OFI Lyophilised Powder Replacement with a

This increase in setting time, led to a delay in strength development at early age, with all cement samples containing OFI additions showing a decrease in strength with respect to the control sample also monitored at an early age. With time, the strength of OFI samples also increased, but it did not surpass the strength of the control sample. Moreover, at 3 and 7 days, the samples containing higher OFI percentages failed and crushed immediately when subjected to compression. This is illustrated in Figure 4 where compressive strength over time is compared.

This increase in setting time shows the potential advantage of using OFI as a retarding agent when casting in high ambient temperatures. Concreting in hot weather presents various challenges and retarding agents reduce the rate of hydration.

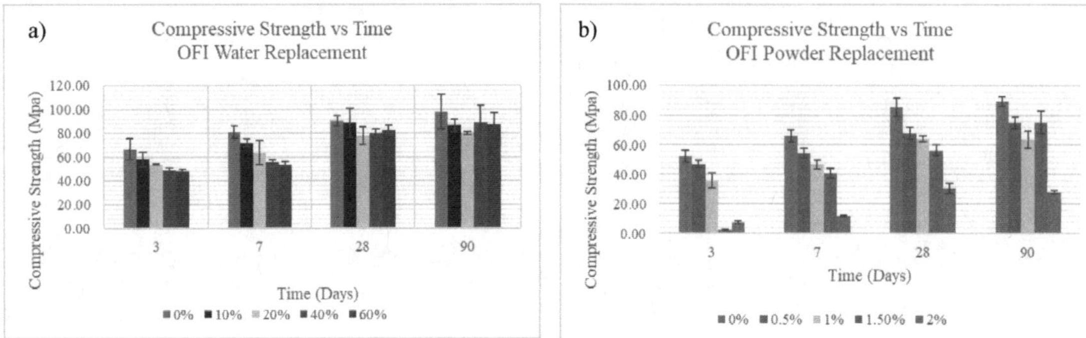

Figure 4. Graphs showing results for cement pastes compressive strength vs time for a) OFI water replacement and b) OFI Lyophilised Powder Replacement

3.2. Mortar Hardened Properties

3.2.1. Ultrasonic Pulse Velocity

Figure 5 presents the Ultrasonic Pulse Velocity(UPV) results obtained at 3,7,28 and 90 days for both types of replacement. Results obtained for the mortars containing OFI water replacement showed a percentage decrease of 7% in UPV in the early stage, with respect to the control samples at 3 days. This UPV decrease was reduced to 4% at 7 days, ultimately leading to an increase of 4.3% in UPV at 28 days. This increase was further maintained at 90 days. The highest increase in UPV was obtained for the highest OFI water replacement (60%).

Samples containing OFI powder replacements, showed similar trends to the OFI water replacements, with an initial decrease in UPV at 3days of testing and an increase in strength at 7 and 28 days of testing. In this case, the initial decrease in UPV was more evident and this is attributed to the higher OFI concentration.

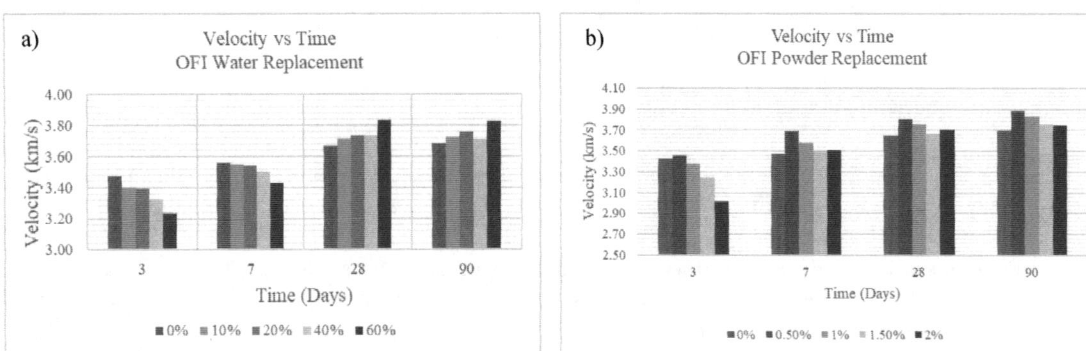

Figure 5. UPV vs time for a) OFI water replacement and b) OFI Lyophilised Powder Replacement

3.2.2. Compressive and Flexural Strength

Compressive strength results shown in Figures 6 obtained from the mortars containing OFI additions exhibited similar trends to the results obtained for Ultrasonic Pulse Velocity. The results obtained for the compression and flexural tests, show an initial reduction in strength for the mortar samples containing the higher percentage of OFI water replacement. The smaller percentage replacement showed equal or higher compressive and flexural strength results even at the early stages and this may be due to the fact that retardation was not much evident in these samples. However, at the 28 day of testing, the highest water percentage replacement (60%) showed the greatest increase in compressive and flexural strength from all the percentages with respect to the control samples.

In the case of the OFI powder replacements, similar results were obtained. In this case the initial decrease in strength was greatest for the highest percentage inclusion. The lowest inclusion of OFI powder (0.5-1%) registered the highest increase in compressive strength throughout the testing period.

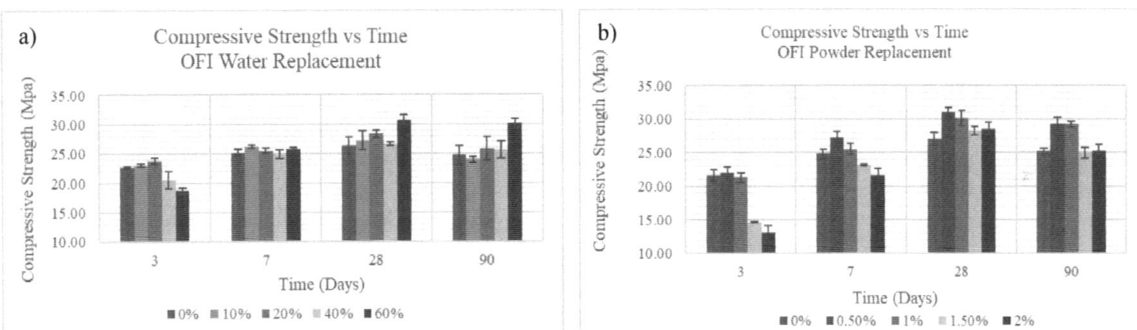

Figure 6. Graphs showing results for mortar samples, compressive strength vs time for a) OFI water replacement and b) OFI Lyophilised Powder Replacement

4. Conclusion

The use of *Opuntia ficus-indica* extracts showed promising results when used in cement pastes both as water and power replacements. Results for setting time indicate that *Opuntia ficus-indica extracts* can be used as retarding additive in both liquid and powder form, where an increase in both the initial and final setting times was recorded. Nevertheless, a relationship between adequate workability and retardation is crucial. Cement pastes with 0.5% and 1% OFI Powder replacements, exhibited prolonged setting time and therefore had a retardation effect, while still maintaining adequate workability. However, this increase in setting time negatively affected the compressive strength gain in cement pastes.

In the case of cement-based mortars, results indicated a general decrease in workability for increasing percentage replacement of both OFI powder and water replacements. Compressive strength results also showed an initial decrease in strength at 3 and 7 days due to the retardation. At 28 days increases in UPV, compressive and flexural strength were recorded for OFI replacements in both powder and liquid form. At certain percentages, this increase in strength surpassed that of the control mortar. This indicates that *opuntia ficus-indica* both in powder and liquid form is a potentially useful addition to enhance the hardened properties of cement-based mortars.

IOP Conf. Series: Materials Science and Engineering **442** (2018) 012016 doi:10.1088/1757-899X/442/1/012016

References

[1] Arreola Sanchez M, Martinez Molina W, Alonso Guzman E, Chavez Garcia H, Torres Acosta A, Bernabe Reyes C, Velazquez Perez, J 2015 Effect of the addition of dried opuntia cactus fibers-setting times and mechanical resistance in portland cement mortar base *2nd International Conference on Green Materials and Environmental Engineering*

[2] Borg R P, Buhagiar J, and Iacobucci C G 2014 Opuntia ficus indica methods of preapration and applications in mortars and concrete *Unpublished manuscript*

[3] Chandra S, Eklund L, and Villarreal R 1998 Use of cactus in mortars and concrete *Cement and Concrete Research* 28 (1) 41-51

[4] European Committee for Standardization 2004 Testing concrete - Part 4: Determination of ultrasonic pulse velocity *EN 12504-4 Brussels*

[5] European Committee for Standardization (2013) Testing for mechanical and physical properties of aggregates - Part 6: Determination of particle density and water absorption *EN 1097-6 Brussels*

[6] Hernández E, Cano-Barrita P d J, and Torres-Acosta A 2016 Influence of cactus mucilage and marine brown algae extract on the compressive strength and durability of concrete *Materiales De Construcción* 66 (321) e074

[7] Hernández-Zaragoza J, Coronado-Márquez A, López-Lara T, Horta-Rangel J, de Querétaro S, and México Q 2008 *Mortar improvement using nopal additive J Pacd* 10 120-125

[8] Martinez-Molina W, Torres-Acosta A, Martínez-Peña G, Guzmán E A, and Mendoza-Pérez I 2015 Cement-based, materials-enhanced durability from opuntia ficus indica mucilage additions *ACI Materials Journal* 112

[9] Martínez-Molina W, Torres-Acosta A A, Celis-Mendoza C E, and Alonso-Guzman E 2015 Physical properties of cement-based paste and mortar with dehydrated cacti additions *International Journal of Architectural Heritage*, 9, 443-452

[10] Monteiro P 2006 *Concrete: Microstructure, properties, and materials* (McGraw-Hill Publishing)

[11] Neville A and Brooks J 1987 *Concrete technology* (Harlow, England: Prentice Hall)

[12] Ramírez-Arellanes S, Cano-Barrita P d J, Julián-Caballero F, and Gómez-Yañez, C (2012) Concrete durability properties and microstructural analysis of cement pastes with nopal cactus mucilage as a natural additive *Materiales De Constucción* 62 327-341

[13] Ribeiro, Erika Maria de Oliveira, Nicácio Henrique da Silva, José Luiz de Lima Filho, Júlio Zoe de Brito, and Maria da Paz Carvalho da Silva. "Study of carbohydrates present in the cladodes of Opuntia ficus-indica (fodder palm), according to age and season." *Food Science and Technology* 30, no. 4 (2010): 933-939.

The effects of water-cement ratio and chemical admixtures on the workability of concrete

Ali M Mansor[1], Ruben Paul Borg[2], Ahmed M M Hamed[2], Mohammed M Gadeem[1], Mansour M Saeed[1]

[1] Faculty of Engineering, University of Tripoli, Libya
[2] Faculty for the Built Environment, University of Malta, Malta

E-mail: ruben.p.borg@um.edu.mt

Abstract. Concrete chemical admixtures are used to improve the fresh and hardened properties of mortar or concrete in different applications. Their addition can allow for the reduction of the water to cement ratio (w/c), without affecting the workability of the mixture. By reducing the amount of water, the cement paste will have higher density. However, the reduction of the water content in a concrete mixture should be done in such a way that complete cement hydration take place and sufficient workability is achieved. In order to maintain the workability of the concrete mix, the dosage of the admixture must be carefully calculated and must be taken into account in the calculation of the w/c ratio. In the present study, three types of chemical admixtures; Type A water-reducing, Type D water-reducing and retarding and Type F high range, water-reducing admixtures conforming to ASTM C 494/C 494M – 04 standards, were used to optimize the percentage of the admixture with respect to the w/c ratio. The suitable time for the addition of the admixture is also investigated. Results show that using 1.5 % of Type A admixture with 0.45 w/c gives 45.6 MPa, 28 day compressive strength with an associated slump of 110 mm, compared to zero slump without admixture. 1.0 % Type A admixture with 0.52 w/c gives 33 MPa, 28 day compressive strength with an associated slump of 95 mm, compared to 35 mm slump achieved without admixture. For Type D admixture, using 1.5 % with 0.5 w/c the slump was 190 mm with 47 MPa, 28 day compressive strength, compared to 15 mm slump achieved without admixture. 1.0% of Type F admixture with 0.35 w/c and 0.45 w/c gave slumps of 25 mm, 225 mm and strength of 63.5MPa and 55.3MPa respectively. For all mixes, the best slump results were achieved when the admixtures were added during the initial mixing process, while maximum compressive strengths were achieved for different times of inclusion of the admixtures.

Keywords: Admixtures, water reducer, concrete slump

1. Introduction

Many important characteristics of concrete are influenced by the ratio of water to cementitious materials (w/c) used in the mixture. By reducing the amount of water, the cement paste will have higher density, which results in higher paste quality and hence higher compressive strength and lower permeability. Reducing the water content in a concrete mixture should be done in such a way so that complete cement hydration may take place and sufficient workability of concrete is maintained for placement and consolidation during construction.

The w/c ratio needed for cement to complete its hydration ranges from 0.22 to 0.25 [1]. The existence of additional water in the mixture is needed for the workability of the concrete. Reducing the water content in a mixture may result in a stiffer mixture, which reduces the workability and increases potential placement problems.

Water reducing admixtures or, as they are more commonly known, plasticisers are admixtures for concrete, which are used to achieve certain properties in concrete more effectively than by other means, and to maintain the quality of concrete during the stages of mixing, transporting, placing, and curing in adverse weather conditions.

Admixtures vary widely in chemical composition, and many perform more than one function. Two basic types of admixtures are available: chemical and mineral admixtures [2]. All admixtures to be used in concrete construction should meet specifications; tests should be made to evaluate how the admixture will affect the properties of the concrete. The effectiveness of an admixture depends upon factors such as type, brand, and amount of cementing materials; water content; aggregate shape, gradation, and proportions; mixing time; slump; and temperature of the concrete.

The objective of this work was to investigate the effect of w/c ratio on the efficiency of the concrete chemical admixtures used. Three types of chemical admixtures, conforming to ASTM C 494/C 494M standard specification, were used: Type A water-reducing, Type D water-reducing and retarding and Type F high range water-reducing admixtures. Different w/c ratios were considered for each type of admixture to investigate its effect on the workability (slump). Fresh and hardened concrete tests were conducted.

2. Materials

Ordinary Portland cement (OPC) type 42.5N conforming to the Libyan standards 340-2009 was used as the main binder. The fine aggregate used is a local Libyan natural sand referred to as Zliten sand, with a specific gravity 2.66, moisture absorption 0.18%, fine materials 2.2% and unit weight 1660 kg/m^3 conforming to the Libyan standards 49-2002. Two different sizes of coarse aggregate were used, 20 mm and 14 mm conforming to the Libyan standards 256 and ASTM C29/129M: 2007. A clean water free from contaminants was used as mixing water for the production of concrete specimens. Three types of the chemical admixtures, conforming to ASTM C 494/C 494M – 04 standards, were used; Type A water-reducing admixtures (Plasticizer), Type D water-reducing and retarding admixtures (Plasticizer) and Type F water-reducing, high range admixtures (Super-plasticizer),

3. Experimental work

Table 1 shows the quantities of materials used for one cubic meter of concrete. The water cement ratio was adjusted for the required initial slump.

Table 1, Concrete mixes.

Cement (kg)	Coarse Agg. (kg)		Fine Agg. (kg)	Water (kg)	w/c %	Admixtures (kg)
	20 mm	14 mm				
360	466.2	299.3	684.5	122.6 – 230.6	0.34 – 0.64	3.6 – 5.4

A sufficient number of mixes were batched and prepared, for each type of admixture and for each w/c ratio selected. Six cubes 150x150x150 mm were prepared for each mix. Ingredients were introduced to the mixer and mixed in a dry form for homogeneity. The mixing water was added first and the batch was mixed for two minutes. Then the assigned portion of the admixture was added and the batch was remixed for another three minutes. The cubes were filled in three layers and demoulded after 24 h. The concrete specimens were water cured in a controlled environment until the time of testing.

4. Results and observations

4.1 Type A admixture

The results and observations of the effect of the w/c ratio with the use of type A admixture on the slump and compressive strength of the concrete mixes are shown in table 2.

Table 2, Results and observations of using Type A chemical admixture with various w/c ratios.

Admixture, % of cement	w/c %	Workability		Slump (mm)		Compressive strength (MPa)	
		Without admixture	With admixture	Without admixture	With admixture	7 days	28 days
1.5	0.3	Very low	Very low	0.0	0.0	7.5	11
	0.4	Very low	Low	0.0	25	31.4	47.6
	0.45	Very low	Medium	0.0	110	41.2	45.6
	0.5	Low	High (with little segregation)	25	195	37	45.7
	0.53	Low	High (with segregation)	50	225	31.2	43
1.0	0.45	Very low	Low	0.0	15	25.6	31.7
	0.5	Very low	Low	10	45	26.8	30.7
	0.52	Low	Medium	35	95	24	33
	0.55	Medium	High (with segregation)	90	210	29.7	39
	0.6	High	High (with segregation)	175	240	27	38.5

Results showed that for 1.5% Type A admixture a slump of 110 mm could be achieved with 0.45 w/c ratio compared to zero slump without admixture. At the same percentages, the 28 days compressive strength was recorded as 45.6 MPa. A higher compressive strength of 47.6 MPa was reported for 1.5% admixture and 0.4% w/c ratio with a low workability. For 1.0% admixture and 0.45 w/c ratio, neither a good workability nor a good compressive strength were achieved, comparing to the 1.5% admixture mix. A slump of 95 mm was reported at 0.52 w/c ratio associated with a compressive strength of 33 MPa for the 1.0% admixture mix. Figure 1, a and b show the relations between w/c ratio and concrete slump for 1.0% and 1.5% Type A admixtures, respectively.

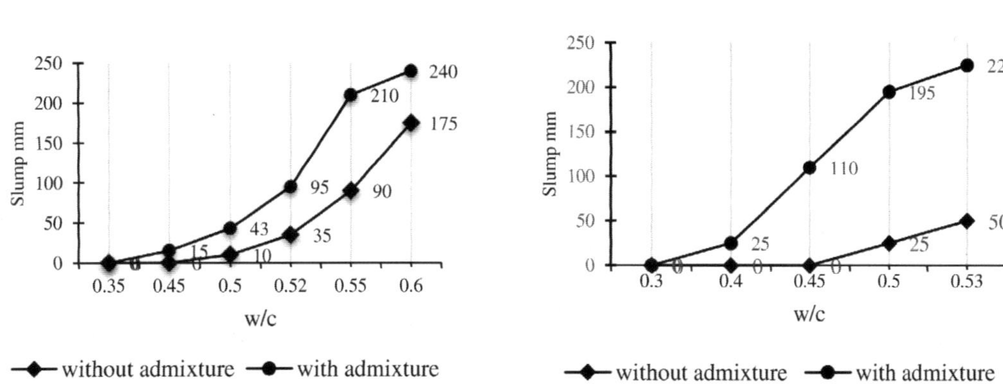

(a): 1.0% admixture

(b): 1.5% admixture

Figure 1. Slump with and without type A admixture

Figure 2, a and b show the relations between w/c ratio and concrete compressive strength for 1.0% and 1.5% Type A admixture, respectively

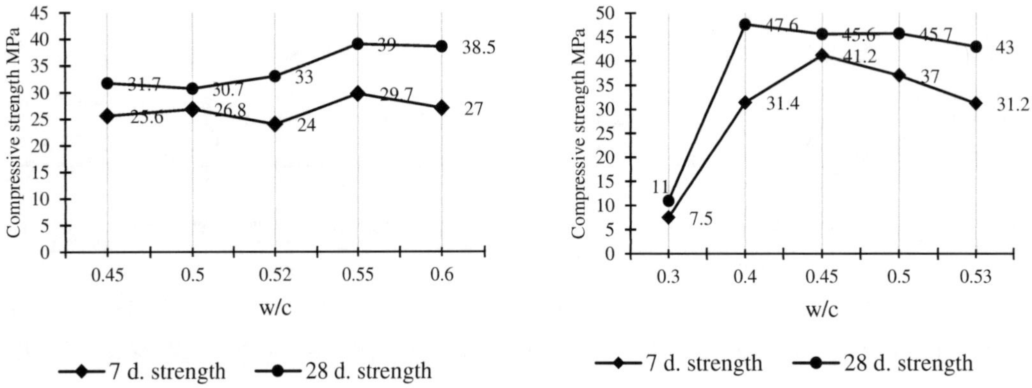

(a): 1.0% admixture

(b): 1.5% admixture

Figure 2. Compressive strength for concrete with and without Type A admixture

4.2 Type D admixture

Table 3 illustrates the results and observations of the effect of the w/c and Type D admixture on the slump and compressive strength of the concrete mixes.

Table 3, Results and observations of using type D chemical admixture with various w/c ratios.

Admixture, % of cement	w/c %	Workability		Slump (mm)		Compressive strength (MPa)	
		Without admixture	With admixture	Without admixture	With admixture	7 days	28 days
1.5	0.35	Very low	Very low	0.0	0.0	24.6	48
	0.45	Very low	Low	0.0	25	39.8	52
	0. 5	low	High	15	190	33.6	47
	0.55	High	High (with segregation)	145	210	28	40. 6

Similar to type A chemical admixture, results showed that for 1.5% type D admixture a slump of 190 mm could be achieved with 0.5 w/c ratio comparing with only 15 mm achieved without admixture. At the same percentages, the 28 days compressive strength was recorded as 47 MPa. A higher compressive strength of 52 MPa was reported at 0.45% w/c ratio with a poor workability. At 0.55 w/c ratio, a slump of 145 mm was reported for the mix without admixture, associated with ingredients segregation and a lower compressive strength of 40.6 MPa when the admixture was introduced. Figures 3 and 4 show the relations between w/c ratio and concrete slump and concrete compressive strength for 1.5% type D admixture, respectively.

IOP Conf. Series: Materials Science and Engineering **442** (2018) 012017 doi:10.1088/1757-899X/442/1/012017

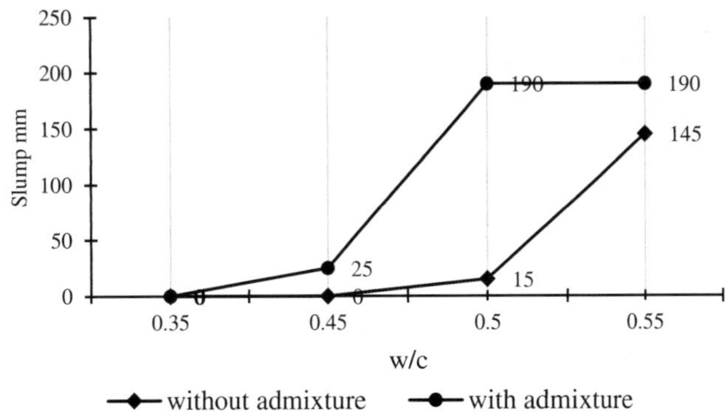

Figure 3. Slump with and without type D admixture

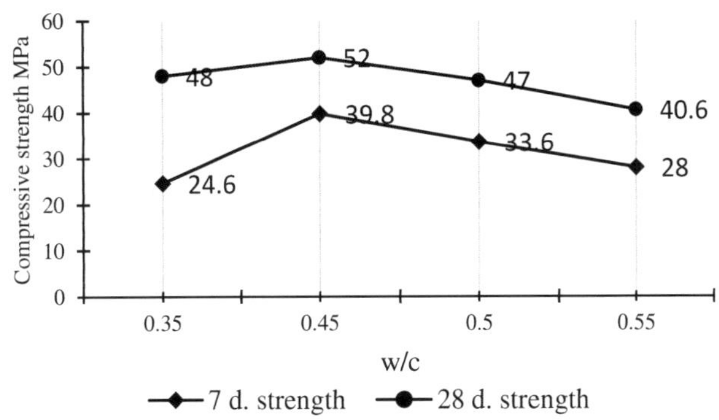

Figure 4. Compressive strength with and without type D admixture

4.3 Type F admixture

Table 4 illustrates the results and observations of the effect of the w/c and type F admixture on the slump and compressive strength of the concrete mixes.

Table 4, Results and observations of using type F chemical admixture with various w/c ratios.

Admixture, % of cement	w/c %	Workability		Slump (mm)		Compressive strength (MPa)	
		Without admixture	With admixture	Without admixture	With admixture	7 days	28 days
1.0	0.3	Very low	Very low	0.0	0.0	53	58.5
	0.35	Very low	Low	0.0	25	50. 6	63.5
	0. 45	Very low	High (with segregation)	0.0	225	39.7	56
	0.5	Very low	High (with segregation)	15	225	31.7	42.7

For type F chemical admixture, as a high range water reducing admixture, results showed that for 1.0% addition, a high workability was reported with a slump of 225 mm at 0.45 w/c ratio, associated with ingredients segregation, comparing with 0.0 mm achieved without admixture. A higher compressive strength of 63.5 MPa was reported at 0.35% w/c ratio with a very low workability.

IOP Conf. Series: Materials Science and Engineering **442** (2018) 012017 doi:10.1088/1757-899X/442/1/012017

Figures 5 and 6 show the relations between w/c ratio and concrete slump and concrete compressive strength for 1.5% type F admixture, respectively.

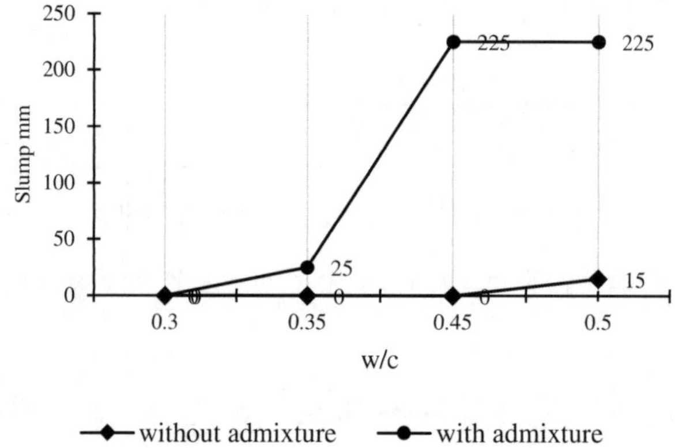

Figure 5. Slump with and without type F admixture

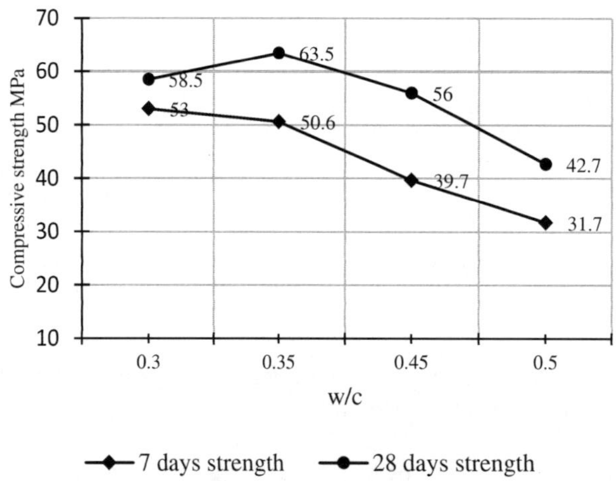

Figure 6. Compressive strength with and without type F

4.4 Effect of time of inclusion of admixture

Different groups of concrete mixes were prepared to investigate the effect of the time of introducing the admixture on the concrete slump and compressive strength.

Table 5 illustrates the results and observations of the effect of the time of inclusion of 1.5% of type A, D and 1.0% of type F chemical admixtures on the slump and compressive strength, respectively. Slump variation with the time of inclusion of the admixture indicates that up to 15 minutes, the admixtures were acting, almost through the same mechanism of dispersing the flocculated cement grains. At the mentioned percentages of type A, D and F admixtures and w/c ratios, the highest compressive strength values were reported at 10 min for type A., at 5 min for type D and at 10 min for type F.

Table 5, Effect of the admixtures inclusion time on the slump and compressive strength.

Type of admixture	Adding time (min.)	w/c	Slump (mm)	7 d. strength	28 d. strength
A	0	0.5	160	28.5	40.4
	5		210	32	45.4
	10		200	34.5	48.5
	15		185	33.5	44.4
D	0	0.49	165	35.4	46.57
	5		200	38.5	55.8
	10		210	38.8	51
	15		220	38.3	48.6
F	0	0.43	220	31.3	39.7
	5		210	39.4	48
	10		210	41.6	50
	15		210	36.6	45.5

5. Conclusions

On the basis of the materials and mixes used, the following conclusions can be drawn from the research presented in this paper:

1. A certain w/c ratio is required for the concrete chemical admixtures to be effective.
2. The efficiency of type A and type D chemical admixtures (plasticizers) starts at about 0.4 w/c ratio.
3. The efficiency of type F chemical admixtures (super-plasticizers) starts at about 0.35 w/c ratio and has no effect on the concrete workability after 0.45 w/c ratio.
4. The time of inclusion of the admixture affects both the workability and the compressive strength.
5. The plasticizing action of water reducers is related to their adsorption and dispersing effects in the cement-water system.
6. For greatest significance and to improve the corresponding performance, the right variety of the concrete chemical admixture should be selected and the dosage and time of inclusion should be determined before.
7. Trial mixtures should be made with the admixture and the materials used for the concrete mix, at temperature and relative humidity anticipated on the job.
8. It is recommended that the dosage of admixture as recommended in the manufacturer's data sheet should be verified through trial mixes.

References

[1] Ramachandran, V.S. 1995, Concrete Admixtures Handbook – Properties, Science, and Technology.
[2] Gidion Turuallo 2004, analysis of the optimum percentage of admixture of sikament-163, which can be added into concrete mixture to achieve the maximum of concrete compression strength, Indonesia, publication/287818287
[3] www.ce.memphis.edu/1101/notes/concrete/PCA.../Chap06.pdf.
[4] Viveka N D, et al, Portland cement hydration in the presence of admixtures - black gram pulse and superplasticizer, India Mat. Res. vol.11 no.4 São Carlos, 2008.
[5] www.theconcreteportal.com/compat.html Cement - Superplasticizer Compatibility - The Concrete Portal.
[6] Junwu Li, et, al, 2017, A Study on the effects of chemical admixtures on the strength of Portland cement, chemical engineering transactions, China, vol. 59.
[7] The Florida Department of Transportation Research Center, 2016, final report effects of chemical and mineral admixtures on performance of Florida structural concrete, Florida, Research.center@dot.state.fl.us.

Calculation of plastic viscosity of concrete mixture using the modified empirical formula

M Daukšys, A Klovas

Faculty of Civil Engineering and Architecture, Kaunas University of Technology, Lithuania

E-mail: mindaugas.dauksys@ktu.lt

Abstract. Nowadays various types of viscometers are being designed to measure the rheological properties of cement paste and fresh concrete. Most of them are very expensive and require specific knowledge about their usage. Then in practice usually the empirical formulas are used to calculate the main rheological properties – yield stress (Pa) and plastic viscosity (Pa·s). Therefore, in the literature, there are some formulas which can't be used to calculate analytically plastic viscosity of concrete mixture, e.g. it cannot estimate the influence of different chemical admixture on plastic viscosity of concrete. The paper focuses on the modification of the empirical formula to calculate the plastic viscosity of fresh concrete mixture. In this paper the way of viscosity formula modification was described. For an empirical formula modification, the coefficients which describe the influence of different chemical admixture on plastic viscosity of concrete mixture were used. Coefficients $K_{calc.}$ were calculated as a ratio between different viscosity values of cement pastes obtained without and with certain chemical admixtures. Viscosity of cement paste was obtained by using rotational viscometer Rheotest RN4. Coefficients describe the change in cement pastes viscosity when adding different type (superplasticizer, viscosity modifying agent, air-entraining agent and air voids removing agent) and amount of chemical admixture.

Keywords: concrete mixture, empirical formula, plastic viscosity

1. Introduction

Concrete is the most popular material which is used in construction field nowadays. Estimation of rheological properties of concrete mixtures and optimization of their compositions are important tasks in the modern construction technology. Rheological parameters according to Bingham model can be calculated according to the equation (1) [1,2]:

$$\tau = \tau_0 + \eta \cdot \bar{\gamma} \qquad (1)$$

where: τ – shear stress, Pa; τ_0 – yield stress, Pa; $\bar{\gamma}$ – shear rate, s^{-1}; η – plastic viscosity, Pa·s.

Various type of viscometers are being used to determine the rheological parameters of fresh concrete They measure the shear stress at varying shear rates [3,4]. However, most of them are very expensive and require specific knowledge to use them. An alternative approach for modelling the rheological behaviour of fresh concrete was proposed. The values of yield stress and plastic viscosity are correlated to the slump value and the flow time, respectively [5-13]. According to scientist [5], the relation between the yield stress τ_0 and the slump flow S does not depend on the mould geometry. The study [12] presented the comparative analysis of various viscosity models including Hu, Ferraris and De Larrard and of Toutou. The calculations were based on changes of plastic viscosity of concrete mixture depending on relative solid concentration (Φ- the solid volume of concrete /Φ*- maximum solid concentration of mixture) of mixtures. According to authors, the proposed models assume that the concrete is divided into two scales: paste (water, cement, additives and superplasticizers) and aggregate (sand and gravel). The model takes in-to the account the effect of additives and the superplasticizers (content and type).

The plastic viscosity is defined as the proportional coefficient between shear stress and shear rate under a state of steady shear, and it is greatly affected by colloidal particle interaction forces including Brownian forces, hydrodynamic forces and viscous forces be-tween particles [11]. The changes in cement paste viscosity have a great effect on the plastic viscosity values of fresh concrete. Due to this assumption, the change of cement paste viscosity adding different type and amount of chemical admixtures was established.

The aim of the study was to determine the possibilities of modification of empirical formula which could estimate the influence of different chemical admixture on the plastic viscosity of fresh concrete.

2. Materials and methodology

The Portland cement CEM II/A-LL 42. R confirming to the European standard EN 197-1 was used for the test. Physical and mechanical properties of Portland cement are presented in Table 1.

Table 1. Physical and mechanical properties of Portland cement

Property	Value
Specific surface area, m^2/kg	410
Particle density, kg/m^3	3.05
Normal consistency of cement paste, %	26.5
Volume stability, mm	0.8
Initial setting time, min.	195
Compressive strength after 2 days / 28 days, MPa	27.1 / 54.0
Loss on ignition, %	5.05
Insoluble materials, %	-
SO_3, %	2.48
Cl^-, %	0.015
Alkalis, calculated by Na_2O equivalent, %	<0.8

Two different sizes of fine aggregate from "Kvesu" quarry were used in this study: sand fraction 0/1with bulk density 1520 kg/m^3, fineness module 1.78 and sand fraction 0/4 with bulk density 1710 kg/m^3, fineness module 2.62. Sieve analysis of the fine aggregate is conducted according to EN 12620 and presented in Table 2. Gravel fraction 4/16 and bulk density 1327 kg/m^3 was used as the coarse aggregate. Sieve analysis of the fine aggregate and coarse aggregate are conducted according to EN 12620 and presented in Table 2.

Cement pastes were mixed in the laboratory in forced type mixer „Automix". Cement was dosed by weight, water and different chemical admixtures were dosed by volume. The plasticizing admixture was mixed with water and poured into the mixer. Remaining admixtures were dozed directly to the cement paste.

Concrete mixtures were mixed for 3 minutes in the laboratory in forced type concrete mixer. Only dry aggregates were used for concrete mixtures. Cement and aggregates were dosed by weight while on the other hand, water and different chemical admixtures were dosed by volume. 90% of water was instantly poured into the mix. The plasticizing admixture was mixed with 10% of remaining water and poured into the mixer. Remaining admixtures were dozed directly to the mix.

Table 2. Sieve analysis of the fine and coarse aggregates

Diameter of the sieve's mesh, mm	The amount of poured out material, %		
	Sand fraction 0/1	Sand fraction 0/4	Gravel fraction 4/16
16.0	100.00	100.00	98.80
8.0	100.00	100.00	42.10
4.0	100.00	95.10	4.30
2.0	99.80	81.80	1.00
1.0	99.10	54.60	0.52
0.500	77.40	12.40	0.44
0.250	2.20	0.70	0.36
0.125	0.50	0.30	0.32
0.000	0.00	0.00	0.00

The description of chemical admixtures which were used in this study are presented in Table 3.

Table 3. Description of chemical admixtures

	Glenium SKY628	Rheomix 880	Rheomatrix 100	Microair G (LP)
Purpose	Superplasticizer	Air voids remover	Viscosity modifier	Air entrainer
Compound-based	Polycarboxylate ether	Propoxylate-etoxylate	Synthetic copolymer	Modified resin
Dosage, %	0.6-1.8	0-0.3	0-1.1	0-0.3
Density, g/cm^3	1.06-1.10	0.97 – 0.02	1.0-1.02	0.98-1.04
Viscosity, mPa·s	-	<600	-	-
pH	-	-	6-9	9-11
Chloride quantity, %	<0.1	-	<0.1	<0.01

The slump of the concrete mixture was evaluated according to standard EN 12350-2 and the density - according to standard EN 12350-6.

The rheological properties of cement pastes were obtained using rotational viscometer Rheotest RN4. The plastic viscosity values of cement paste are presented in Table 4. Viscometer consists of measuring drive with quick acting coupling rotor or cone and measuring cup coupling, connecting cable for control unit connection and stand with a temperature-controlled plate guide. For the test was used the cylinder measuring system which consists of the measuring cup (assembled) with coupling and cylinder rotor. General technical data of equipment: torque 0.1 to 150 mNm; speed range 0 to 1000 rpm; frequency: 0.001...10 Hz; viscosity 1…107 mPas; share rate 0.13...1300 s^{-1}.

Concrete mixture`s rheological property yield stress was evaluated according to equation (2). The yield stress of non-vibrated concrete mixture was calculated according to its fresh technological parameters: density and slump [13]. This equation was adjusted according to the standard cone which is used to evaluate the slump of the concrete mixture.

$$\tau_0 = \frac{0.00815 \cdot \rho_m}{\left(\sqrt{\frac{0.498}{30-SL}} - 0.001724 - 0.024\right)^2}; \tag{2}$$

where: τ_0 – yield stress of concrete mixture, Pa; ρ_m – density of concrete mixture, kg/m^3; SL – slump of concrete mixture, cm.

Second mixture`s rheological property – plastic viscosity was evaluated according to equation (3). Plastic viscosity is established by the concentration of aggregates which are used in the concrete mixture [13].

$$\eta_b = \eta_v \cdot \exp\left(\frac{a_c \cdot \rho_v}{\rho_v + \frac{V}{C}\cdot\rho_c - b_c \cdot \rho_v} + \frac{a_{sm}\left(1-\varphi_{st}-\varphi_o-\frac{V}{\rho_v}-\frac{C}{\rho_c}\right)}{1-\varphi_{st}-b_{sm}\left(1-\varphi_{st}-\varphi_o-\frac{V}{\rho_v}-\frac{C}{\rho_c}\right)} + \frac{a_{st}\cdot\varphi_{st}}{1-b_{st}\cdot\varphi_{st}} \right); \tag{3}$$

where: η_v – dynamic viscosity of water in 20 °C, Pa·s; φ_o – air content of concrete mixture, %; V, C – water and cement quantities in concrete mixture (1 m^3); ρ_v, ρ_c – densities of water and cement, kg/m^3; φ_{st} – volume concentration of coarse aggregate in concrete mixture; a_c, a_{sm}, a_{st} – factors describing the form of cement, fine and coarse aggregate`s particles ($a_c = 2.6$; $a_{sm} = 2.5$; $a_{st} = 2.6$); b_c, b_{sm}, b_{st} – factors describing the density distribution of cement, fine and coarse aggregate respectively in cement paste, mortar and concrete mixture.

The empirical formula (3) was modified by a coefficient $K_{calc.}$ that describes the influence of different chemical admixture on the plastic viscosity of concrete mixture (4).

$$\eta_b = \eta_v \cdot \exp\left(\frac{a_c \cdot \rho_v}{\rho_v + \frac{V}{C}\cdot\rho_c - b_c \cdot \rho_v} + \frac{a_{sm}\left(1-\varphi_{st}-\varphi_o-\frac{V}{\rho_v}-\frac{C}{\rho_c}\right)}{1-\varphi_{st}-b_{sm}\left(1-\varphi_{st}-\varphi_o-\frac{V}{\rho_v}-\frac{C}{\rho_c}\right)} + \frac{a_{st}\cdot\varphi_{st}}{1-b_{st}\cdot\varphi_{st}} \right) \cdot K_{calc.} \tag{4}$$

where: $K_{calc.}$ – the coefficient which describes the change in mixture`s plastic viscosity when using different type and content of chemical admixture.

3. Results and discussions

Coefficient $K_{calc.}$ was calculated as a ratio between different viscosity values of cement pastes obtained without and with certain chemical admixtures. Table 4 shows the calculated coefficient`s $K_{calc.}$ values. Rheological properties of the yield stress (Eq. 2) and the plastic viscosity (Eq. 4) with different type and content of chemical admixture were analysed.

The content of chemical admixture was gradually increased. The dosage of superplasticizing admixture based on polycarboxylate ether varied in the range from 0.6 to 1.4% of cement mass, the dosage of air voids removing admixture based on propoxylate-etoxylate – from 0.05 to 0.3%, dosage of viscosity modifying admixture based on synthetic copolymer – from 0.1 to 1.1% and dosage of air entraining admixture based on modified resin – from 0.05 to 0.3% (Table 4). In addition, the content of superplasticizing admixture was constant (1.4% of cement) and content of air voids removing admixture, viscosity modifying admixture and the air-entraining admixture was gradually increased, respectively. During the research, W/C ratio was constant 0.47.

Table 4. Coefficients describing the change in cement paste viscosity when adding different type and amount of chemical admixture

No.	Chemical admixture content, %	Plastic viscosity of cement paste, Pa·s	Calculated coefficient, $K_{calc.}$
Superplasticizing admixture alone			
1	0	0.506	-
2	0.6	0.093	0.184
3	0.8	0.093	0.183
4	1.0	0.043	0.085
5	1.2	0.039	0.077
6	1.4	0.039	0.077
Superplasticizing admixture (1.4 %) and air voids removing admixture			
1	0	0.039	0.077
2	0.05	0.043	0.085
3	0.10	0.036	0.071
4	0.15	0.043	0.085
5	0.20	0.043	0.085
6	0.25	0.043	0.085
7	0.30	0.032	0.063
Superplasticizing admixture (1.4 %) and viscosity modifying admixture			
1	0	0.039	0.077
2	0.1	0.032	0.063
3	0.3	0.032	0.063
4	0.5	0.032	0.063
5	0.7	0.032	0.063
6	0.9	0.034	0.067
7	1.1	0.046	0.091
Superplasticizing admixture (1.4 %) and air entraining admixture			
1	0	0.039	0.077
2	0.05	0.029	0.057
3	0.10	0.029	0.057
4	0.15	0.029	0.057
5	0.20	0.029	0.057
6	0.25	0.030	0.059
7	0.30	0.032	0.063

Concrete mixture compositions are given in Table 5.

Table 5. Compositions of the concrete mixture.
Amount of materials for 1m^3 of concrete mixture

Materials	Unit	Concrete mixture compositions						
		BA4-0	BA4-1	BA4-2	BA4-3	BA4-4	-	-
Portland cement	kg	380	380	380	380	380	-	-
Water	l	178	178	178	178	178	-	-
Gravel fraction 4/16	kg	986	986	986	986	986	-	-
Sand fraction 0/4	kg	574	574	574	574	574	-	-
Sand fraction 0/1	kg	286	286	286	286	286	-	-
Superplasticizer	l	2.15	2.87	4.03	4.30	5.02	-	-
W/C	–	0.47	0.47	0.47	0.47	0.47	-	-
		BA5-0	BA5-1	BA5-2	BA5-3	BA5-4	BA5-5	BA5-6
Portland cement	kg	380	380	380	380	380	380	380
Water	l	178	178	178	178	178	178	178
Gravel fraction 4/16	kg	986	986	986	986	986	986	986
Sand fraction 0/4	kg	574	574	574	574	574	574	574
Sand fraction 0/1	kg	286	286	286	286	286	286	286
Superplasticizer	l	5.02	5.02	5.02	5.02	5.02	5.02	5.02
Air voids remover	l	-	0.20	0.39	0.59	0.78	0.98	1.18
W/C	–	0.47	0.47	0.47	0.47	0.47	0.47	0.47
		BA6-0	BA6-1	BA6-2	BA6-3	BA6-4	BA6-5	BA6-6
Portland cement	kg	380	380	380	380	380	380	380
Water	l	178	178	178	178	178	178	178
Gravel fraction 4/16	kg	986	986	986	986	986	986	986
Sand fraction 0/4	kg	574	574	574	574	574	574	574
Sand fraction 0/1	kg	286	286	286	286	286	286	286
Superplasticizer	l	5.02	5.02	5.02	5.02	5.02	5.02	5.02
Viscosity modifier	l	-	0.38	1.14	1.90	2.66	3.42	4.18
W/C	–	0.47	0.47	0.47	0.47	0.47	0.47	0.47
		BA7-0	BA7-1	BA7-2	BA7-3	BA7-4	BA7-5	BA7-6
Portland cement	kg	380	380	380	380	380	380	380
Water	l	178	178	178	178	178	178	178
Gravel fraction 4/16	kg	986	986	986	986	986	986	986
Sand fraction 0/4	kg	574	574	574	574	574	574	574
Sand fraction 0/1	kg	286	286	286	286	286	286	286
Superplasticizer	l	5.02	5.02	5.02	5.02	5.02	5.02	5.02
Air entrainer	l	-	0.19	0.38	0.57	0.76	0.95	1.14
W/C	–	0.47	0.47	0.47	0.47	0.47	0.47	0.47

As shown at Table 6, the slump of concrete mixture with different type and content of chemical admixture varied in the range from 220 to 260 mm and the density of concrete mixture – from 2360 to 2400 kg/m^3. The consistency of concrete mixture varied in the S5 (≥220±30 mm) class range according to standard EN 12350-2.

Table 6. Technological properties of the concrete mixtures

Technological property	Unit	Concrete mixture compositions						
		BA4-0	BA4-1	BA4-2	BA4-3	BA4-4	-	-
Density	kg/m^3	2360	2390	2360	2400	2400	-	-
Slump	mm	240	240	220	260	250	-	-
		BA5-0	BA5-1	BA5-2	BA5-3	BA5-4	BA5-5	BA5-6
Density	kg/m^3	2400	2390	2390	2380	2390	2390	2380
Slump	mm	250	220	250	250	240	250	250
		BA6-0	BA6-1	BA6-2	BA6-3	BA6-4	BA6-5	BA6-6
Density	kg/m^3	2400	2390	2390	2380	2370	2380	2360
Slump	mm	250	240	250	240	230	230	230
		BA7-0	BA7-1	BA7-2	BA7-3	BA7-4	BA7-5	BA7-6
Density	kg/m^3	2400	2390	2400	2390	2370	2370	2380
Slump	mm	250	250	250	260	240	250	260

The yield stresses of different compositions of concrete mixtures were calculated according to their fresh technological parameters: density and slump (Table 6) using equation (2). The plastic viscosity values of concrete mixtures with different type and content of chemical admixture were evaluated according to modified equation (4). Figure 1 – Figure 4 shows the influence of different type and content of chemical admixture on yield stress and plastic viscosity of the concrete mixture.

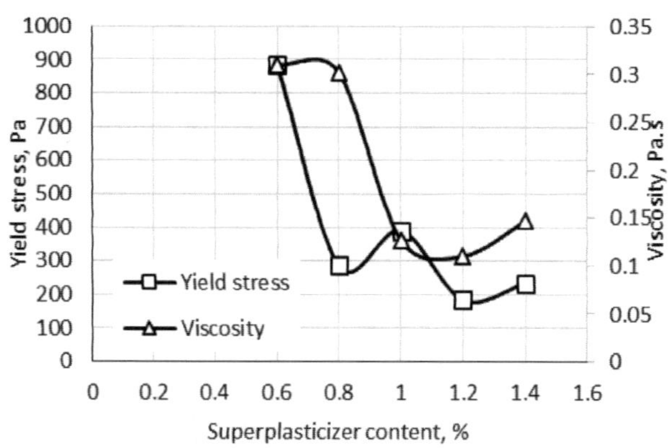

Figure 1. The dependence between superplasticizer content and mixture`s (BA4) yield stress and plastic viscosity

As can be obtained from Figure 1, the increase of superplasticizer content from 0.6 to 0.8% of cement mass in mixture composition decreased the yield stresses of fresh concrete from about 884 to 286 Pa (about 3.1 times). Further increase of superplasticizer content from 0.8 to 1.4% of cement mass in mixture composition showed that the yield stresses decreased about 1.2 times. The increase of superplasticizer content from 0.6 to 1.4% of cement mass in mixture composition gradually increased the plastic viscosity of fresh concrete from about 1.4 to 1.9 Pa.s (about 1.4 times). The superplasticizer based on polycarboxylate ether had a significant influence on fresh concrete rheological properties. This could be due to its adsorption effect on cement particles surface and dispersants to avoid particle segregation.

Figure 2 illustrates the relation between air voids removing admixture content and mixture`s yield stress and plastic viscosity. The increase of air voids removing admixture content from 0 to 0.3% of cement mass in mixture composition hadn't any significant influence on yield stresses of fresh concrete – yield stresses varied in the same range from about 234 to 233 Pa. The addition of air voids removing admixture of 0.3% of cement in mixture composition resulted in the decrease of plastic viscosity about 12% compared to viscosity without this admixture. Complex of chemical admixtures – superplasticizer (constant content) based on polycarboxylate ether and air voids removing admixture based on propoxylate-etoxylate hadn't any significant effect on yield stress and plastic viscosity values of fresh concrete.

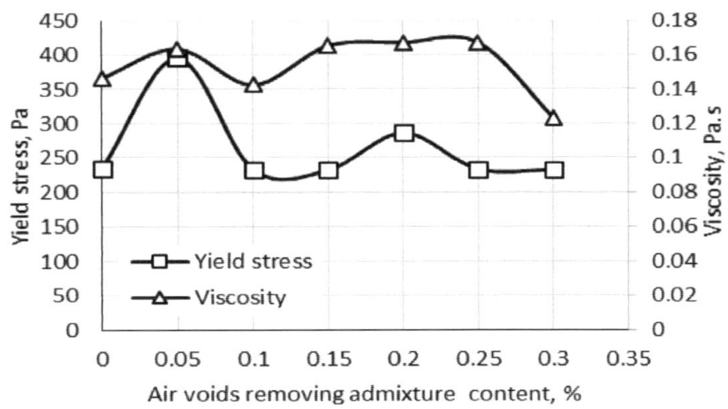

Figure 2. The dependence between air voids removing admixture content and mixture`s (BA5) yield stress and plastic viscosity

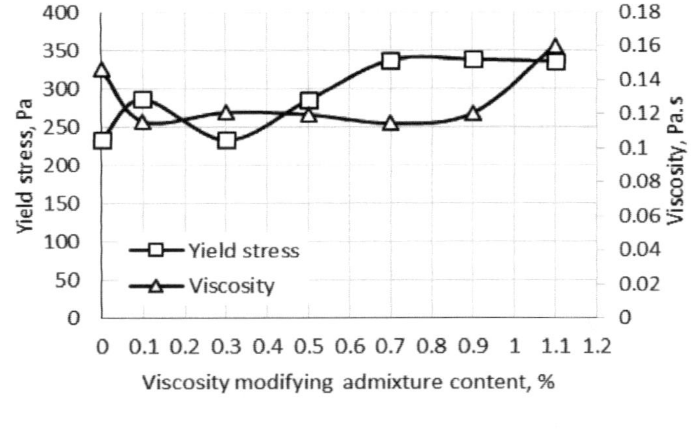

Figure 3. The dependence between viscosity modifying admixture content and mixture`s (BA6) yield stress and plastic viscosity

As can be seen from Figure 3, the increase of viscosity modifying admixture content from 0 to 1.1% of cement mass in mixture composition increased both the yield stresses of fresh concrete from about 234 to 335 Pa (about 1.4 times) and plastic viscosity of fresh concrete from about 0.14 to 0.16 Pa·s (about 0.9 times).

The complex of chemical admixtures – superplasticizer (constant content) based on polycarboxylate ether and viscosity modifying admixture based on synthetic copolymer increased the yield stress and plastic viscosity values of fresh concrete.

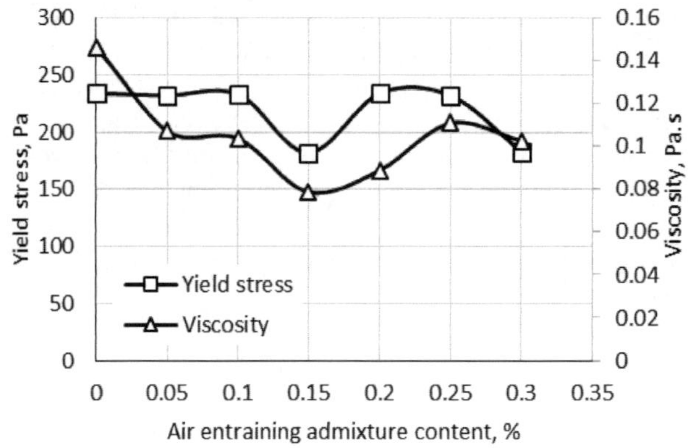

Figure 4. The dependence between air entraining admixture content and mixture`s (BA7) yield stress and plastic viscosity

Figure 4 illustrates the dependence between the air-entraining admixture`s content and mixture`s yield stress and plastic viscosity. The increase of air entraining admixture content from 0 to 0.3% of cement in mixture composition decreased both the yield stresses of fresh concrete from about 234 to 183 Pa (about 1.3 times) and plastic viscosity of fresh concrete from about 0.15 to 0.10 Pa·s (about 1.5 times). The complex of chemical admixtures – superplasticizer (constant content) based on polycarboxylate ether and air-entraining admixture based on modified resin decreased the yield stress and plastic viscosity values of fresh concrete.

To verify the obtained results, the modified empirical formula must be examined in future works and additional experiments are needed to be performed with other concrete mixtures compositions and different types of chemical admixtures.

4. Conclusions

Based on the modified empirical formula, plastic viscosity values of the fresh concrete mixture were calculated using different type and amount of chemical admixture. Coefficients $K_{calc.}$ were calculated as a ratio between different viscosity values of cement pastes obtained without and with certain chemical admixtures. Coefficients describing the change in cement paste viscosity adding the different type of solutions including superplasticizer, viscosity modifying, air-entraining and air voids removing agent were calculated. The change in cement paste viscosity adding different type and amount of chemical admixture was equated to the change in concrete mixture viscosity. To verify the obtained results, the modified empirical formula must be examined in future works and additional experiments are needed to be performed with other concrete mix-compositions and different types of chemical admixtures.

IOP Conf. Series: Materials Science and Engineering **442** (2018) 012018 doi:10.1088/1757-899X/442/1/012018

References

[1] González-Taboada I, González-Fonteboa B, Martínez-Abella F, Carro-López D 2017 Self-compacting recycled concrete: Relationships between empirical and rheological parameters and proposal of a workability box *Constr. Build. Mater* 143 537-546.

[2] Feys D, Wallevik J E, Yahia A, Khayat K H, Wallevik O H 2013 Extension of the ReinereRiwlin equation to determine modified Bingham parameters measured in coaxial cylinders rheometers *Mater. Struc.* 46 289–310.

[3] Domone P 1998 The slump flow test for high-workability concrete *Cem. Concr. Res.* 28 177–180.

[4] Ferraris Ch F 1999 Measurement of the Rheological Properties of High Performance Concrete: State of the Art Report *Jour. Res. Nat. Ins. Stan. Tech.* 104 461–480.

[5] Murata J, Kukawa H 1992 Viscosity equation for fresh concrete *ACI Mater. Jour.* 89 230-237.

[6] Chidiac S E, Maadani O, Razaqpur A G, Mailvaganam N P 2000 Controlling the quality of fresh concrete – a new approach *Mag. Concr. Res.* 52 353–363.

[7] Wallevik J E 2006 Relationship between the Bingham parameters and slump *Cem. Concr. Res.* 36 1214–1221.

[8] Chidiac S E, Mahmoodzadeh F 2009 Plastic viscosity of fresh concrete – A critical review of predictions methods *Cem. Conc. Comp.* 31 535–544.

[9] Zerbino R, Barragán B, Garcia T, Agulló L, Gettu R 2009 Workability tests and rheological parameters in self-compacting concrete *Mater. Struc.* 42 947–960.

[10] Mahmoodzadeh F, Chidiac S E 2013 Rheological models for predicting plastic viscosity and yield stress of fresh concrete *Cem. Concr. Res.* 49 1–9.

[11] Dengwu J, Caijun Sh, Qiang Y, Xiaopeng A, Yu L, Huang L 2017 Effect of constituents on rheological properties of fresh concrete – A review *Cem. Concr. Comp.* 83 146–159.

[12] Soualhi H, Kadri E H, Bouvet A, Ngo T T, Cussigh F, Belaidi A S E 2017 New model to estimate plastic viscosity of eco-friendly and conventional concrete *Constr. Build. Mater.* 135 323–334.

[13] Skripkiūnas G 1993 Optimization of concrete macrostructure according to technological and performance properties and raw material resources. *Doctor. Disser.* (Lithuanian).

Matching safety and sustainability: the SAFESUST approach

Alessio Caverzan, Marco Lamperti Tornaghi and Paolo Negro

European Commission, Joint Research Centre, Ispra, Italy

E-mail: paolo.negro@ec.europa.eu

Abstract. The construction industry, as a main energy consumer and a foremost contributor to greenhouse gas emissions, has been undergoing a "green revolution" in the recent years. Sustainability has become a prominent issue, and environmental methods such as footprint schemes and Life Cycle Analysis approaches are being considered in the design and in the rehabilitation of buildings. However, all environmental assessment methods are applied at a later design stage, to provide a final indication of the life cycle environmental performance. A more effective way would be to consider the environmental issues in the early design stage, along with structural reliability and safety, and the global performance indicator should be expressed in economic terms. The need for an integrated design approach, to tackle safety and sustainability together has been the object of a recent workshop, in which the acronym SAFESUST was introduced. SAFESUST is an acronym to mean SAFEty and SUSTainability. It also identifies a research work-package on impact of sustainability and energy efficiency requirements on building design and retrofit, being conducted by the European Commission - Joint Research Centre, as a part of the project Safe and Cleaner Technologies for Construction and Buildings. The SAFESUST approach has been implemented into a Sustainable Structural Design (SSD) method, which considers both environmental and structural parameters in a life cycle perspective. The integration of environmental data in the structural performance is the focus of the method. Structural performances are considered in a probabilistic approach, through the introduction of a simplified Performance Based Assessment method.

Keywords: Building Sustainability; Sustainable Structural Design; Performance Based Design; Building Rehabilitation

1. The SAFESUST approach

The global population now exceeds seven billion. This means that during the past 250 years or so, it has increased tenfold that of the Industrial Revolution in the mid-18th century, which is believed to have been 700 million. Eighty percent of the global population lives in developing regions, which means that the consumption of resources and energy will increase enormously in the future. Resources and energy are some of the most fundamental elements for the daily life of humankind. In recent years, it has been recognised that increasing fossil energy consumption could even change the global climate. It is anticipated that global warming will cause extremely serious problems in the future, in fact, climate change driven by global warming has already increased the intensity and frequency of weather action such as typhoons/hurricanes and torrential rainfalls, causing enormous damage; in addition, the frequency or intensity of heavy precipitation events has likely increased in North America and Europe [1]. On the other hand, developed countries, such as EU, have accumulated a huge amount of infrastructure and building over a long time. It means that these structures have to be properly maintained by taking cost, natural resources consumption, and more severe loading and environment into consideration. In other words, it is very important how to incorporate sustainability concepts into construction industry.

SAFESUST is an acronym to mean SAFEty and SUSTainability. It identifies a research work-package on Impact of sustainability and energy efficiency requirements on building design and retrofit, being conducted by the European Commission - Joint Research Centre as a part of the project: Safe and Cleaner Technologies for Construction and Buildings. The acronym appeared in the title of the workshop: A roadmap for the improvement of earthquake resistance and eco-efficiency of existing buildings and cities, and the word soon during the discussions which took place, became a neologism. Expressions such as "SAFESUST concept", "SAFESUST problem" and "SAFESUST approach" were commonly used. For this reason, in this freshly created word, possibly remains the essence of the roadmap which has been started to be drawn at the workshop: SAFEty and SUSTainability.

There is sufficient evidence of the fact that there could be not anymore safety without sustainability. This evidence underpins actions to mitigate what could be the most difficult problem mankind have faced since ever: saving our only planet and this was not to be discussed at the workshop. What became soon evident during the discussion is that there can be no sustainability without safety, in tackling the improvement of the existing building heritage.

2. Why a Roadmap?

Today, more so than ever before, global issues such as the climate change are overlapped to local crises like the shrinkage of the construction sector, which followed the economic and financial crisis in EU [2]. On one hand this scenario is a nightmare for stakeholders involved in the building process, but - on the other hand - it represents a good chance to create the technical conditions for a deep and lasting urban transformation. This is a stimulating challenge that the construction chain will face in the coming decades.

The destinations of this route are smart, resilient and sustainable cities and the starting point is the present building stock. The path is made by the innovation of products and processes and the basic tools to have a coherent and safe travel are a compass and a map to guide the action. Without a roadmap, all efforts would lack direction and thus could easily be inefficient. The age profile analysis of the EU's building heritage [3] reveals that only the minority of these 27 billion m^2 is recent, namely built after 1991: the main part of the stock was built between 1961 and 1990 and a significant percentage before 1960. This means that those buildings are likely to have poor thermal and environmental performances and, in addition, they were built without modern structural design codes and often without seismic prescriptions.

Any actions aimed at improving energy and environmental efficiency without addressing safety at the same time is bound to failure. No seismic provisions were considered in the construction of very old buildings, and those enforced at the time of the construction of more recent ones are typically considered to be insufficient. The problem of seismic safety is being considered also in those regions which were not considered as affected by the earthquakes in the past and, sometimes, it is technically more severe there. The improved knowledge of the seismicity of Europe has significantly increased the areas for which at least some seismic provisions should be enforced.

Other dynamic actions, e.g., traffic disturbances, industrial or mining activities, pose similar problems. In more general terms, requirements for service loads are nowadays more stringent than those considered in the design and construction of old buildings, as well as for the deformability requirements. Ageing of materials, poor maintenance and corrosion also affect the resulting structural safety of existing buildings.

The need to pursue is the integrated renovation/reconstruction with the SAFESUST approach: to tackle the problem of the improvement of the structural safety at the same time of the energy and environmental performance in a Life Cycle (LC) perspective should then be kept in mind.

3. Who's game?

In defining a roadmap, those who will have to follow it have to be identified. To this extent, the experience of the SAFESUST workshop has been instrumental. Many individuals from the scientific committee have expressed doubts and concern about the format of the workshop. The idea of mixing traditional sessions based on oral presentations, open discussions and take-home assignments to be left with the rapporteurs was seen as rather uncommon, however, it worked well. What was seen as rather bizarre, and possibly unprecedented as for this topic, was to ask for the participation of experts from differing disciplines. The workshop was intended to bring in the same room experts on structures, architecture and city planning, materials, energy and finance, so that they could learn from each other, discuss and seek synergies and possible agreed priorities. From one side it might have been difficult to identify the leading experts from all disciplines, and convince them that it might have been worth playing that game. From the other, somebody was preoccupied because the disciplines were seen as too far apart to speak the same language, or too close to reach a common view.

The workshop was unanimously seen as a success by the participants, and the main lesson learnt from this success is profound: a solution to the problem of the improvement of safety and eco-efficiency of existing buildings can be found only in a multidisciplinary perspective. Defining the rules of the game calls for the participation of all technical experts.

Any solution conceived having in mind only one aspect of the problem is bound to failure. The need for new, interdisciplinary, expertise was expressed. Different roles will continue to be necessary; at the same time, a new approach calls for a common language as well as shared rules. This is best expressed by the need for new taxonomy, semantic and metric. Moreover, it became evident from the presentations, as well as during the discussions, that not all the players were indeed in the room: other actors have to be involved, such as local authorities and communities, owners and investors; to that extent, another issue has to be included to the roadmap. The SAFESUST approach will from now on also refer to the involvement of all experts and stakeholders, sharing the new common language.

4. The circle game

Circular economy: Cradle to grave (or Cradle to cradle?). Prevent, reduce, reuse, recycle... Where is the starting point?

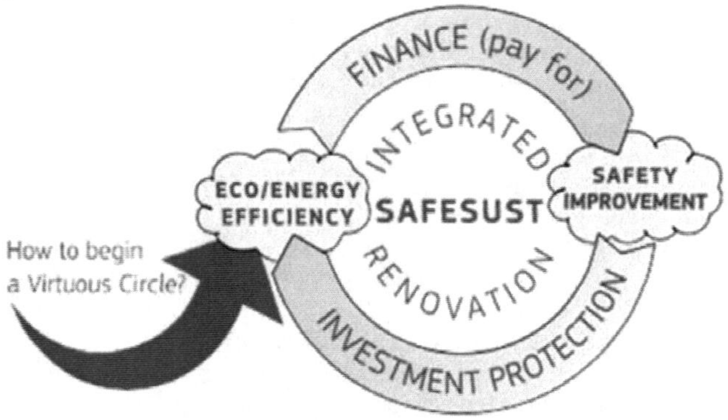

Figure 1. The virtuous circle.

Everybody would agree that there is no starting point in circle. There has been a strong agreement at the workshop: it is not effective to just check the safety requirements of a satisfactory design in terms of eco-efficiency, as well as it is not effective to check the level of eco-efficiency of a compliant safety design. Safety and eco-efficiency should be addressed together, at the same time, and this should be done in the design phase.

At the moment, design norms just relate to safety performance, and eco-efficiency can just be checked afterwards. Design norms should be extended to include eco-efficiency. Speaking about safety, the neighbors' safety is equally important than ours. In case of accident, especially in historic town centres, the fire propagation, or the seismic interactions among buildings, can affect also inherently safe buildings. In addition, to preserve one property is not enough if the surrounding is injured. For all these reasons, an innovative approach must be developed at urban scale to improve the city resilience.

5. The rules

He/she who moves first loses. There seem to be a first move advantage in chess, for sure not in addressing the improvement of safety and eco-efficiency of buildings. If one tries to improve the energy efficiency of a building, without making adequate structural safety, loses. No point in having your old car repainted, if the engine has not been fixed!

Who holds the starting pistol? Typically, owner/investor and architect initiate the game. However, in a circle game, moving first does not make sense if the others remain still, therefore each actor should be encouraged to move. The architect is often the first technical interface with the owners/investors, his/her role is crucial at the beginning, to collect and represent the requirements from the other experts and stakeholders since the first project stage. The architect must define, together with his/her counterparts the main aspects of the building in terms of cost and general layout.

A rational and collaborative pre-design analysis should prevent jeopardising the integrity of the project at later stages. For this purpose, rules should be known in advance, so that whoever moves does that without preventing the others from setting their objectives and applying their methods. SAFESUST objectives should be put forward from the beginning, so that they can be defined in economic terms and be considered by all the experts who intervene in the design. A specific training might be required, and possibly a new coordinated expertise should take the lead: a SAFESUST expert.

6. Good and bad players

Eco-efficiency is a challenge, but it can also be an opportunity. When the opportunity is addressed in economic terms, players try to demonstrate that their products are superior to the others'. There are no doubts that industrial competition harvests innovation; however, a general consensus was soon reached about the fact that there is no good, or bad, material. The use of structural materials is responsible for a huge fraction of the global CO_2 production; however, there is no alternative to the current wide use of steel and concrete. No material is *per se* better performing in environmental terms, the best solution can be identified only when the environmental performances, together with the safety performances, are defined for the building, in a life cycle perspective: the SAFESUST approach.

Technical advances could help to give the start. For instance, it is much easier to recycle materials which have been designed for recycling, whereas reusing normal scrap might be difficult or even impossible. To that extent, new technologies to ensure the possibility to completely recycle the construction materials have been developed. It is for instance possible to produce concrete from which aggregates can be fully recovered at the end of its life. Adopting such technologies would for sure help in starting the game.

7. Pay to play

It's a long way. In each technical session, many needs were identified as for research and innovation, awareness increase, education, training, barrier breaking, integration... The main obstacle to enter the virtuous cycle is economic.

From one side, it should become evident that the cost increase for buying a better performing, more efficient and safer building corresponds to an investment with a high return. The investor should be made fully aware of this, as well as of the importance of preserving the resources and ameliorating the quality of his life and the life of his community. Moreover, the figures shown at the workshop seem to indicate that the cost for design is a small percentage of the total cost of the building (and it might become negligible if compared to the cost of operating the building). It was also shown that in recent years, as an effect of energy efficiency regulations, the building's price index has increased, but it has much less than constructions cost: the owner should realise that investing in safety and eco-efficiency is a good business. Education and public awareness play an important role, to this extent.

On the other hand, the fact that the increase in the costs associated to new eco-efficient construction is not reflected by the increase in retailing prices of new buildings might not be good news for the construction industry. The advantages for the community should govern over those of the owner or of the single investor, and this could be enforced by means of regulations and incentives.

Figure 2. The economy circle.

In a roadmap, the most urgent need should be identified, and a consensus was soon reached about the importance of adequate investments in research and innovation. And also in this case, a specific, SAFESUST, approach was said to be needed: the current lines of funding for research and innovation are focused on eco-efficiency, but seem to forget the problem of the improvement of safety of existing buildings.

In the last resort, citizens pay for the game. They do it indirectly – as taxpayers – fostering studies/research/policies, and they do directly, incurring higher prices to buy or rent buildings with better performances. If people are not aware of these technical advances, and even more if safety and sustainability are not recognised as crucial for their lives, how is it possible to call for resources on this? In other words: who will pay for a show no one has ever heard of?

8. Sustainable Structural Design

To reach all the targets stated above, a new way to conceive structures is needed. The design process should concern the whole life cycle of structures and their components, in order to reduce environmental impacts and increase the durability of structures. In this light, the *Sustainable Structural Design* (SSD) method hereinafter presented could be the solution to design constructions, considering all buildings requirements in a holistic way.

The SSD method is conceived as a supporting tool for the building design process. It takes into account technical-structural aspects along with environmental ones during the entire LC. It tends to optimize the building design in terms of structural and environmental performance, configuring a design method both for safety and sustainability [4] [5].

The framework of the SSD method is based on three main pillars, corresponding to the three evaluation steps: *Energy Performance Assessment; Life Cycle Assessment and Structural Performance Assessment*. The following sections describe the details of each component. Finally, a *Global Assessment Parameter*, expressed in monetary terms, is obtained, to make the necessary design decisions.

9. Energy performance assessment

The operational Life Cycle Energy Assessment (LCEA), which formally should be part of the Life Cycle Assessment (LCA), corresponds to the evaluation of the total energy during the operation phase only. It is performed independently from the other LCEA components. Even though the output of this step, the operating energy E_O, corresponds to an environmental impact, it is more convenient to treat it independently from the LCA.

The main reason for this decision is the fact that the cost of energy typically includes a tax that is intended to represent the environmental impact corresponding to the production and use of the energy, from whatever source the energy is produced. Such tax, which is fixed by the relevant authorities, is in principle accounting for the full environmental impact and therefore it should not be included in the LCA.

10. Life cycle assessment

The life cycle assessment is performed according to the standard cradle to grave approach. The contribution of the operating phase will be determined using the results of the first step and the output of this phase will be expressed in terms of total equivalent CO_2 emissions.

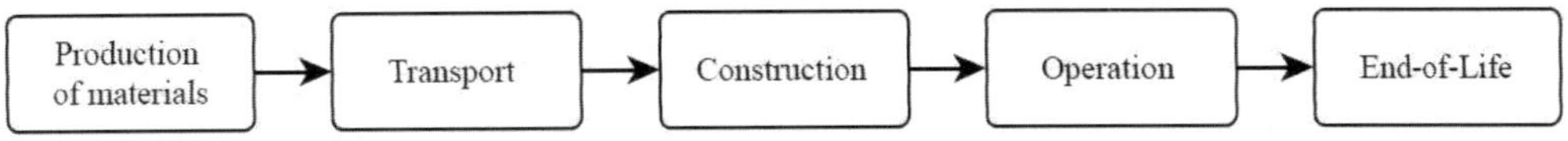

Figure 3. Flowchart of the LCA of a building.

Introducing the Life Cycle Impact Assessment (LCIA) in a LCA, the emissions and resources consumed that can be attributed to a specific product are compiled and documented in a LC Inventory. An impact assessment is then performed, considering human health, the natural environment, and issues related to natural resource use. Impacts considered in a LCIA include climate change, ozone depletion, eutrophication, acidification, human toxicity (cancer and non-cancer related) respiratory inorganics, ionizing radiation, ecotoxicity, photochemical ozone formation, land use, and resource depletion. Different emissions and resources consumed, as well as different product options, can then be cross-compared in terms of the impact indicators.

11. Structural performance assessment

The third step of the SSD method deals with the structural design using a Performance-Based Assessment (PBA). In other terms, the design process should not be seen in the sole aspect of structural response but also in the aspect of structural performance [6] is expressed in predefined design targets that the structure is required to meet over its design life [7].

The structural PBA consists in the implementation of probabilistic scenarios that can occur during the lifespan of the structure in terms of uncertainties [8]. Therefore, not only loads imposed on the structure but also uncertainty and probabilistic response should be taken into account in the structural analysis [9]. The uncertainties are grouped into three main categories namely hazard uncertainty (e.g. earthquakes, winds), structural uncertainty (e.g. stiffness, material properties) and interaction mechanism uncertainty (e.g. pressure duration) [8] [10]. To this regard, the PBA allows structural systems to be designed in a way that meets performance targets in terms of capacity, safety and quality [7].

The final results of the structural PBA are presented in economic context by evaluating all the costs associated with a structural solution, as well as the expected losses that may occur to the building during its life for all the design targets.

The development of PBA methods is gaining big interest in the field of structural engineering. This interest originates from the successful implementation of the Performance-Based Earthquake Engineering (PBEE) method from the Pacific Earthquake Engineering Research (PEER) Center [11]. The PEER's PBEE method has been fundamental in addressing the importance of integrating loss-assessment within structural design. However, such method seems too complicated to be applied to ordinary projects due to complex probabilistic relations and high number of parameters [4] [5]. Considering the latter, a simplified Performance-Based Assessment (sPBA) method has been introduced by Safety & Security of Buildings Unit of the Directorate Space Security & Migration – European Commission, Joint Research Centre (JRC) [4] [12]. The framework of the sPBA method aims at reducing the complexity and the amount of data needed as well as simplifying the procedure of estimating losses due to uncertainties. The method has been developed for seismic actions, but the same approach could, *mutati mutandis*, be applied to other actions on the structure.

The output of the analysis determines the cost of the structure together with expected losses for each defined limit state, corresponding to different Peak Ground Acceleration (PGA) and Inter-storey Drift Ratio (IDR) [13]. The steps of the sPBA method are detailed in the following.

11.1. Limit States Definition

The limit states are defined in terms of damageability and the expected costs for each limit state are calculated. The damage (limit) states that can be defined in a building are low damage, heavy damage, severe structural damage and loss of the building/collapse. The Engineering Demand Parameter (EDP) that measures the structural damage is the Inter-storey Drift Ratio, thus the IDR values are calculated for each damage level by using fragility curves.

11.2. Structural Analysis

The structural analysis step consists in the calculation of the PGA values that cause the IDR values defined in the previous step. This correlation is defined through skeleton curves that may be obtained from the Incremental Dynamic Analysis (IDA), according to the FEMA-350 guidelines [14] or simply from a standard pushover analysis. After performing this analysis to a structural system, the PGA versus IDR can be defined for each configuration of damage state.

11.3. Hazard Analysis

In the hazard analysis, the output of the structural analysis is used to estimate the probability of exceedance. Modern seismic codes provide the relation between the Return Period (T_R) and the PGA [15]. For example, the Italian seismic code [16] provides a set of values of PGAs for nine return periods (30, 50, 72, 101, 240, 201, 475, 975 and 2475 years) along with the interpolation formula for values in between:

$$log(a_g) = log(a_{g1}) + log\left(\frac{a_{g2}}{a_{g1}}\right) \cdot log\left(\frac{T_R}{T_{R1}}\right) \cdot \left[log\left(\frac{T_{R2}}{T_{R1}}\right)\right]^{-1} \quad (1)$$

where:

a_g is the PGA calculated for a defined damage state,
T_R is the return period which corresponds to that state,
a_{gi} are the intermediary values of PGA taken from the seismic map,
T_{Ri} are the return periods corresponding to a_{gi}.

After determining a_g from the previous step, the return period T_R can be defined for each damage state. The probability of exceeding R_N in N years is expressed from the following equation [16]:

$$R_N = 1 - \left(1 - \frac{1}{T_R}\right)^N \quad (2)$$

11.4. Cost Analysis

The total cost of the building C_{TOT} is the sum of the initial construction cost (investment: I) and the expected total loss (L) over the project's life cycle. While the initial cost includes the expenses related to the initial establishment of the facility, the estimation of expected total loss is more complex and involves different stakeholder categories [4] [5].

$$C_{TOT} = I + L \quad (3)$$

More specifically, the contractor estimates in advance the time needed to repair the damages of each limit state. This information about time, needed for each limit state, is associated with the downtime loss. Downtime refers to the period of time in which the system fails to provide its primary function and therefore downtime loss expresses the amount of money that will be spent (lost) while the building is not used. Later, for each state, the structural engineer calculates the cost of repair of the damages [4] [5]. Therefore, the expected loss C_i for each limit state i is expressed as the sum of monetary loss (the amount of money needed to repair the damaged building) and downtime loss (the amount of money spent during the repairing actions e.g. for rent, removal, etc...).

$$C_i = E(Loss_{repair} \vee \text{IM}) + E(Loss_{downtime} \vee \text{IM}) \quad (4)$$

$$L = \sum_{i=1} C_i \cdot (R_i - R_{i+1}) \quad (5)$$

12. Global Assessment Parameter

The outputs of the previous phases are expressed in terms of different units of measurements: energy for the Energy Performance Assessment, mass of the equivalent carbon dioxide emissions for the Life Cycle Analysis and costs for the Structural Performance Assessment. These quantities cannot be summed up to obtain a single global parameter. Therefore outputs of the energy and environmental impact will be converted into monetary units (costs) as it will be explained in the following. At this stage, all the financial issues related to discount costs and benefits over the time, to calculate the net present value, may be performed following common financial procedures.

12.1. Energy performance into monetary units

The calculated amount of energy corresponds to energy consumed during the use phase and usually it is expressed in kilowatt-hour (kWh) or in cubic metres of natural gas (m^3 gas) [17]. The price for gas and electricity in Europe can be determined using the data of the Statistical Office of the European Union (Eurostat) [18] for each member state or the average price of all the Union in the household or industrial category. Therefore the total energy price $R_{E(Energy)}$ can be calculated for a specific country or using the average price by the following equation:

$$R_{E(Energy)} = Q_{(Energy)} \cdot P_{(Energy)} \tag{6}$$

where:
Q_{Energy} is the amount of the energy consumption (kWh or m^3 gas),
P_{Energy} is the price of one kWh or m3 of gas (€/kWh or €/m^3 gas).

12.2. Environmental performance into monetary units

Until today in Europe there are market prices only for the carbon dioxide (CO_2), therefore only the environmental impacts associated with the global warming potential (carbon footprint) can be converted into economic costs [4] [5]. To this end, ILCD is a crucial element to convert all the considered burdens into a CO_2 equivalent amount, and from this calculate an equivalent economic impact [19].

Carbon footprint is defined as the total amount of greenhouse gases (GHG) emissions caused by building life cycle phases, usually expressed in equivalent tonnes of carbon dioxide (CO_{2eq}). Otherwise, the embodied energy includes the energy consumed by processes associated with the production, use and demolition of the building, usually expressed in MegaJoule (MJ), kiloWatt-hour (kWh) or cubic metres of natural gas (m^3 gas) [17]. The conversion of the results into monetary unit can be done following the EU directives.

The carbon dioxide price per tonne is linked with the European Union Emissions Trading System (EU ETS). According to the EU ETS, each member state agrees on maximum national emission limit that should be approved by the European Commission. Then the Union countries allocate allowance values to their industrial operators, who are able to buy or sell such allowances named European Emission Allowances (EUA) [20]. The total number of permits issued, either auctioned or allocated, defines the price per carbon which is therefore, determined by stock exchange rules.

Considering the carbon prices deriving from EU ETS, the monetary cost of the environmental impact referring carbon footprint $R_{E(CO2)}$ can be expressed as follows:

$$R_{E(CO_2)} = Q_{(CO_2)} \cdot P_{(CO_2)} \tag{7}$$

where:
$Q_{(CO_2)}$ is the amount of CO_{2eq} emissions calculated from the analysis (tonne),
$P_{(CO_2)}$ is the price of one tonne of CO_2 according to the EUA (€/tonne).

12.3. Equation of the Global Assessment Parameter R_{SSD}

After converting environmental impacts into monetary unit, the Global Assessment Parameter R_{SSD} of the SSD method can be expressed as the total sum of environmental and structural impact as follows:

$$R_{SSD} = R_{E(CO_2)} + R_{E(Energy)} + C_{TOT} \tag{8}$$

where $C_{TOT} = I + L$.

The Global Assessment Parameter, which is the main outcome of the procedure, represents a direct evolution of the more traditional design process, based on the (initial) Price of the building, which indeed represents the greater part of R_{SSD}. The procedure is scalable to be extended to other processes: building renovation or retrofitting, even to compare different solutions including demolition; infrastructure projects...; for this purposes it remains open to include assessment of other risks, such as fire, wind, floods, etc.

13. Conclusions

The outcome of a recent workshop on the renovation of existing building heritage has been summarized. The main conclusion of the workshop, in which experts from many disciplines (structural engineering, architecture, city planning, finance) participated was that in the renovation of existing

buildings, as well as in the design of new ones, a method to account for safety and sustainability in recent years is needed. The acronym SAFESUST (SAFEty and SUSTainability) was introduced.
A possible way to implement the SAFESUST requirements is represented by the Sustainable Structural Design method. The method is based on the independent evaluation of Energy, Life Cycle and Structural performances, which are finally combined into a Global Assessment Parameter, expressed in economic terms.

References

[1] IPCC, AR. "Intergovernmental panel on climate change." Climate change 2007: Synthesis report (2007).
[2] European Commission (2013): Energy-Efficient Buildings: Multi-annual roadmap for the contractual PPP under Horizon 2020. *European Construction Technology Platform*, Brussels, Belgium.
[3] Buildings Performance Institute Europe (2011): Europe's buildings under the microscope: A country-by-country review of the energy performance of buildings. *Buildings Performance Institute Europe (BPIE)*, Brussels, Belgium
[4] Tsimplokoukou K, Lamperti Tornaghi M, Negro P (2014): Building Design for Safety and Sustainability. *JRC Science and Policy Report*, Joint Research Centre, Ispra, Italy.
[5] Lamperti Tornaghi M, Loli A, Negro P (2018): Balanced Evaluation of Structural and Environmental Performances in Building Design. *Buildings*, 8(4), 52
[6] Fragiadakis M, Vamvatsikos D, Karlaftis MG, Lagaros ND, Papadrakakis M (2015): Seismic assessment of structures and lifelines. *Journal of Sound and Vibration*, 334, 29–56
[7] Negro P, Mola E (2006): Performance-Based Engineering Concepts: Past, Present and Future. *Proceedings of the First European Conference on Earthquake Engineering and Seismology*, 1-10, Geneva, Switzerland
[8] Olmati P, Petrini F, Gkoumas K (2014): Fragility analysis for the Performance-Based Design of cladding wall panels subjected to blast load. *Engineering Structures*, 78, 112-120
[9] Müller HS, Haist M, Vogel M (2014): Assessment of the sustainability potential of concrete and concrete structures considering their environmental impact, performance and lifetime. *Construction and Building Materials*, 67, 321-337
[10] Spence SM, Gioffrè M (2012): Large scale reliability-based design optimization of wind excited tall buildings. *Probabilistic Engineering Mechanics*, 28, 206-215.
[11] Deierlein, G.; Krawinkler, H.; Cornell, C. *A framework for performance-based earthquake engineering*. Pacific conference on earthquake engineering, 2003, pp. 1–8.
[12] Negro, P.; Mola, E. *A performance based approach for the seismic assessment and rehabilitation of existing RC buildings*. Bulletin of Earthquake Engineering 2017, 15, 3349–3364.
[13] Romano E, Negro P, Taucer F (2014): Seismic performance assessment addressing sustainability and energy efficiency. *JRC Science and Policy Report*, Joint Research Centre, Ispra, Italy
[14] U.S. Federal Emergency Management Agency. *Recommended Seismic Design Criteria for New Steel Moment-Frame Buildings* (FEMA 350). 2000.
[15] European Committee for Standardization, EN1998-1 (2005): General rules, seismic actions and rules for buildings, *CEN (European Committee for Standardisation)*, Brussels, Belgium.
[16] D.M. 26/01/2010 (2010): Aggiornamento decreto 11 marzo 2008 in materia di riqualificazione energetica degli edifici
[17] Biswas WK (2014): Carbon footprint and embodied energy consumption assessment of building construction works in Western Australia. *International Journal of Sustainable Built Environment*, 3 (2), 179-186
[18] Statistical Office of the European Union (Eurostat).
[19] Wolf, M.A.; Pant, R.; Chomkhamsri, K.; Sala, S.; Pennington, D. T*he International Reference Life Cycle Data System (ILCD) Handbook* - JRC Reference Reports. European Commission JRC, Ispra, Italy, 2012.
[20] European Energy Exchange (EEX).

Chloride ion detection through the voltage response of a galvanic pair

Stephen Sammut[1], Edward Gatt[1], Ruben Paul Borg[2]

[1] Department of Micro & Nano Electronics, Faculty of ICT, University of Malta
[2] Faculty for the Built Environment, University of Malta, Malta

E-mail: stephen.sammut.00@um.edu.mt, ruben.p.borg@um.edu.mt

Abstract. Chloride ion ingress into reinforced concrete structures is one of the primary causes of concrete degradation. Real-time monitoring of concrete structures for the ingress of chloride ions is therefore of paramount importance towards the achievement of durable and sustainable structures. It is therefore important that the chloride ion detection method used must be accurate and reliable. This paper will endeavour to show how embedded 316L stainless steel and EN3b mild steel galvanic electrode pairs will react to chloride ions present in the concrete pore solution. Through practical experimentation the correlation between the chloride ion concentration in the pore solution and galvanic voltage measurements between the electrode pairs was explored. An experimental setup was proposed, based on concrete samples with in-built fluid containment vessels holding sodium chloride solution, which was allowed to permeate through the pore structure of the concrete over a defined period of time. At particular time intervals core samples were taken at different depths in concrete samples, and the chloride ion content was determined with depth of penetration through titration on the extracted material. The chloride ion content results from titration tests were compared to the measured galvanic voltages. Within certain limits there was good correlation between the chloride ion concentration and the measured voltage readings across the galvanic electrode pair. Galvanic voltage measurements provide a relevant technique for the detection of chloride ions in concrete and in the design of a chloride ion sensory system. Validation testing was also carried out, where samples with known salt concentrations were prepared against which the results obtained could be compared.

Keywords: chloride ion penetration; galvanic corrosion; concrete degradation

1. Introduction

Chloride ions, Cl^-, as well as species such as O_2, and SO_4^{2-} are able to penetrate the concrete pores through various mechanisms. One of the principal methods of penetration is diffusion which occurs when a concentration gradient is established. This gradient causes the species to travel from the surface of the structure, into the pores and then through the pores into the internal regions of the concrete. Once in the internal regions, the chloride ions come into contact with the rebar structure and cause the initiation of corrosion [1].

1.1. Theoretical basis

The pore solution in concrete is highly alkaline, with a pH of around 12.5 to 13.5. This pH value is due to the substantial amounts of the hydroxides of calcium, sodium and potassium which are dissolved in the pore solution. The highly alkaline environment of the pore solution passivates the steel, which forms a layer of gamma ferric oxide, thereby protecting the steel and preventing corrosion [2].

Chloride ions dissolved in the pore solution, result in subsequent localised corrosion, in specific areas which are not protected by passive film, thereby resulting in active zones which act as anodes. The surrounding areas on the other hand act as the cathodes in which the reduction of oxygen occurs.

The anodic and cathodic half-cell equations which tie together the galvanic process are represented below [3]:

Anodic half-cell reactions

$$Fe \rightarrow Fe^{++} + 2e^- \quad \textbf{(1)}$$

$$2Fe^{++} + 4OH^- \rightarrow 2Fe(OH)_2 \quad \textbf{(2)}$$

$$2Fe(OH)_2 + {}^1/_2O_2 \rightarrow 2FeOOH + H_2O \quad \textbf{(3)}$$

Cathodic half-cell reactions

$$2H_2O + O_2 + 4e^- \rightarrow 4(OH)^- \quad \textbf{(4)}$$

2. Methodology

2.1. Experimental procedure

The procedure followed during the experimental programme, was based on ASTM C1556-11a. This is the standard test method used for determining the apparent chloride diffusion coefficient of cementitious mixtures by bulk diffusion [4].

The test samples designed and built for this experimental process are shown in Figure 1. The procedure to construct the test samples started by preparing the electrodes required to form the galvanic pairs. The two metals used were EN3B Mild Steel, which act as the anodes and 316L Stainless Steel, which act as the cathodes. The edges of the electrodes were rounded to remove all pointed edges thereby ensuring that contact area is maximised, with solvents being used to remove any organic residues.

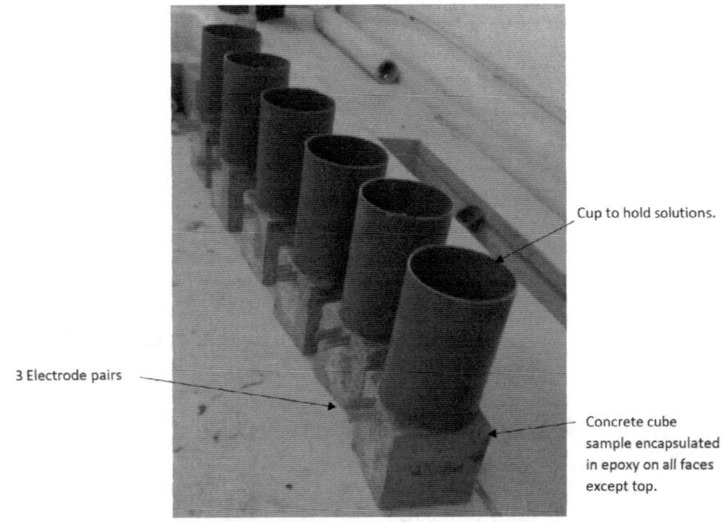

Figure 1. Concrete blocks with fluid containment vessels attached.

The electrode pairs which were spaced as shown in Figure 2 were set in moulds intended for 100mm cubic samples of mortar. The cement used was CEM 1 52.5N, together with crushed rock 0/4 aggregate sand, sourced from a Coralline Limestone quarry in Malta. The water/cement ratio was set at 0.5 with a sand/cement ratio of 2.7.

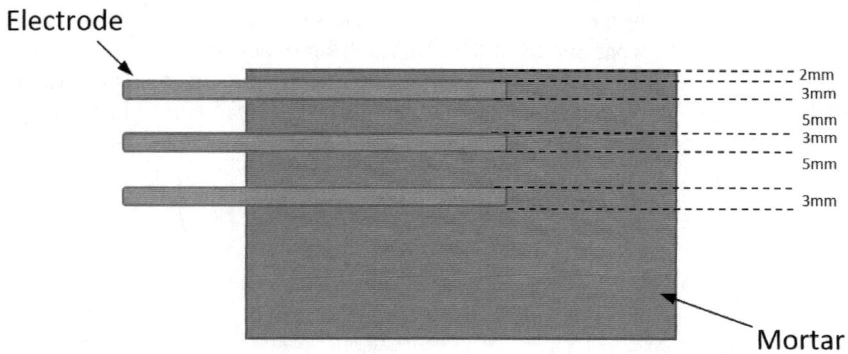

Figure 2. Diagram showing the spacing between electrodes in the mortar cube (not to scale).

A PVC fluid containment vessel was then attached with sealing adhesive to each concrete block, in order to ensure a watertight seal. This vessel allows the testing fluids to be contained and also allows galvanic voltages across the electrodes to be studied while the actual diffusion is occurring.

The exposed concrete surfaces were then coated with an epoxy structural system sealant. This sealant was necessary to minimise the possibility of fluid egress through the exposed surfaces thereby ensuring that the chloride ions move through the pore solution predominantly by diffusion rather than other bulk transport methods.

In line with the method as proposed in the ASTM C1556-11a the next step was to expose the samples to a solution of saturated calcium hydroxide $Ca(OH)_2$ which was placed in the fluid containment vessels fitted on the concrete samples. The solutions were kept replenished up to the maximum mark in the fluid containment vessels. The exposure to the calcium hydroxide continued until the concrete mass did not increase by more than 0.1 mass% per 24 hours.

After the exposure to the calcium hydroxide, the samples` fluid containment tanks were emptied, cleaned with distilled water and then filled with the concentrated brine exposure liquid. The exposure liquid was prepared by mixing 165g ±1g of pure NaCl per dm^3 solution with distilled water. The exposure to the solution was continued for a number of days with the fluid containment vessels on all the samples being kept filled and replenished with solution to ensure that hydrostatic pressure on the surface of the concrete remains constant.

3. Results

The samples were studied over two time periods namely, Period 1 being between 1 and 19 days and Period 2 between 20 and 120 days. The first time period, Period 1, was monitored to study the correlation between the galvanic voltage readings and the chloride ion concentration in early age concrete before the pore solution starts to dry out. The latter time period, Period 2, was studied to see how galvanic voltage varies as the concrete matures in controlled ambient conditions.

Throughout the experimental period the voltages across each of the three electrode pairs was measured. For each measurement three spot readings was taken and the average was then calculated, with all samples giving results which were very similar. The average values were plotted as presented in Figure 3. With reference to Figure 2, Electrode set 1 was the galvanic pair at the top next to the fluid containment vessel, electrode set 2 was the middle galvanic pair set and electrode set 3 was the lowermost galvanic pair.

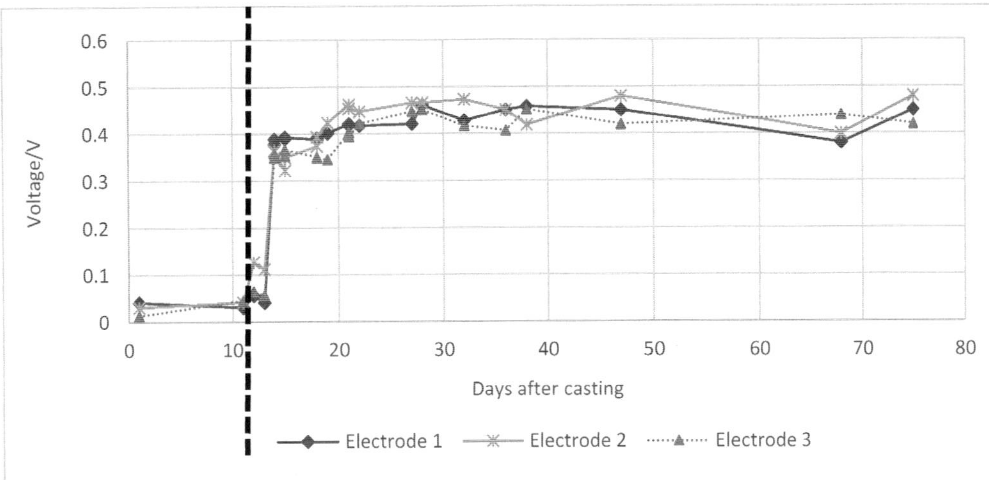

Figure 3. Galvanic voltage measured across electrodes for sample A1. Thirteen days (marked by dotted line) after casting exposure to chloride ions occurs. Each of the series shows the voltage values across an electrode pair.

As can be seen in Figure 3, the galvanic voltages measured before the exposure to chloride ions occurred, were all below the 0.3V threshold (actually all values measured were below 0.1V). As soon as exposure to chloride ions occurred the galvanic voltage increased instantly to values in the range between 0.35V and 0.45V. This meant that the presence of chloride ions and the voltage increase occurred at exactly the same time. It was also observed that the voltage increased at exactly the same time as when the chloride ion exposure occurred, in all the samples. Moreover the voltage across the galvanic pair instantly increased by more than a factor of four as soon as there is chloride ion contact with the metals.

The fact that the voltage increased in all the samples at the moment in which chloride ion exposure occurred, indicates that there is strong correlation between the galvanic voltage and the chloride ion concentration.

3.1. Titration analysis.

To calibrate the voltage increase with the chloride ion concentration values, the chloride concentration in the pore solution at the electrode positions was studied chemically using a volumetric procedure.

The volumetric procedure used was based on an argentometric titration process which involved the use of silver salts to determine the concentration of chloride ions present. The procedure used was based on BS 1881-124:1988 [5].

The concrete sample was cored at accurately measured depths and dust samples were collected for analysis. The dust samples were taken at 2mm depth intervals up to a depth of 14mm from the surface of the concrete block. Further samples were taken at depths of 25, 35, 45 and 50mm with around 8g of dust being taken for each depth.

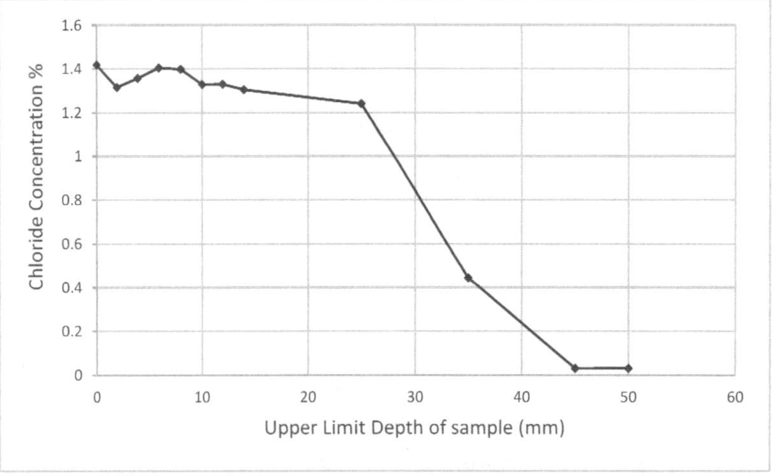

Figure 4. Chloride Concentration vs Depth of dust sample collection (After 23 days).

Figure 4 shows the profile of the chloride concentration changing with depth as obtained from the volumetric chemical analysis. It can be seen that as the sample depth increases, the chloride concentration decreases. This decrease becomes pronounced after the 25mm mark from the surface is exceeded. The deepest of the three electrode sets is situated at around 21mm from the concrete surface. The titration results therefore show that all the 3 electrode pairs, the lowest of which is set at 21mm from the sample surface, are situated in an area which reaches a chloride concentration of between 1.2% and 1.4% of the cement weight. The results therefore give an indication that the galvanic voltage across the electrode pair increases when the chloride ion concentration increases above 1.2% of the cement weight.

3.2. Validation Testing.

To confirm threshold chloride limits above which galvanic voltages start to increase above the 0.3V threshold, three samples with known sodium chloride concentrations added to the mortar mix were prepared as shown in Figure 5. These samples consisted of three mortar cubes with electrode sets A, B and C embedded. Each of the samples had a different sodium chloride salt concentration in the mortar mix.

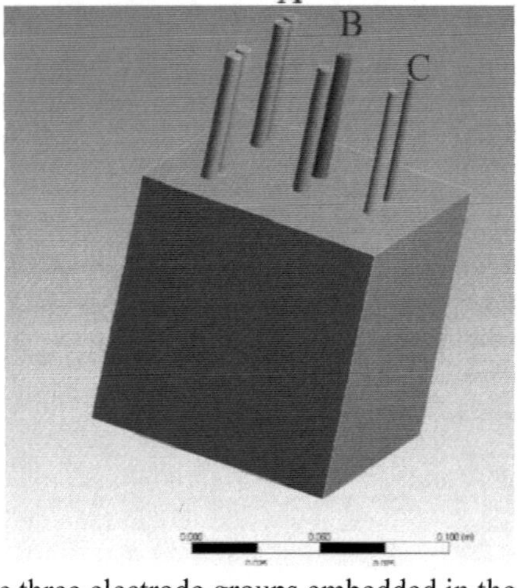

Figure 5. The three electrode groups embedded in the mortar.

One of the mortar cubes was made with mortar only i.e. 0% salt, the other one had salt equivalent to 1% of the cement weight added to the mortar mix, while the third sample had salt equivalent to 5% of the cement weight added. The material used and ratios (apart from the added salt concentrations) were the same as that used previously for experiments discussed in this paper. With reference to Figure 5 the electrode sets B and C were used in this experimental procedure to form the galvanic pairs. Electrode set A on the other hand formed by a set of four 316L stainless steel electrodes through which resistivity changes in the concrete with varying salt concentrations could be monitored through the use of the Wenner probe principle.

Electrode set B consisted of a 6mm diameter galvanic pair of electrodes, one of which was made of 316L stainless steel and the other EN3b mild steel. Electrode set C consisted of a galvanic pair made of the same metals as set B but with the electrodes being of a smaller 3mm diameter. The galvanic voltages measured between the electrodes pairs of each of the sets labelled B and C were recorded over a period of 115 days and the results were tabulated in Table 1.

Table 1. Galvanic voltages on the two stainless steel/mild steel galvanic pairs 5, 6 (6mm diameter) & 7, 8 (3mm diameter).

	Galvanic Voltage 6mm V			Galvanic Voltage 3mm V		
	6mm	**6mm**	**6mm**	**3mm**	**3mm**	**3mm**
Days after casting	**0%**	**1%**	**5%**	**0%**	**1%**	**5%**
4	0.01	0.15	0.36	0.03	0.20	0.37
7	0.03	0.21	0.38	0.04	0.23	0.38
10	0.03	0.21	0.38	0.04	0.23	0.38
14	0.08	0.21	0.40	0.08	0.12	0.34
19	0.05	0.14	0.40	0.08	0.12	0.37
38	0.01	0.11	0.39	0.02	0.06	0.27
49	0.01	0.10	0.33	0.02	0.05	0.21
59	0.03	0.17	0.28	0.05	0.06	0.28
68	0.08	0.19	0.37	0.07	0.06	0.25
88	0.05	0.09	0.27	0.08	0.12	0.21
95	0.03	0.07	0.24	0.06	0.08	0.17
115	0.05	0.10	0.25	0.03	0.02	0.18

The results shown in Table 1 over the first 19 days show that with a salt content of 1% the galvanic voltage between the galvanic pair of electrodes is higher than that noticed in the sample with 0% salt content. In the sample with the 5% salt content the galvanic voltage presented the high values similar to those measured in the experimental procedures as presented in section 2.1 when exposure to sodium chloride occurred. The values obtained in this sample, exceeded the correlated directly with the galvanic voltage values shown in Figure 3 thus confirming that chloride ion concentration above threshold causes galvanic voltage to instantly rise thereby validating the results of the chloride ion concentrations measured through the titration as shown in Figure 4.

The standard deviations for the voltages obtained in the period between day 4 and day 19 were calculated to study how closely grouped the results were when measurements were taken across a time period. The graph in Figure 6 shows tightly grouped voltage results, thereby indicating the suitability of the use of this method towards the construction of chloride ion sensory systems.

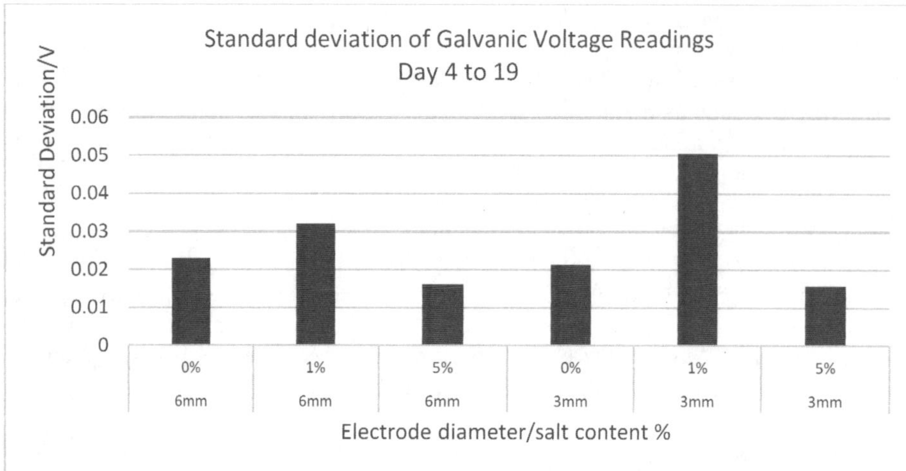

Figure 6. Standard deviation values of Galvanic Voltages obtained in the period between Day 4 to day 19 post molding.

The three samples were air dried in a controlled environment at an ambient temperature of 21°C and RH of 60%. It can be seen from Figure 7 that as the mortar matures, the galvanic voltage trend line for the 5% salt sample decreases and tends to converge with the trend lines of the 1% and 0% salt concentrations.

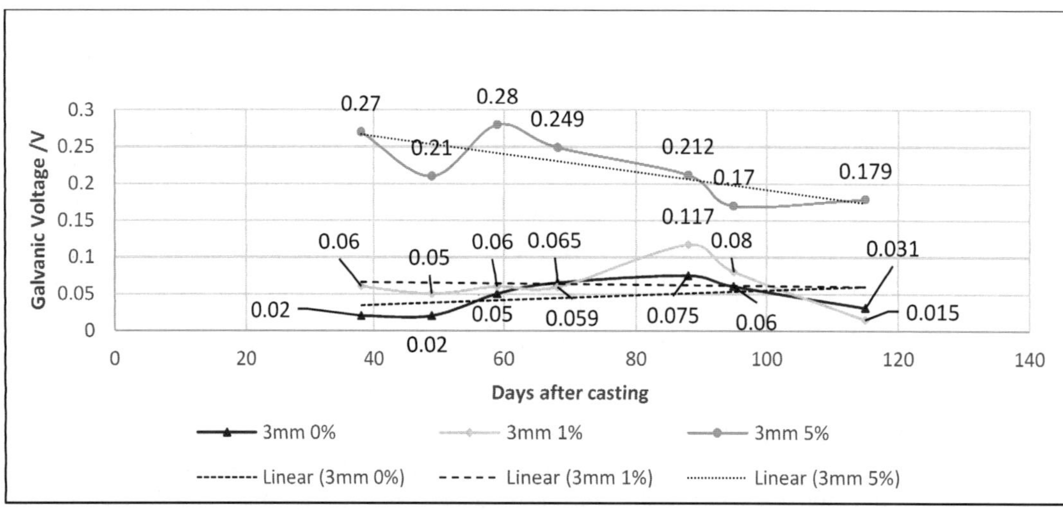

Figure 7. Galvanic voltages for the 3mm electrodes plotted against time.

There can be more than one reason for this reduction in galvanic activity as the sample matures. The main reason relates to the actual mechanism of ionic movement through the pore solution. For corrosion to occur there needs to be a path for ion transfer between the two electrodes. In the case of the selected galvanic pair there needs to be a liquid medium, through which the hydroxide ions formed at the cathode can move towards the anode, where the oxidation reaction with the iron occurs [2] [6].

Another reason for the reduction in galvanic activity with maturity of the sample, is the increase of the concrete's resistivity as the concrete dries. Testing using electrode set A, discussed in previous literature, confirmed that while resistivity is inversely proportional to ionic concentration, drying of the sample causes an increase in the concrete's resistivity [7] [8].

4. Conclusions

The results presented and discussed in this paper show that chloride concentrations in excess of circa 1.2% cause a galvanic electrode pair to present an increase in the voltage across it. This voltage increase may be measured and used to indicate the presence of chloride ions in the pore solution. Repeat testing has presented reliable results throughout.

A system based on a 316L stainless steel and En3B mild steel pair is therefore an effective solution which can be used to detect chloride ions above a threshold concentration of circa 1.2%. However the results also indicated that as the concrete dries up the galvanic voltage becomes a less reliable means of measuring chloride ion concentration. Methods of moisture retention in the sensor's active zone must therefore be explored for an effective sensory system to be developed.

Looking at the validation of results presented in section 3.2 it can also be seen that such a system has potential for future development and lends itself well towards effective miniaturisation. If one compares the first 20 days for both the 6mm and 3mm electrodes, very similar responses can be seen. This in the sense that for 5% salt content, both 3mm and 6mm diameter electrodes immediately exhibited a galvanic voltage above the 0.3V threshold. The other two samples (1% and 0% salt concentrations) never reached this voltage level across either of their electrode pairs. It can therefore be deduced that the electrode′s surface area does not noticeably affect the open circuit voltage measurement.

The galvanic voltages obtained in this paper are similar to those achieved in other experimental procedures as found in prior literature which also recorded voltages of between 0.5 and 0.6V when the galvanic pairs were exposed to chloride ion solutions. Such literature also indicated that the oxygen level has a predominantly important effect on the galvanic activity and therefore the open circuit voltages measured. Tests in literature have shown that with increased oxygen concentration in solution the galvanic voltage can reach 0.87V in a mild steel/stainless steel galvanic [9]. Hence oxygen concentration enhancement methods, will be another area which would need to be studied and explored to be integrated in a sensory system in other future works carried out.

References

[1] L. Bertolini, "Corrosion of Steel in Concrete- Prevention, Diagnosis, Repair, Second Completely revised and enlarge edition.," Milan, Wiley, 2013, pp. 1-250.

[2] The Electrochemical Society, "Uhlig′s Corrosion Handbook, Third edition," Pennington, Wiley, 2011, pp. 94-600.

[3] American Concrete Institute, "Protection of Metals in Concrete against Corrosion," USA, 2001, pp. 222R-1 - 222R-28.

[4] ASTM, "Standard test method for determining the apparent chloride diffusion coefficient of cementitious mixtures by bulk diffusion," ASTM, 2013.

[5] BSI, "BS 1881-124:1988 Testing concrete. Methods for analysis of hardened concrete," London, 1988.

[6] B. Nimmo and G. Hinds, "Beginners Guide to Corrosion," USA, National Physics Laboratory, 2003, pp. 1-10.

[7] R.P.Borg, E.Gatt and S.Sammut "Chloride Ion Detection in Concrete through Galvanic and Resistivity methods," *Jordan Journal of Civil Engineering,*, vol. 11, no. 4, 2017.

[8] P. Spiesz and H. Brouwers, "Application of the conductivity sensor as a chloride detector in concrete," in *1st International Conference on the Chemistry of Construction Materials pp351-354*, Eindhoven.

[9] D. Qu, S. Qian and B. Baldock, "Effects of galvanic coupling between carbon steel and stainless steel reinforcements," in *NACE Northern Area 2003 Conference pp 1-26*, Ontario, Canada, 2003.

IOP Conf. Series: Materials Science and Engineering **442** (2018) 012021 doi:10.1088/1757-899X/442/1/012021

Sustainable structural rehabilitation and strengthening of the "Ponte delle Grazie" bridge in Faenza, Italy

M Micheloni[1], **M La Monica**[1], **D Parmeggiani**[2], **P Barchi**[2]

[1] Studio Micheloni, Florence, Italy
[2] Unione Romagna Faentina – Settore Lavori Pubblici, Faenza, Italy

E-mail: michelangelo@micheloni.org

Abstract. The "Ponte delle Grazie" is a three span bridge, with a total length of 72 meters, built between 1948 and 1952 in Faenza, Italy. The reinforced concrete main beams of the deck have undergone a strong deterioration over the years. In detail atmospheric agents and chemical aggressions caused a strong deterioration of the concrete, up to the point of making the structure not accessible and at risk of collapse. In fact, the five main beams were heavily damaged, as well as the concrete bearings were strongly compromised. So, an urgent intervention was necessary to save the structure of this historic bridge. A delicate restoration has allowed to remove the deteriorated concrete and to restore the resistant sections with new materials compatible with the old remaining structures. In particular, a specific rehabilitation procedure was studied using fiber-reinforced cement mortar with low elastic modulus, that is shrinkage compensated, in combination with composite materials reinforcements. Without modifying the structural behaviour of the bridge, the deteriorated concrete was restored and reinforced, in a sustainable way, in order to make the structure safe and usable again.

Keywords: bridge, infrastructure, rehabilitation, concrete, FRP

1. History and description of "Ponte delle Grazie"

The "Ponte delle Grazie" is a three span bridge, with a total length of 72 meters, designed and built between 1948 and 1952 in Faenza, Italy. The bridge has a historical value linked not only to its age but also to its role as a link between the two oldest parts of the city: the historic center of Faenza and the Borgo Durbecco.

In the last years a strong deterioration of concrete was find in the main beams of the lateral spans, in the Gerber half joints of the central span and in the reinforced concrete bearings on the breast walls of abutment. The deterioration of the concrete was extended up to the point of making the structure not accessible and at risk of collapse and an urgent intervention was necessary to save the structure of this historic bridge, following the procedure described in the following.

Figure 1. The "Ponte delle Grazie" bridge located on the Lamone River in an original postcard.

1.1. Description of the bridge

The bridge belongs to the category of "Gerber Bridge" (isostatic) with three spans with "internal half joints".

The type in question was usually used when it was desired to have an advantageous distribution of the stresses in the structure (less positive moments in exchange for negative moments on the supports), without renouncing the advantages of the isostaticity. The disadvantage of this structural typology is that the detail in the beam seat region represents a potentially sensitive structural detail.

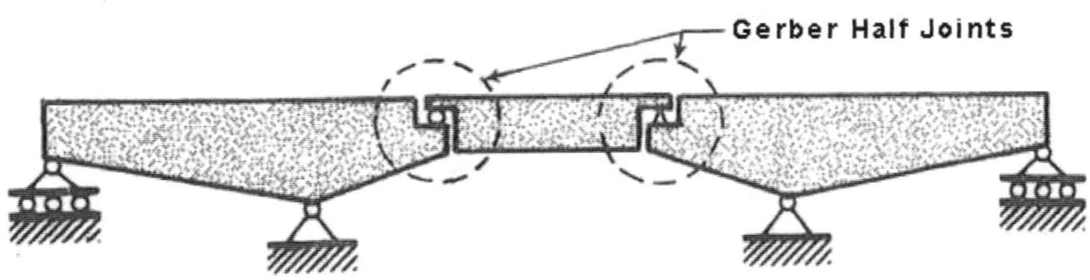

Figure 2. The structural scheme of the bridge with two Gerber half joints inside the central span

The side spans have a length of 21 meters, while the central span has a total length of 30 meters. The part of the bridge between the two piers is made by two 6 meters cantilever spans with a suspended span of 18 meters between them. The connection between cantilever and central span is made with Gerber half joints. The total length of the bridge is 72 meters.

The cross section of the deck is made of 5 beams of variable height in reinforced concrete with 240cm wheelbase. and connected with stringers. The total deck width is 13 meters.

1.2. History of the bridge

The bridge was designed in 1949 and built between 1950 and 1952. The original project of the bridge and many documents regarding its construction are conserved at "Archivio di Stato di Faenza" [2], and so it was possible to analyze and study these documents to increase the knowledge of the bridge.

Figure 3. The elevation of "Ponte delle Grazie" in the original drawings from "Archivio di Stato di Faenza" [2]

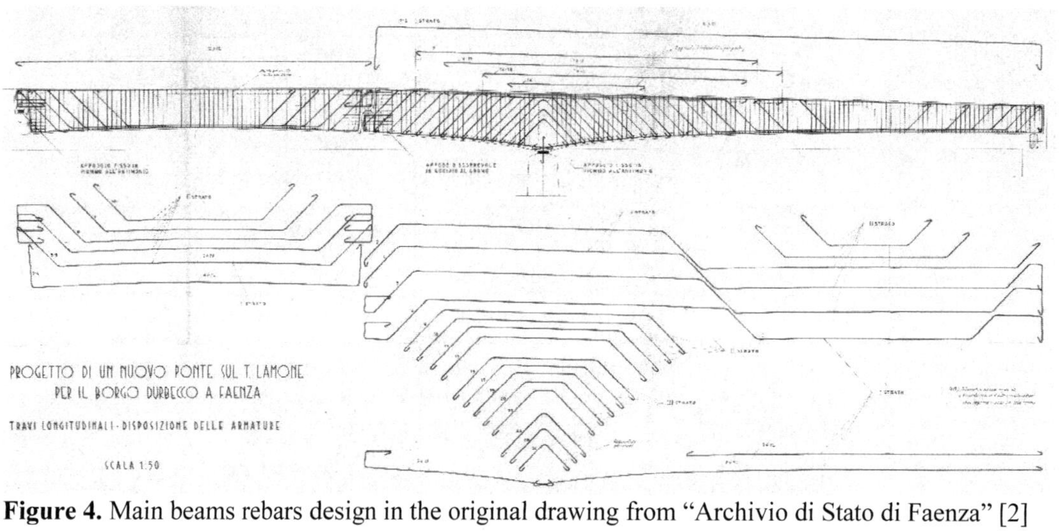

Figure 4. Main beams rebars design in the original drawing from "Archivio di Stato di Faenza" [2]

2. The knowledge process

The knowledge of the structure and its state of conservation goes through the analysis of the following steps.

- Analysis of the deterioration of the concrete and reinforcement steel
- Analysis of materials
- Analysis of the structure and its service level

3. The deterioration of the concrete and reinforcement steel

The first task to be carried out was the identification of the critical aspects of the structure and its deterioration.

The most critical deterioration was found in the following parts:

- Main beams of the side spans
- Gerber half joints of the central span
- Reinforced concrete bearings on the breast walls of abutment

3.1. Main beams deterioration

The main beams of the side spans presented a strong deterioration, especially near the abutment, as demonstrate by visual inspections and laboratory tests on materials.

Figure 5. Beams deterioration on the side span

Figure 6. Beams deterioration on the side span

The dangerous state of the main beams in the area near to the abutments was evident. In fact not only the concrete is strongly damaged, but also the steel bars are deeply corroded and the stirrups are exploded.

3.2. Gerber half joints deterioration

A disadvantage inherent in "Gerber bridge" type of construction is that there are problems associated with leakage through the half joint. This enables moisture to accumulate at the beam seats thereby increasing the propensity for the deterioration of the concrete and reinforcement steel.

Furthermore, together with water leakage and moisture, there is also an accumulation of chlorides, dangerous for concrete and steel bars.

The situation is adversely compounded by limited access to the bearing seat joint, which leads to both inspection and maintenance issues.

So, it is very important to analyze the detailing deficiencies and deterioration.

IOP Conf. Series: Materials Science and Engineering **442** (2018) 012021 doi:10.1088/1757-899X/442/1/012021

Figure 7. Gerber half joint at actual state

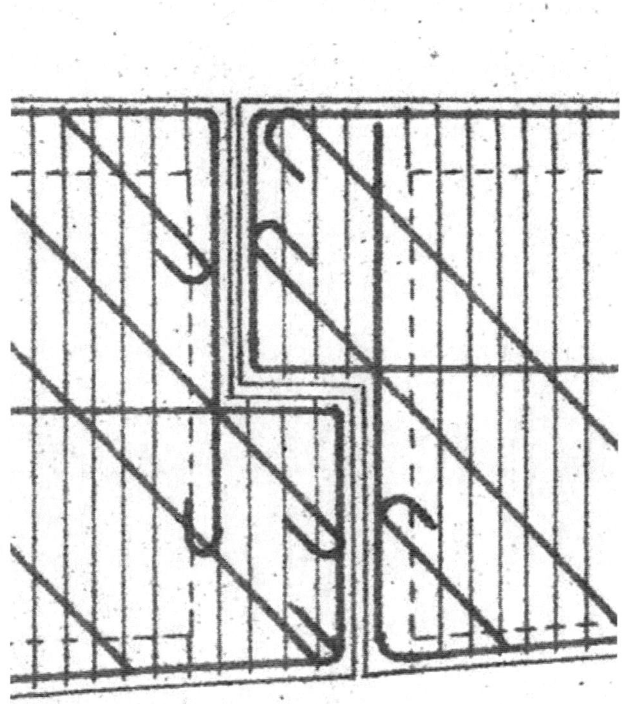

Figure 8. Gerber half jpint in the original drawing from "Archivio di Stato di Faenza" [2]

3.3. Bearing deterioration

The main beams of the side spans rest on the breast walls with reinforced concrete bearings. This kind of "pendulum bearing" is made with reinforced concrete and they represent a very sensitive detail.

Figure 9. Deterioration and cracks on the concrete pendulum bearing

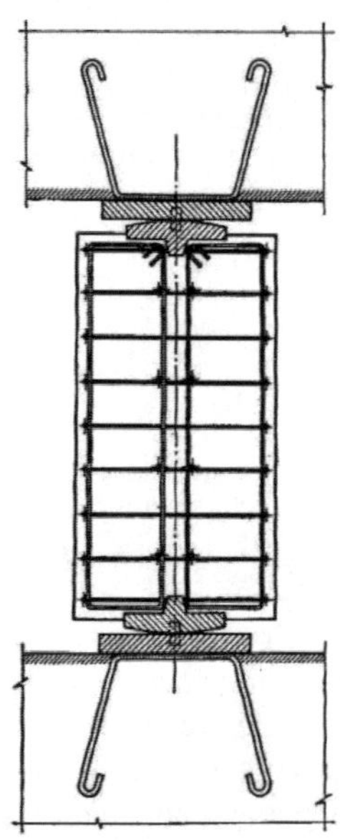

Figure 10. Concrete pendulum bearing typical scheme [1]

In detail this concrete bearings exhibit a strong degradation phenomena, with evident breaking lesions in place, as evidenced by the following images.

The dangerous state of the supports is evident and it has become necessary to take urgent action.

4. Analysis of materials

The first task to be carried out was the identification of the critical aspects of the structure and its deterioration.

A detailed test program was performed on the existing concrete to determinate the mechanical properties:

- Compression Test On Concrete Cores
- Sclerometric Hammer Test
- Son-Reb Test

and to determine the concrete deterioration:

- Carbonation Test
- Chlorides Test

4.1. Compression Test On Concrete Cores

Sixteen concrete cores were taken from various parts of the structure and subjected to compression tests in the laboratory.

Although it is possible to average the values of the compressive strength of the 16 concrete cores and although the average values obtained are in line with the requirements of the original project, it is clear that some cores have given very low values compared to the others. It is therefore essential to investigate in more detail with an approach that divides concrete cores according to the structural elements that form the bridge.

In fact, the structure has been realized in 3 parts of works and therefore each of these phases has been characterized by different concrete supply. Furthermore, the exposure of the concrete is also different for the various structural elements and therefore also the deterioration of the characteristics over time may have happened differently.

The samples are then divided according to the structural elements stated in the following table 1.

Table 1. Properties of concrete

Bridge parts	Average compression strenght - f_m	Concrete class
Pier (1st part of works during 1948)	18,6 MPa	C10/13
Abutment (2nd part of works during 1949)	31,8 MPa	C23/28
Deck slabs (3th part of works during 1950)	41,9 MPa	C33/41
Longitudinal girders of central span (3th part of works during 1950	27,5 MPa	C19/23
Longitudinal girders of side span (3th part of works during 1950	17,9 MPa	C9/11
Whole bridge	24,4 MPa	C16/20

These results are confirmed also by Sclerometric Hammer and Sonreb site tests.

IOP Conf. Series: Materials Science and Engineering **442** (2018) 012021 doi:10.1088/1757-899X/442/1/012021

4.2. Test for the determination of carbonation

Following the codes UNI 9944:1992, all the samples of concrete cores, after breaking, were also subjected to carbonation tests in the laboratory.

The test allows the detection of the state of conservation of structural elements in concrete through the chemical process of carbonation, which provides for the action of carbon dioxide that is present in the air to neutralize the lime present in the concrete.

The average value of the carbonation depth on the 15 cores is about 6cm. This is symptomatic of a strong phenomenon of carbonation and deterioration of the concrete in progress.

4.3. Test for the determination of chlorides

Following the codes UNI 9944:1992 and UNI EN 206-1, some samples of concrete cores have been subjected to this test that allows to determine the percentage of water-soluble chlorides present in a mass of concrete, so as to be able to give a judgment on the risk of corrosion by chlorides that can reach a critical level on the surface of the concrete, with consequent verification of localized breakdown of the film of passivity of the reinforcements and subsequent priming of the corrosion mechanism.

From the tests carried out, there was a low percentage content of chlorides in the concrete of the slabs and of the internal parts of the beams, lower than the indicated range of 0.2-0.4% in UNI EN 206-1. Instead, in the first 4 cm of the edge beams a high risk of chloride corrosion was found with a percentage of chlorides above the maximum limit of the regulatory range, i.e. 0.6% compared to the mass of the cement.

Table 2. Depth of chlorides

Depth	percentage of chlorides in concrete mass - %
4cm	0,61
11 cm	0,09
16 cm	0,07

The strong attack of the chlorides in the concrete cover of the lateral beams is therefore evident (first 4cm) confirming the high corrosion caused by the reinforcement bars.

5. Analysis of the structure and its service level

The results of the test on materials are used to develop a finite element model that represents the real geometries and mechanical characteristics of the bridge. The software SAP2000 is used to generate the FEM of the bridge.

Three different types of loading scheme are applied to the model:

1. The loading scheme of current Italian regulations (D.M. 14.01.2008) [4], that is a scheme with on lane no. 1 a load per axis Q_{1k} = 240 kN and a distributed load q_{1k} = 7.2 kN / m^2 and on lane no. 2 a load per axis Q_{1k} = 200 kN and a distributed load q1k = 2.5 kN/m^2 and on the remaining area q_{Rk} = 2.5 kN/m^2.
2. The loading scheme of the original design of the bridge of 1949 (R.D.L.16.11.1939 n. 2229) [5], that is, a pattern of civilian loads consisting of an indefinite column of 12ton trucks with 8ton axes and 4ton alternating and equidistant of 3m.
3. A loading scheme with reduced loads, that is a scheme of two lanes, each composed of an indefinite column of 6 ton vehicles. (60 kN) with a length of 5 meters, inside which there is a 20 ton vehicle. (200 kN) with length 8 meters. In combination with this group of actions a load of 2.5 kN/m^2 is considered along the carriageway and on the sidewalks.

Considering the results of the materials test, the safety level of the bridge is not satisfied for any of the three loading scheme.

This critical situation is confirmed also by a site load test with a column of 5 ton vehicles. During this test a very high deformation is recorded with unusual residual deformation.

All these facts s imposed the closure of the bridge and the urgent start of consolidation works.

6. The structural rehabilitation and reinforcement process

The structural rehabilitation and reinforcement follow two approaches:

- Repair of concrete and reinforcement of the bearings on the abutment and of the side span of the main beams.
- Install of new steel structures to protect the bearings on the abutment and the Gerber half joints.

In fact, the desire to maintain the original scheme of the bridge, in a sustainable way, requires not only to repair and reinforce the concrete, but also to insert new structures to protect the most delicate parts. The main beams of side spans and the concrete pendulum bearing were repaired and reinforced to comply with the requirements of the loading scheme 3, described in the previous paragraph 5. Moreover the Gerber half joint and the concrete pendulum bearing are very critical structural details, so, even if the structural check is satisfied, a new "protection" structures are placed on these parts.

6.1. Main beams rehabilitation and reinforcement

The main beams of the side spans are repaired with a deep cleaning and restoration of concrete and reinforced with a wrapping of carbon fiber textiles. The bearing strengthening in this way is verified considering the loading scheme n. 3 described in the previous section 5.

The phases of these works are the following:

- Deep cleaning of concrete surfaces with high pressure hydro-scarification (2000 bar)
- Reconstruction of the concrete section with a two-component thixotropic, fiber-reinforced cement mortar with low elastic modulus that is shrinkage compensated.
- Wrapping with carbon fiber textiles.

In the areas where the cognitive tests on the concrete had given the lowest results, a large quantity of deteriorated concrete was removed. In these points the 40cm wide beam has been reduced to only 18cm.

Figure 11. The main beams after the cleaning with hydro-scarification.

Figure 12. 3D laser scanner of the main beams after hydro-scarification

The reconstruction phase was very critical, because it was very important to guarantee the perfect adhesion of the new cement mortar to the existing concrete. In the mean time the new material must have the same low elastic modulus of existing concrete. To comply with these requirements a specific two-component thixotropic, fiber-reinforced cement mortar was selected. Some pull-off tests are performed to check the adhesion between the new and old concrete.
Inside the new part of concrete reconstruction, a steel net and some steel bars are placed to increase the reinforcement bar cross-section, that was lost due to corrosion.
After the reconstruction of the beam section a wrapping with carbon fiber textiles of 1000gr/sqm was necessary to satisfy the verification for the shear loads, in accordance with the CNR-DT code [7].
A specific detail is designed to wrap the FRP textiles around the beam cross-section. In the lower part of the beam the FRP textile can be laid on the beam surface, but to pass true the deck slab, it is necessary to place the FRP textile true holes with 30mm diameter.

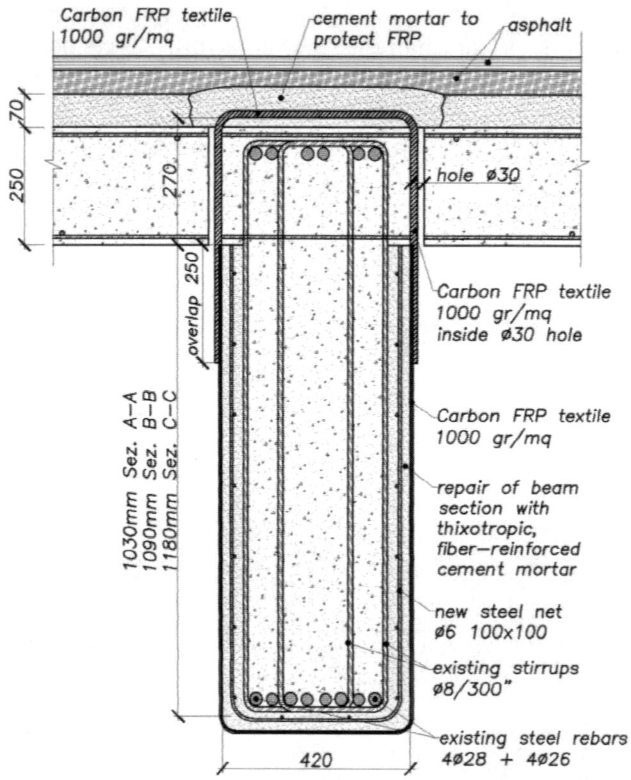

Figure 13. Scheme of reconstruction and strengthening of main beams

6.2. Gerber half joints protections

The Gerber half joints on the central span are verified considering the results of the test on the materials and the loading scheme n. 3 described in the previous paragraph 5. However it is important to consider the limited access to the bearing seat joint which leads to both inspection and maintenance issues. So there is no guarantee that the actual situation can get worse without the possibility of intervention.

For these reasons a protective metal structure, that comes into operation only in case of collapse of the half joint and prevents the collapse of the bridge central span, was considered.

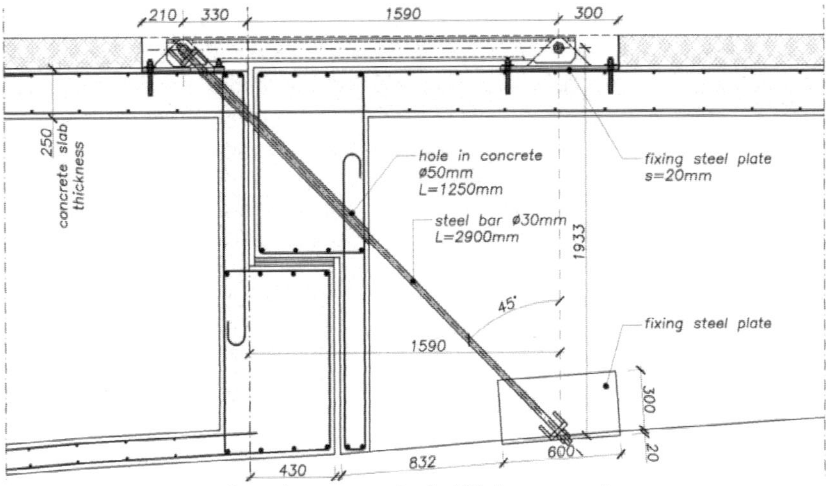

Figure 14. Scheme of steel structure for half joint protection

The steel structure works with a strut and tie scheme, where an upper strut is placed under asphalt and a tie rod tilted through the existing structure up to the lower face of the beam. This steel structure does not modify the working scheme of half joint, but it sustains the central span in case of half joint collapse.

6.3. Reinforced concrete bearings rehabilitation and reinforcement

The reinforced concrete pendulum bearings are repaired with a soft cleaning and restoration of concrete and reinforced with a wrapping of carbon fiber textiles. In this way the bearing strengthening is verified considering the loading scheme n. 1 described in the previous paragraph 5. However, it is important to consider the particularity and delicacy of this kind of bearing. So there is no guarantee that the actual situation can get worse without the possibility of intervention.

For these reasons some steel columns are put in place as a structure of defense, that comes into operation only in case of collapse of the concrete bearings.

Figure 15. Reinforced concrete bearing after concrete reconstruction.

Figure 16. Concrete bearing after wrapping with FRP and the steel protections.

7. Final tests and conclusions

The quality of the work was confirmed by a complete series of final test on the materials and on the structure. In particular a specific set of static and dynamic test with vehicle loads were performed on the entire bridge structure.

In detail the static tests demonstrate the safety of the structure and the correct execution of the works, but it is very interesting to underline the results of dynamic testing through microwave interferometry [3]. In fact, as a result of the reinforcement works, there was an increasing of the natural frequency of the structure of about 0,15Hz. This value is synonymous with a substantial increase in bridge deck stiffness.

At the end it is possible to conclude that the delicate restoration work with specific materials and procedures, has allowed to reinforce the bridge, in a sustainable way, in order to make the structure safe and usable again.

References

[1] Pozzati P. Ceccoli 1977 Teoria e tecnica delle strutture vol 2, tomo 2, fig. 3.38 p 223
[2] Original project of "Ponte delle Grazie" placed in "Archivio di Stato di Faenza" - fondo b.51/1952
[3] Pieraccini M, Parrini F, Fratini M, Atzeni C, Spinelli P, Micheloni M 2007 Static and dynamic testing of bridges through microwave interferometry *NDT&E International 40* pp 208–214.
[4] Italian regulations at the time of rehabilitation project: NTC2008 - D.M. Infrastrutture del 14 gennaio 2008 "Nuove norme tecniche per le costruzioni" e Circolare 2 febbraio 2009 n. 617 Consiglio Superiore Lavori Pubblici "Istruzioni per l'applicazione delle <Nuove norme tecniche per le costruzioni> di cui al DM 14 gennaio 2008"
[5] Italian regulation at the time of the bridge project (1949): "Regio Decreto Legge del 16 novembre 1939 n. 2229 Norme per l'esecuzione delle opere in conglomerato cementizio semplice ed armato" and "Normale n. 6081 del 9 giugno 1945 del Ministero dei Lavori Pubblici per le strade di grande traffico con importanza militare".
[6] UNI EN 1992-1-1 - Eurocodice 2 - Progettazione delle strutture di calcestruzzo - Parte 1-1: Regole generali e regole per gli edifici.
[7] CNR-DT 200R1/2013 "Istruzioni per la Progettazione, l'Esecuzione ed il Collaudo di Interventi di Consolidamento Statico mediante l'utilizzo di Compositi Fibrorinforzati" – Consiglio Nazionale delle Ricerche Italiano - 10 ottobre 2013 – versione del 15 maggio 2014.

Concrete Composite – Sustainable Material for Floating Islands

I Broukalová and A Kohoutková

Department of Concrete and Masonry Structures, Faculty of Civil Engineering, Czech Technical University in Prague

E-mail: iva.broukalova@fsv.cvut.cz

Abstract. The aim of the paper is presenting of a nature-friendly achievement of an artificial floating island made from concrete composite with dispersed reinforcement (fibre reinforced concrete). Fibre reinforced concrete is referred to as a material contributing to sustainable construction, as utilisation of fibre reinforced composites enhances durability and resiliency of structures and structural elements. The paper describes the floating island design, its shape and manufacturing process. The focus is also on the composition of fibre reinforced concrete and testing of its material characteristics. The successful application of fibre reinforced concrete in the floating island approved efficiency of fibre reinforced concrete as a material with high potential to meet demands of sustainable building.

Keywords: sustainable concrete use, floating island, fibre reinforced concrete, mechanical properties

1. Introduction

Concrete has a reputation of material with a high environmental impact. Unjustifiably. In comparison with other structural materials, the assessment usually shows better result for concrete. The pure public reputation of concrete is probably caused by global numbers; concrete is the most commonly used structural material and its production volume is the largest of all building materials.

2. Minimizing of concrete production impacts

Particular concrete components play a varying role in the environmental impacts of concrete. In the analysis of reduction of concrete production impacts, the influence of concrete components – aggregate, cement, water shall be considered.

The extraction of aggregates influences the appearance of the landscape, affects the underground and surface waters. Mining may adversely affect hydrogeological conditions. Environmental impacts also include the further processing of extracted stone, especially crushing.

Cement production accounts for 80% of the total energy consumption embodied in concrete. It is the most demanding process also regarding assessment of the carbon footprint, i.e. CO_2 production. In production of a tonne of Portland clinker, more than 800 kg of CO_2 is generated. 55% of CO_2 is released during calcination of limestone, and the remaining 40% is caused by the combustion of fuel. Therefore, many research projects focus on reducing the amount of cement, and substitution binders are used that partly or fully replace cement in the concrete mix.

Water for concrete production should be clean. Drinking water is commonly used because its use does not require any testing. In the so-called wasteless technology, recycled water is used. Recycled water contains an increased proportion of fine particles (cement, aggregate and admixtures), which is not uniform and can vary over time. The quality and properties of recycled water must be taken into account in the design of the concrete mixture to reach the required consistency of fresh concrete, what may complicate the concrete production.

In the effort to minimize concrete production impacts, not only individual concrete components´ impact should be focused on. Also, other principles should be considered, such as choosing the appropriate shape of the structure / element, which corresponds with the straining and stresses in the element, correct design and efficient use of the material, based on knowledge and understanding of the material features. Special types of concrete – cement composites with controlled properties, fibre reinforced concrete, high performance concrete, etc. may be used if appropriate [1]. Along with effective use and economic profit, the application of special cement composites may bring additional architectural and aesthetic benefits. A high durability of the material is recommended. Durable structures require less maintenance during the life cycle and less repairs. In case of more durable structures, it will take longer to reach the stage, when demolition and construction of a new building are needed. The environmental impacts and economic disadvantages associated with construction are then offset by longer operation of the structure without repairs and remedies. High initial costs and lack of information on the new concrete composites are often a reason of refusing the modern and more suitable material. The modern concrete composites need to be promoted by engaging and motivating exploitations.

3. Promoting application of concrete fibre reinforced concrete

The story combines two aspects of sustainable construction. The first aspect is the exemplary approach of a producer of the concrete mixture component to the protection of environment, the other is the application of progressive material and the demonstration / proof of the advantages and the beneficial properties of the advanced concrete composite.

Natural aggregates for concrete production, are necessarily linked to excavation activities. Excavation influences the nature of the landscape; in the extreme case the excavation may create a moonscape. The related company has a responsible approach to the environment and the areas affected by the excavation works are a challenge for them to create unique natural habitat. In the area where gravel is excavated from lakes, a pleasant landscape is created with a variety of plants and animals bound to water sources. Despite mining in progress, the surrounding landscape is not adversely affected. The revitalization of the entire territory is taken already during the excavation process. The afforestation of shores of the excavation lakes is provided by natural species indigenous to the region, as this leads to the creation of more stable, natural and more environmentally friendly communities.

An interesting achievement in the field of land rehabilitation is the care of strictly protected birds - common terns (Sterna hirundo), which are protected under the EU Council Directive 79/409 / EEC on the conservation of wild birds. The artificial nesting sites are created on sand gravel lakes. Initially, wooden floating islands were made. After verifying that this nesting method is convenient for birds, the decision was made, to create a floating island from durable material – concrete.

The project was divided into several sub-stages. Firstly, a draft of the island shape and construction was prepared and requirements on concrete properties were specified. The concrete mixture was proposed, and properties of concrete were tested in laboratory. The production technology was thought out and a prototype was elaborated. Finally, the prototype was tested in real conditions.

3.1 Design of the artificial floating island

For the design of artificial floating islands, the principal requirements were specified:

- the artificial island shall float in stable position;
- the shape of the island shall enable connection of several islands together;
- the upper deck shall be drained;

- the dimensions of the island must enable its transportation on a commonly used lorry (oversize transport shall be avoided);
- the weight of the island must enable its placing on water by a commonly used small trucks crane.

Material chosen for manufacturing of the island was fibre reinforced concrete (FRC). Fibre reinforced concrete (FRC) is an advanced modern material. Compared to common concrete it is more tough, more durable, more ductile. Cracking in FRC elements has favourable layout; the width of cracks is smaller. FRC influences the failure mode of members; the rupture may change from the brittle failure mode to the ductile failure, what tributes to reliability of structures.

If FRC is used, the thickness of the element can be reduced, while maintaining the same resistance as a conventional concrete element. FRC members are more durable than concrete members. That´s why FRC material can be classified more environmentally friendly than conventional concrete.

3.2 Properties of the fibre reinforced concrete

The mixture design proposed high-strength fibre reinforced concrete with synthetic (polymeric) fibres. The volume content of fibres is 0.55%. Fine aggregates were used with max grain size 8 mm. The proposed FRC is easily compactable and it has very low permeability. The mixture composition is in the table 1.

Table 1. Composition of the fibre reinforced concrete

Component	**Description / value**
Aggregate	mined fine aggregates - fraction 0 / 4
	mined coarse aggregates - fraction 4 / 8
Cement	CEM I 42,5 R
Admixture	microsilica compacted
Plasticizer	two types of carboxylate based plasticizers
Water / cement ratio	0.48
Super-plasticizing admixture	PCE with a long-term consistence
Fibre reinforcement	high-modulus synthetic macro-fibres

A test wall with thickness of 60 mm was concreted to verify that concrete in the thin element will be sufficiently dense without air voids and non-compacted areas. The test demonstrated demanded density of FRC (see figure 1).

Figure 1. Test specimen for verification of concrete density in a thin wall

Conventional tests were used to determine material characteristics.

For the fresh mixture testing, five batches of 48 litres volume were mixed. The slump flow test detected diameter of the concrete spread ranging from 560 mm to 650 mm; the average from all measurements was 600 mm. The density of fresh FRC mixture was 2290 kg/m^3 and the air content was 1.6%.

Material characteristics of the hardened concrete were tested on cubes 150/150/150 mm (compressive strength and tensile splitting strength), cylinders 150/300 mm (modulus of elasticity) and prisms 150/150/700 (flexural strengths). Results of the test are shown in the table 2 and figure 2.

Table 2. Tests of hardened fibre reinforced concrete

Test	**Minimal value**	**Maximal value**	**Average**
Compressive strength at age 24 hours [MPa]			1.5
Compressive strength at age 3 days [MPa]	39.4	40.5	40.0
Compressive strength at age 7 days [MPa]	53.3	54.6	54.0
Compressive strength at age 28 days [MPa]	70.0	78.0	72.8
Compressive strength at age 90 days [MPa]	72.5	81.4	76.4
Watertightness at age 28 days [mm]	6	11	8
De-icing agents - 150 cycles	39.2	58.8	48.4
Bending tensile strength [MPa]	5.71	6.26	6.02
Tensile splitting strength 7 days [MPa]	11.2	11.9	11.5
Tensile splitting strength 28 days [MPa]	11.4	15.5	13.1
Modulus of elasticity [GPa]	31.5	34.0	32.8

The effect of fibres on the toughness and ductility of the composite is proven in flexural tests performed with prisms 150/150/700 mm in a four-point bending test.

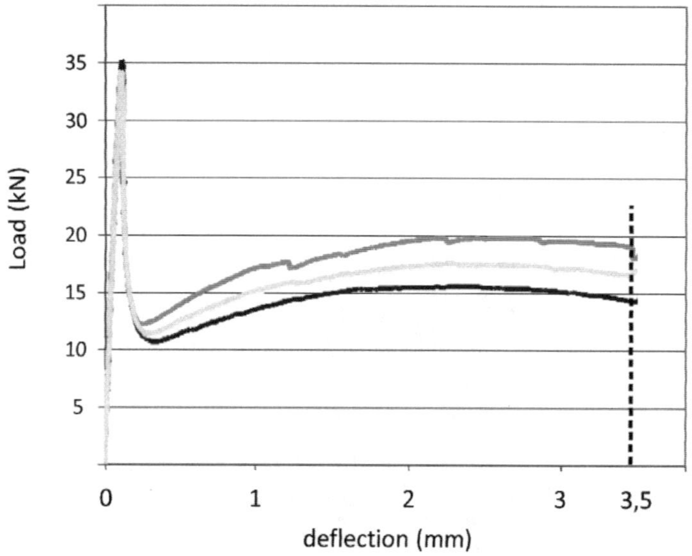

Figure 2. Results of flexural tests (Load – deflection curve)

3.3 Shape of the artificial floating island

The shape of the element results from above mentioned demands and the need to resist water pressure, eventually pressure of ice on the water level.

The basic shape is a prism with hexagonal base. The shape is both suitable for carrying of water pressure and allows simple joining of the elements to larger units in a form of a honeycomb pattern (see figure 3).

Figure 3. Assembling of artificial floating island - three elements joint together

Utilisation of FRC allowed to minimize thickness of walls of the element. The stability in floating is based on theoretical analyses; it is provided by optimizing the distribution of mass in the element, especially the solid heavy bottom. Above the water level, there is a drained deck with a raised edge and drain holes. A five-centimetre concrete layer separates the upper deck and the internal cavity, which is filled with hardened polystyrene. The height of the element is 1.7 meters, the dimensions in plan were derived so that the hexagon is inscribed in a circle with a diameter of 2 meters. The total weight is less than three and a half tonnes. Thus, the element can be transported on conventional trucks.

A reusable formwork was used for casting of the element. The top deck with the raised edge was formed with a polystyrene insert. Plastic pipes for draining of the deck are inserted into the edge. The walls and bottom floor are concreted after a technology break. Concrete rebars are embedded in the work joint to provide connection of the upper and lower parts of the element.

In the bottom, lifting hooks are fixed for manipulation and transport, and for anchoring the island in the lake. The element is manufactured bottom up and in this position, it is also transported (figure 4).

The first artificial island was launched in the water in July 2009. The element was carried on the water level by the truck crane bottom up (figure 5) and in the water, the island slowly turned to stable position bottom down. Since then, more than 40 artificial islands have been manufactured. They are installed on water levels in the Czech Republic and surrounding countries.

The floating island can serve as an artificial island for habitation by water birds and other species and growing of plants, for recreation, for landing of small boats and other purposes, whether sporting or agricultural. The advantage of the proposed solution are favourable shape, dimensions and weight, which enable simple transport of the element and creating of larger area by joining of elements together.

Figure 4. FRC element for artificial floating islands after demoulding

Figure 5. Transporting of the element by the truck crane

4. Conclusions

The advances in concrete mix design, production technology, inventions in advanced materials, new use of traditional materials, new perspectives in the designing concrete structures should lead to a change of understanding concrete as a material hostile to the environment.

Concrete composites with dispersed reinforcement (fibre reinforced concretes) have a great potential to meet the requirements of "sustainable" building. These advanced materials enable reduction of the element dimensions, which brings decrease in the consumption of the material. The fibre reinforced concrete is durable, so the need for repairs is reduced and the service life of the structure is prolonged.

A positive example of utilization of the advanced material in an interesting application where the benefits of material are effectively exploited can help to extended use of this composite and contribute to sustainable constructions. The presented achievement was prepared in cooperation of nature conservationists from Ornithological Society, concrete and concrete components producers and members of scientific institution.

Acknowledgements

This outcome has been achieved with support of the Ministry of Education, Youth and Sports of the Czech Republic, project No. SGS17/123/OHK1/2T/11 “Durability of concrete composites with recycled materials” and Technology Agency of the Czech Republic, project TH02030649 “Environmentally Efficient Construction and Demolition Waste for Structures”.

References

[1] Bílý, P.; Fládr, J.; Ryjáček, P.; Vodička, J.: Development of special fibre reinforced concrete for exposed concrete pavements on bridges, In: *Fibre Concrete 2015 - Technology, Design, Application.* Praha: České vysoké učení technické v Praze, Fakulta stavební, 2015. ISBN 978-80-01-05683-7.

[2] Broukalová, I.; Kohoutková, A.: Interesting Utilisation of Fibre Concrete as a Contribution to More Environmentally Friendly Production of Concrete, In: *Fibre Concrete - Technology, Design, Application.* Praha: České vysoké učení technické v Praze, Fakulta stavební, 2009. pp. 61-64. ISSN 2336-338X. ISBN 978-80-01-04381-3.

[3] Šeps, K.; Fládr, J.: Deformation properties and testing of synthetic fibres, In: *Fibre Concrete 2013 - Technology, Design, Application.* Praha: České vysoké učení technické v Praze, Fakulta stavební, 2013. pp. 431-438. ISSN 2336-338X. ISBN 978-80-01-05240-2.

[4] Trejbal, J.; Prošek, Z.; Fládr, J.; Tesárek, P.: Utilization of Surface-Modified Polymer and Glass Micro-Fibers as Reinforcement in Cement Composites, In: *Special Concrete and Composites 2017.* Zürich: Transtech Publications, 2018. pp. 225-230. Key Engineering Materials. vol. 760. ISSN 1013-9826. ISBN 978-3-0357-1240-7.

[5] Veselý, V.V.; Broukalová, I.; Vodička, J.; Lorek, K.: Innovative Usage of Fibre-Concrete, In: *Fibre Concrete - Technology, Design, Application.* Praha: České vysoké učení technické v Praze, Fakulta stavební, 2009. pp. 253-261. ISSN 2336-338X. ISBN 978-80-01-04381-3.

IOP Conf. Series: Materials Science and Engineering **442** (2018) 012023 doi:10.1088/1757-899X/442/1/012023

Effects of climate change on structures; analysis of carbonation-induced corrosion in Reinforced Concrete Structures in Malta

Bernice Mizzi[1], Ying Wang[1] and Ruben Paul Borg[2]

[1] Faculty of Engineering and Physical Sciences, University of Surrey, Guildford, United Kingdom

[2] Faculty for the Built Environment, University of Malta, Malta

E-mail: mizzibernice@gmail.com, ying.wang@surrey.ac.uk, ruben.p.borg@um.edu.mt

Abstract. The various elements that are affecting the Earth's climate have brought climate change to the top of the priority list amongst scientists and policy-makers. Expected changes to local climatic conditions impact directly on the surrounding environment and potentially lead to changes in the degradation processes of building materials, affecting the durability and service life of infrastructures. The aim of this paper is to investigate the effects of future climate projections on concrete structures in Malta, in particular on carbonation-induced corrosion resulting from increasing temperatures and CO_2 concentrations. Thirteen reinforced concrete structures in Malta were chosen for a retrospective analysis in order to validate two carbonation depth prediction models. The validated prediction models were subsequently used to evaluate the varying climate change scenarios in order to determine the effects on concrete carbonation depth for several concrete grades. The age of the structures used for the retrospective analysis ranged from 10 to 60 years. The field data verified the validity of both prediction models for structures with carbonation depths less than 50mm. Although both models proved valid for the retrospective analysis, a difference was noted between the models with regards to the predicted carbonation depth in relation to different climatic scenarios. An increase in carbonation depth of up to 40% is being predicted, by 2070, when considering the worst case climatic scenario. The findings prove that climate change plays a major role on the carbonation depth of concrete, which in turn reduces the service life of concrete structures.

Keywords: reinforced concrete structures, climate change, carbonation-induced corrosion

1. Introduction

Reinforced concrete forms an essential part of our world's built environment and its durability issues have been widely researched. The main deterioration process which highly degrades the durability of reinforced concrete is corrosion [1]. Reactions between atmospheric Carbon Dioxide (CO_2) and the hydrated phases of concrete lead to a decrease in the concrete potential hydrogen (pH) level, which in turn causes de-passivation of the embedded rebar and results in corrosion. This phenomenon is referred to as carbonation-induced corrosion of concrete [1,2]. Variations in concrete mixes and exposure to different environments result in different carbonation rates [3]. Climate change, has a direct impact on the surrounding environment. Therefore variations in precipitation amounts and relative humidity levels coupled with temperature rise and air pollution intensification may lead to changes in the degradation processes, which in turn have a direct bearing on the safety, durability and serviceability of infrastructures [4].

Climate change is a major international problem which humankind is currently enduring. Greenhouse gas (GHG) emissions, including CO_2 generated by anthropogenic activities are affecting the Earth's climate [5,6]. In 2014, the Intergovernmental panel on climate change (IPCC) 5th assessment report (AR5), declared that anthropogenic GHG emissions were the highest in history from 2000 to 2010 [7]. Furthermore, the Mediterranean region has been identified as a 'hot spot' region for climate change [8–10]. In consequence to Malta's strategic location within this region, the Maltese archipelago is experiencing pronounced effects due to change in climate. From 1951 till 2010, the maximum and minimum temperatures have increased by +1.2°C and +1.1°C respectively [11].

Albeit the common consensus amongst researchers with regards to the changing climate, it is only till recently that the effects of climate change on structures is being studied. A simplified model proposed by Yoon et al [12], was one of the first researches which confirmed that carbonation depths shall increase due to the effect of CO_2 concentrations resulting from a changing climate.

Subsequent studies of Stewart et al [13–15] built up on the work of Yoon et al [12] where the effect of temperature on the CO_2 diffusion coefficient was being taken into consideration. Furthermore, advanced probabilistic and reliability-based methods were developed which considered time to crack initiation, crack propagation and failure due to reinforcement corrosion. These studies found that in particular regions and depending on different scenarios of Australian cities, the risk due to carbonation-induced corrosion can increase by 400% by the year 2100. However, the model used in these researches is an approximation, as it does not consider the time-dependent effect of CO_2 concentration and other parameters such as temperature and humidity.

The numerical model presented in Zhou [16], which is based on Fib Bulletin 34 [17], has been evaluated using the special report on emissions scenarios (SRES) climate change prediction data. A comparison between other predicted carbonation models and the new model was investigated where it resulted that the variations in predictions escalate as the exposure time and extent of climate change increase.

Talukdar et al [16–18] argued that there is the need for a model which predicts reliably the depth of carbonation as a function of time. A numerical model which includes the effects of changing variables such as humidity, porosity, temperature and atmospheric CO_2 concentrations was developed [16–18]. The model involves simultaneous solutions of chemical reaction equations of CO_2 and Calcium hydroxides ($Ca(OH)_2$) based on past research studies [21,22]. Talukdar et al [19], refined the one-dimensional numerical diffusion equation, which is used to forecast the diffusion coefficient of CO_2 in concrete as a continuation to the work described in Talukdar et al [16–18]. The refined model together with the latest climate forecasts obtained from the AR5 of the IPCC were used to forecast carbonation depths for a series of urban environments in the United States. The results confirmed that climate change will affect carbonation depth and the magnitude of increase of such carbonation depth is dependent on the Representative Concentration Pathways (RCP) scenario and geographic location. Furthermore, such an increase in depth will also have a bearing on the concrete cover proposed by the current building codes.

Likewise, the latest RCP emission scenarios were used in the recent study of Peng et al [23], in order to investigate the carbonation-induced deterioration in three typical Chinese cities. In 2100, the mean carbonation depths of reinforced concrete structures in China are expected to increase up to 45% as a result of changing parameters of atmospheric CO_2, temperature and relative humidity. Considering that climate change is a global phenomenon, similar implications on concrete infrastructure is expected for other countries. This study therefore explored the predicted effects of climate change on structures in Malta up until 2070. Being the first study of its kind to be conducted in the Maltese Islands, the salient points earmarked in this paper may be considered as a tool for local structural designers and construction enterprises to embrace the right adaptation measures to rectify the effects of future climate change on local structures.

IOP Conf. Series: Materials Science and Engineering **442** (2018) 012023 doi:10.1088/1757-899X/442/1/012023

2. Carbonation models and climate scenarios

2.1. Carbonation Models

Carbonation, having a direct bearing on the service life of reinforced concrete structures, has brought about the need to develop a model for estimation of the depth of carbonation as a function of different variables in order to quantify this phenomenon. Despite the broad knowledge on this subject, a numerical model which accurately predicts carbonation depth is still not available [24,25]. This is due to the fact that integrating all the various factors affecting the carbonation process into a prediction model results in a complex process [24,25].

Following extensive research on the various available existing models, an empirical-mathematical model based on Fick's first law and a statistical model, developed from multiple regression of various parameters, were chosen for this research study. Potential practical applicability and unequivocal functionality were the two main criteria used for choosing these models. Both of the carbonation models do not include corrosion propagation time to cracking and therefore determine only the time to steel de-passivation.

2.1.1. Statistical model of carbonation – Model 01. The statistical model of carbonation proposed by Silva et al [26] was selected for this research. Being a rudimental model, it combines both CO_2 concentrations and compressive strength of concrete. Albeit its simplicity, the model was developed from a sample comprising 964 case studies through multiple linear regression analysis. Furthermore, this statistical technique is an innovative approach in the field of carbonation modelling. The carbonation coefficient, which is defined as the dependent variable, varies according to concrete characteristics, curing conditions and environmental exposure, and is given as:

$$h = k_w . t^{1/2} \quad \text{for RH} > 70\% \tag{1}$$

and
$$k_w = 3.355c - 0.019C - 0.042f_c + 10.83 \tag{2}$$

Where:
h – carbonation depth [mm]
k_w – carbonation coefficients [mm/year$^{0.5}$]
t – CO_2 exposure time [years]
c – CO_2 content [%]
f_c – concrete compressive strength [MPa]
C – clinker content [kg/m^3]

2.1.2. Square root of time model of carbonation – Model 02. The proposed abridged second carbonation model used in this study is based on Fick's first and second law of diffusion. Proposed in CEB [27] and modified further in Yoon et al [12], Equation 03 expresses carbonation depth as a function of the CO_2 diffusion coefficient and CO_2 binding capacity. The CO_2 diffusion coefficient (D_{CO_2}) is based on the empirical equation proposed by Papadakis [22] and further modified in order to cater for both temperature effects and relative humidity.

$$X_c = \left[\frac{2D_{CO_2}}{a} C_{CO_2} t\right]^{1/2} \tag{3}$$

Where:
X_c – carbonation depth (cm)
D_{CO_2} – CO_2 effective diffusion coefficient [cm^{2}s^{-1}]
t – time [s]
C_{CO_2} – atmospheric CO_2 concentration [gcm^{-3}]
a – an amount of CO_2 for complete carbonation [kgm^{-3}]

2.2. Climate Scenarios
RCPs are emission scenarios, based on GHG concentration trajectories which span until the year 2100 [28]. Though four different scenarios are available, the stringent mitigation scenario (RCP 2.6) and the worst GHG emissions scenario (RCP 8.5) were considered in this paper. Apart from representing two divergent scenarios, these opposing scenarios were chosen because no equivalent scenario similar to RCP 2.6 has been used in previous assessments such as SRES.

The CO_2 concentrations in parts per million (ppm) were converted into mass concentrations in metric units by using the ideal gas law behaviour and considering ambient temperatures [12]. Based on a 30-year climatic period, the monthly mean relative atmospheric humidity over the Maltese Islands was in the range of 70% and 80% [29].

3. Experimental Work

The initial part of the work performed was to verify the validity of two carbonation deterioration models through field data. For the purpose of carbonation model application, the field data was obtained from already-conducted site investigations. Durability assessment was done on thirteen (13) ageing concrete structures of 10 to 60 years consisting of major bridges and buildings in Malta as listed in Table 1. The data was obtained from a local accredited independent materials laboratory. The buildings selected to inspect carbonation and concrete strength included both public and private buildings, tallying to circa 300 cores. The destructive site investigation tests, through which the real values of the depth of carbonation and compressive strengths were obtained, were executed during the period from 2005 till 2017.

Table 1. Local structures used for the retrospective analysis.

Building code	Date of Construction	Building Use
01	1970	Commercial
02	1990	Industrial
03	1980	Commercial
04	1960	Industrial
05	1970	Industrial
06	1980	Water Retaining Structure
07	1960	Residential
08	2005	Commercial
09	1960	Commercial
10	1994	Commercial
11	1970	Commercial
12	1986	Industrial
13	1970	Infrastructure

The data was grouped according to date of construction, location related to exposure and building element. Following which, the statistical variability for durability design parameters, including compressive strength, carbonation depth and carbonation rate, was established and compared to published literature. The accuracy and suitability of the two chosen carbonation models was examined by verifying whether the models can reliably replicate the carbonation depths of the structures determined from the field carbonation results. The predicted calculations from the carbonation depth models were subsequently compared with actual carbonation depths, by plotting a line of equality.

Following validation of the carbonation models, based on the parameters included in equations 1,2 and 3, the projected CO_2 concentrations together with the relative humidity and varying temperatures corresponding to the coming years were inserted in the carbonation models. Predicted carbonation depths were calculated and plotted for different structures having varying concrete grades, assuming that 2017 was the build year of the structures. The concrete grades together with their constituent material quantities were retained identical to those used for the retrospective analysis for the validation of the two carbonation depth prediction models.

4. Results and discussion

The coefficient of variation for compressive strength and carbonation depth of the field results that were used for the retrospective analysis were found to be 0.15 and 0.38 respectively, which resulted in being compliant with other values used in published literatures [30,31]. For both carbonation depth models, it was observed that there is a deficient correlation between the predicted values of carbonation depths for buildings that had an actual carbonation depth in the range of 50mm or more. Contrarily, with a mean difference of circa +/- 7mm between the actual and predicted results, a strong correlation was attained for both the statistical model and the numerical model of carbonation for structures with actual carbonation depths less than 50mm. Despite minor discrepancies between the predicted and the actual carbonation depths were present, both models gave realistic predictions and therefore both carbonation models were used for the prediction of carbonation depths of buildings in future years under varying conditions of climate change.

However, though both models of carbonation have been validated as being accurate and suitable to predict carbonation depth, it is evident that that model 01 does not simulate the influence of varying future climatic factors, as illustrated in table 2 and 3. No major difference in the ultimate carbonation depth was visible when comparing the controlled climatic scenario (RCP 2.6) carbonation depth against the worst scenario (RCP 8.5) carbonation depth for the different grades in question. The subtle percentage difference between the two climatic scenarios for the different concrete grades was between 1.48% and 6.6%. This observation highlights the fact that apart from CO_2 concentration, which is a major climatic variable, temperature (which is a varying parameter not included in model 01) also has a direct bearing on the process of carbonation as indicated by model 02. This is also in agreement with published literature, where an increase in temperature resulted in additional risks of carbonation-induced damage by 2100 [13,14].

Table 2. Model 01: Predicted carbonation depths for different scenarios

Concrete Grade	Projected Carbonation Depth as per MODEL 01 and RCP 2.6 [mm]		Projected Carbonation Depth as per MODEL 01 and RCP 8.5 [mm]	
	2050	2070	2050	2070
C15	38.72	49.05	38.92	49.78
C20	34.41	43.58	34.60	44.31
C25	27.50	34.82	27.69	35.56
C30	22.14	28.04	22.34	28.77
C35	14.72	18.63	14.91	19.36
C40	12.99	16.44	13.18	17.17
C45	8.67	10.97	8.87	11.70

Table 3. Model 02: Predicted carbonation depths for different scenarios

Concrete Grade	Projected Carbonation Depth as per MODEL 02 and RCP 2.6 [mm]		Projected Carbonation Depth as per MODEL 02 and RCP 8.5 [mm]	
	2050	2070	2050	2070
C15	39.17	49.10	44.20	67.59
C20	35.29	44.23	39.82	60.88
C25	29.73	37.26	33.54	51.29
C30	26.85	33.65	30.29	46.32
C35	23.22	29.11	26.21	40.07
C40	23.06	28.91	26.02	39.79
C45	19.69	24.68	22.22	33.97

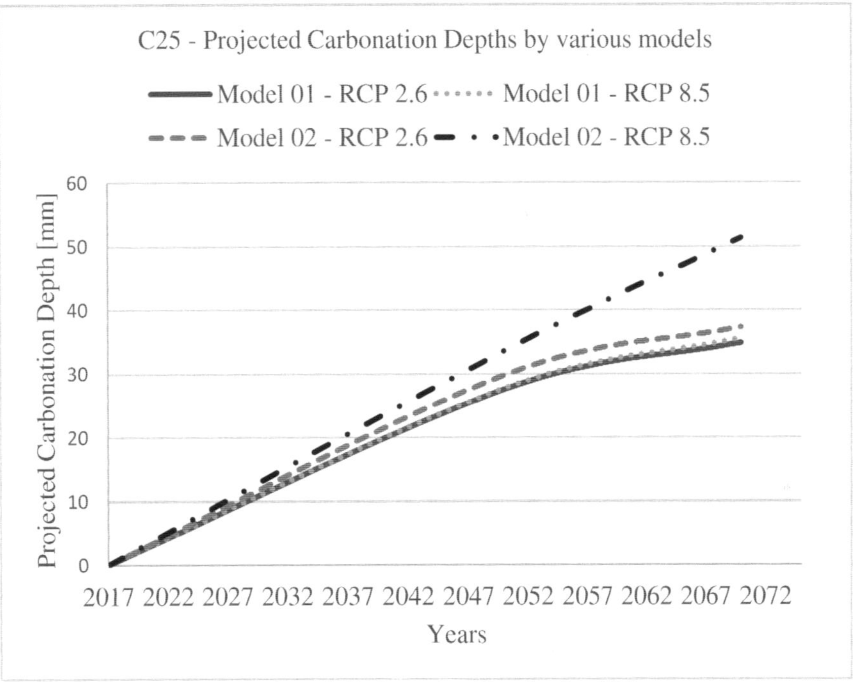

Figure 1. Projected Carbonation Depths for C25 concrete

Concrete structures in 2070, which are constructed from C25, are expected to have a carbonation depth of circa 50mm for the worst case climatic scenario as illustrated in figure 1. Therefore, the cover thickness chosen for such structures should ensure that the carbonation front does not reach the reinforcement during the designed service life of the structure. Environmental exposures in Malta are classified by the Maltese Codes for Design of concrete structures that are based on Eurocode 2 [32]. The Code in general specifies a concrete cover that ranges between 10 to 65mm, depending on for example concrete grade, maximum aggregate size, fire performance and exposure classes. For normal grades of concrete exposed to ordinary environments, a typical value of 25mm (+/-10mm) is generally adopted locally as concrete cover. Therefore, in this paper it is being shown that due to varying climatic variables, carbonation-induced corrosion would begin as early as year 2050 for a structure that is constructed today with a 25mm cover. This is clearly indicated in the predicted carbonation depth results obtained for model 02 that are in the range of or more than 25mm for all concrete grades. This is further highlighted in year 2070 where higher carbonation depths are projected.

5. Conclusions

This research shows that climate change could play a role on the carbonation depth of concrete that will in turn reduce the service life of the structures. It is shown that circa 40% increase in the carbonation depth is expected when comparing the worst case climatic scenario RCP 8.5 to the controlled climatic scenario RCP 2.6. Although the models used in this research are an approximation due to their simplicity, the findings from this research should act as a basis for development of climate adaptation through the improved design of concrete structures locally. A thorough sensitivity analysis could be conducted in order to investigate the effects of random variables (such as varying concrete cover) on the results obtained. Such an analysis can further confirm which concrete parameters could effectively diminish the climate change effects on degradation of concrete structures. This can be extended further by conducting an economic valuation of adaptation with numerous variables being taken into consideration, including increased concrete mix durability, galvanised reinforcement and surface treatments.

References

[1] Tuutti K 1982 Corrosion of steel in concrete (Stockholm: Swedish Cement and Concrete Research Institute)
[2] Lagerblad B 2005 *Carbon dioxide uptake during concrete life cycle: State of the art* (Stockholm: Swedish Cement and Concrete Research Institute)
[3] Neves R, Branco F A and de Brito J 2012 A method for the use of accelerated carbonation tests in durability design *Constr. Build. Mater.* 36 585–91
[4] Nijland T G, Adan O C, van Hees R P and van Etten B D 2009 Evaluation of the effects of expected climate change on the durability of building materials with suggestions for adaptation *Heron* 54 37–48
[5] Thorpe A J 2005 *Climate change prediction: a challenging scientific problem* (Institute of Physics)
[6] Stocker T F, Qin D, Plattner G-K, Tignor M, Allen S K, Boschung J, Nauels A, Xia Y, Bex V and Midgley P M 2013 *Climate Change 2013: The Physical Science Basis. Contribution of Working Group I to the Fifth Assessment Report of the Intergovernmental Panel on Climate Change, 1535 pp* (Cambridge Univ. Press, Cambridge, UK, and New York)
[7] Pachauri R K, Allen M R, Barros V R, Broome J, Cramer W, Christ R, Church J A, Clarke L, Dahe Q, Dasgupta P and others 2014 *Climate change 2014: synthesis report. Contribution of Working Groups I, II and III to the fifth assessment report of the Intergovernmental Panel on Climate Change* (IPCC)
[8] Giorgi F and Lionello P 2008 Climate change projections for the Mediterranean region *Glob. Planet. Change* 63 90–104
[9] Turp M T, Öztürk T, Türkeş M and Kurnaz M L 2014 Assessment of projected changes in air temperature and precipitation over the Mediterranean region via multi-model ensemble mean of CMIP5 models *BLACK SEAMEDITERRANEAN Environ.* 93
[10] Guemas V, García-Serrano J, Mariotti A, Doblas-Reyes F and Caron L-P 2015 Prospects for decadal climate prediction in the Mediterranean region *Q. J. R. Meteorol. Soc.* 141 580–97
[11] Galdies C 2012 Temperature trends in Malta (central Mediterranean) from 1951 to 2010 *Meteorol. Atmospheric Phys.* 117 135–43
[12] Yoon I-S, Çopuroğlu O and Park K-B 2007 Effect of global climatic change on carbonation progress of concrete *Atmos. Environ.* 41 7274–85
[13] Stewart M G, Wang X and Nguyen M N 2012 Climate change adaptation for corrosion control of concrete infrastructure *Struct. Saf.* 35 29–39
[14] Stewart M G, Wang X and Nguyen M N 2011 Climate change impact and risks of concrete infrastructure deterioration *Eng. Struct.* 33 1326–37
[15] Stewart M G and Peng J 2010 Life-cycle cost assessment of climate change adaptation measures to minimise carbonation-induced corrosion risks
[16] Zhou S 2016 A numerical model for concrete carbonation under gradually changing climate condition in future *Concr. Aust.* Vol 42 No 1 45–52
[17] International Federation for Structural Concrete 2006 *Model code for service life design* (Lausanne: fib)
[18] Talukdar S, Banthia N and Grace J R 2012 Carbonation in concrete infrastructure in the context of global climate change – Part 1: Experimental results and model development *Cem. Concr. Compos.* 34 924–30

[19] Talukdar S and Banthia N 2016 Carbonation in Concrete Infrastructure in the Context of Global Climate Change: Model Refinement and Representative Concentration Pathway Scenario Evaluation *J. Mater. Civ. Eng.* 28 04015178
[20] Talukdar S and Banthia N 2013 Carbonation in concrete infrastructure in the context of global climate change: Development of a service lifespan model *Constr. Build. Mater.* 40 775–82
[21] Papadakis V G, Vayenas C G and Fardis M N 1991 Experimental investigation and mathematical modeling of the concrete carbonation problem *Chem. Eng. Sci.* 46 1333–1338
[22] Vagelis G. Papadakis C G V and Michael N Fardis 1991 Physical and Chemical Characteristics Affecting the Durability of Concrete *Mater. J.* 88
[23] Peng L and Stewart M G 2016 Climate change and corrosion damage risks for reinforced concrete infrastructure in China *Struct. Infrastruct. Eng.* 12 499–516
[24] Parrott L J 1987 A review of carbonation in reinforced concrete
[25] Sagues A A, Moreno E I, Morris W and Andrade C 1997 Carbonation in concrete and effect on steel corrosion
[26] Silva A, Neves R and de Brito J 2014 Statistical modelling of carbonation in reinforced concrete *Cem. Concr. Compos.* 50 73–81
[27] Schießl P 1997 *New approach to durability design: an example for carbonation induced corrosion* (Comité Euro-Internat. du Béton)
[28] Moss R H, Edmonds J A, Hibbard K A, Manning M R, Rose S K, van Vuuren D P, Carter T R, Emori S, Kainuma M, Kram T, Meehl G A, Mitchell J F B, Nakicenovic N, Riahi K, Smith S J, Stouffer R J, Thomson A M, Weyant J P and Wilbanks T J 2010 The next generation of scenarios for climate change research and assessment *Nature* 463 747–56
[29] Galdies C 2011 *The Climate of Malta: statistics, trends and analysis* (National Statistics Office)
[30] Sarja A and Vesikari E 1996 *Durability Design of Concrete Structures.* (London: Spon Press)
[31] Ekolu O S 2010 Model validation and characteristics of the service life of Johannesburg concrete structures
[32] Code P 2005 Eurocode 2: Design of concrete structures - Part DESIGN OF CONCRETE -: General rules and rules for buildings

An holistic approach to a sustainable future in concrete construction

Coppola L[1,2], Coffetti D[1,2], Crotti E[1,2]

[1] Department of Engineering and Applied Sciences, University of Bergamo (Italy)
[2] Consorzio INSTM, UdR "Materials and Corrosion", Florence (Italy)

E-mail: luigi.coppola@unibg.it

Abstract. The main challenge for concrete industry is to serve the two major needs of human society, the protection of the environment, on one hand, and - on the other - the requirements of buildings and infrastructures by the world growing population. In the past, concrete industry has satisfied these needs well. However, for a variety of reasons the situation has changed dramatically in the last years. First of all, the concrete industry is the largest consumer of natural resources. Secondly, Portland cement, the binder of modern concrete mixtures, is not as environmentally friendly. The world's cement production, in fact, contributes to the earth's atmosphere about 7% of the total CO2 emissions, CO2 being one of the primary greenhouse gases responsible for global warming and climate change. As a consequence, concrete industry in the future has to face two antithetically needs. In other words, how the concrete industry can feed the growing population needs being – at the same time - sustainable? Sustainability in construction industry can be achieved through three different main routes based on the "3R – Green Strategy": Reduction in consumption of gross energy for construction materials production, Reduction in polluting emissions and Reduction in consuming not renewable natural resources.

Keywords: Sustainability in Construction Materials, 3-R Strategy, Alternative Binders, Waste Management

1. Introduction

With a production of more than 10 billion tons, concrete is the most widely used building material in the world, especially in areas affected by intense economic and demographic growth, such as China and India. Due to these huge volumes, concrete industry – and, in particular, the cement sector – has a very strong environmental impact in terms of greenhouse gases emissions, energy requirement and consumption of natural resources. In fact, it is well known that cement manufacturing is responsible for 5-7% of anthropogenic greenhouse gas emissions [1,2], including the CO_2 released in clinker production (industrial process CO_2: 520 kg CO_2/t of clinker) and by fuel combustion (energy-use CO_2: 350 kg CO_2/t of clinker). Furthermore, despite the high energy demand in clinker production, it has been significantly reduced over the last years since a modern cement plant in Europe nowadays requires about 2900 to 3300 MJ for each ton of clinker produced [3]. Moreover, on average 1.53 ton of raw materials (about 1.22 ton of limestone and 0.31 ton of clay) are required to produce 1 ton of OPC (Ordinary Portland Cement [4]. Finally, it has been estimated that about 44% of construction sand and gravel mined globally are used as concrete aggregates [5]. Thus, cement and concrete industry is under pressure to reduce both energy used and greenhouse gases emissions as well as natural resources consumption, in other words to became more "sustainable" [6,7]. The task is particularly complicated since population is expected to growth up to ten billion in 2050. As a consequence of this, the main challenge for the concrete industry is how to support the increasing demand of buildings and infrastructures of the growing population being at the same time sustainable. The answer to this hard task is achievable by means of the "3R Green-Strategy": Reduce energy-Reduce pollutant emissions-Reduce consumption of natural resources (Figure 1).

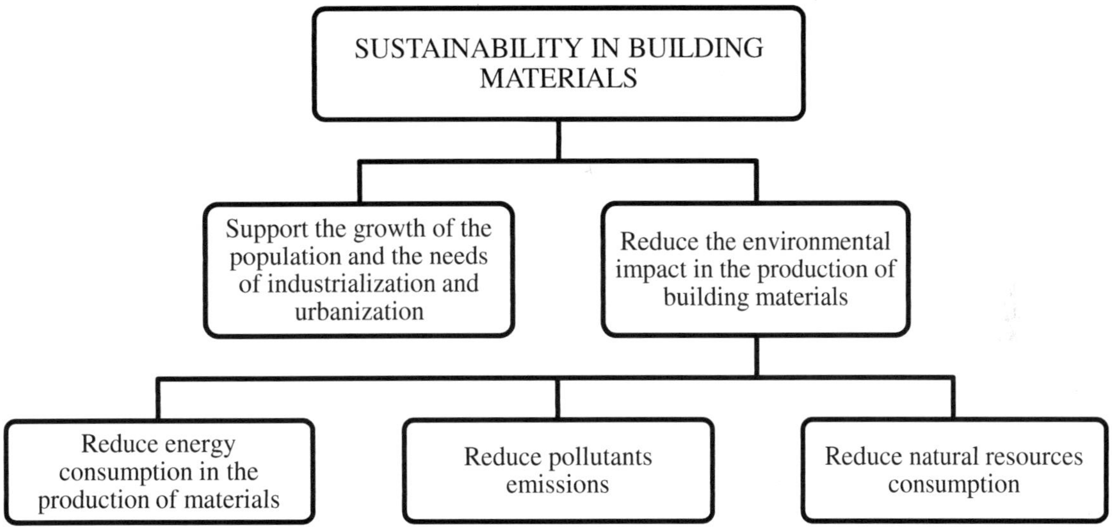

Figure 1: Main strategies to make concrete sector more environmentally friendly

2 Reduction in energy consumption and pollutant emissions in production of building materials

The first two steps of the virtuous path of "3-R Green Strategy" are represented by a strong effort in reducing energy consumption and pollutant emissions by means of (Figure 2):

- a "low energy clinker production process" replacing a large amount of alite-based clinker by belite cements [8–11]. Production of belite cements, in fact, takes place at a lower kiln temperature reducing both fuel consumption and carbon dioxide released into the atmosphere;
- a "limitation of the clinker factor in cements" by replacing this high energy-consuming material with low-carbon supplementary cementitious constituents, such as fly ash [8,12], ground granulated blast furnace slag [9], metakaolin [10] and natural pozzolans [11]. Moreover, supplementary cementitious materials (such as natural pozzolans, fly ashes, blast furnace slag, silica fume, etc.) can be used directly in ready-mix concrete plants to manufacture cementitious mixtures for massive structures (dams, foundations, etc.).) and elements where a slow strength gain is required;
- "alternative Portland-free low-carbon binders" such as alkaline activated materials and geopolymers [13–15]. AAMs and geopolymers are silico-alumina materials (called precursors) with reduced calcium content mixed with strong dosages of alkaline activators. In AAMs and geopolymers the process of hardening is promoted by the dissolution of silica favored by the alkaline activators which generally consist of sodium or potassium silicate and/or the corresponding hydroxides. Therefore, the geopolymers or in general the alkaline

activated binders can be considered "environmentally friendly" since it is not necessary (except for the metakaolin) to burn materials used as precursors. From this point of view, it has been highligthed how a ton of geopolymers produced with fly ash and alkaline activators releases an amount of CO_2 (Global Warming Potential: GWP) equal to 20% than that emitted during Portland cement production;

- "sulfoaluminate cements" whose production requires a lower consumption of primary energy deriving from both lower kiln temperature and grinding of the lower hardness CSA clinker (Table 1). Consequently, the production of sulfoaluminate cement [16,17] is also characterized by lower CO_2 emissions, estimated at about 25% less than that of Portland cement clinker;

Table 1: Global Warming Potential of different binders used in construction materials

Binder	**Global warming potential: GWP ($kgCO_{2eq}/kg$)**
Portland cement	**0,98**
Fly ash based geopolymer	**0,25**
Solfoaluminate cement	**0,74**
Gypsum	**0,025**

- "binary or ternary sulphoaluminate Portland-free binders" where OPC is replaced by supplementary cementitious materials [18,19]. These blends, thanks to the elimination of Portland cement and the use of gypsum (0.025 $kgCO_{2eq}/kg$: Table 1) in ternary mixtures, allow to reduce both energy consumption and polluting emissions up to 40%;
- "a reduction of transport distances of building materials" using both the local natural raw resources and by-products available in each country. With this in mind, use of alternative binders to those based on Portland cement clinker must take place avoiding special materials produced only in technologically advanced countries. Their use, in fact, would lead to an excessive consumption of primary energy and polluting emissions not only for production, but also for transportation;
- "a strong limitation of organic polymers and Volatile Organic Compounds (VOCs)" in pre-packed special products for renders, screeds and repair of existing masonry buildings and reinforced concrete structures [20]. This objective is achieved by using binders characterized by a high dimensional stability so as not to require the use of organic polymers to reduce hydraulic shrinkage. In special products, currently marketed on Portland cement, in fact, to face some intrinsic shortcomings of cementitious materials, such as high shrinkage, poor adhesion and low tensile strength, it necessary to use organic polymers. By replacing the traditional Portland cement with binder having high dimensional stability, therefore, it would make it unnecessary to use shrinkage reducing admixture, water retainers, adhesion enhancers and viscosity modifiers. The lower consumption of organic polymers, in essence, would have the undeniable advantage of reducing materials characterized both by a high consumption of primary energy and by high amounts of CO_2 emitted into the atmosphere. Take into account, for example, that for the production of a rheology modifier polymer, a quantity of primary energy equal to about five times than that for the production of Portland cement is required, while GWP is almost double. The decrease in the consumption of organic substances has the further advantage of limiting the emission of pollutants such as Volatile Organic Compounds (VOC) contributing to make healthier the environments of the building site. Moreover this improves workers' safety, reducing indoor pollution for the benefit of the occupants;

- "materials capable of reducing energy consumption" for the cooling/heating of homes. In fact, data indicate that in "Western Countries" energy consumption for heating and cooling of houses and apartments is about 40% of the total energy consumed. The achievement of these objectives is possible thanks to the use of thermal energy storage techniques based on systems that allow the transfer of heat to the storage medium during the charging period and release the same amount of energy at a later time. These systems exploit Phase Change Materials (PCM) that absorb energy while the phase transition occurs and release it during the return to the original conditions. PCMs for thermal regulation in buildings can be integrated into traditional binders by direct mixing. Specifically, the application in plasters could be the most technologically viable and advantageous one [21].

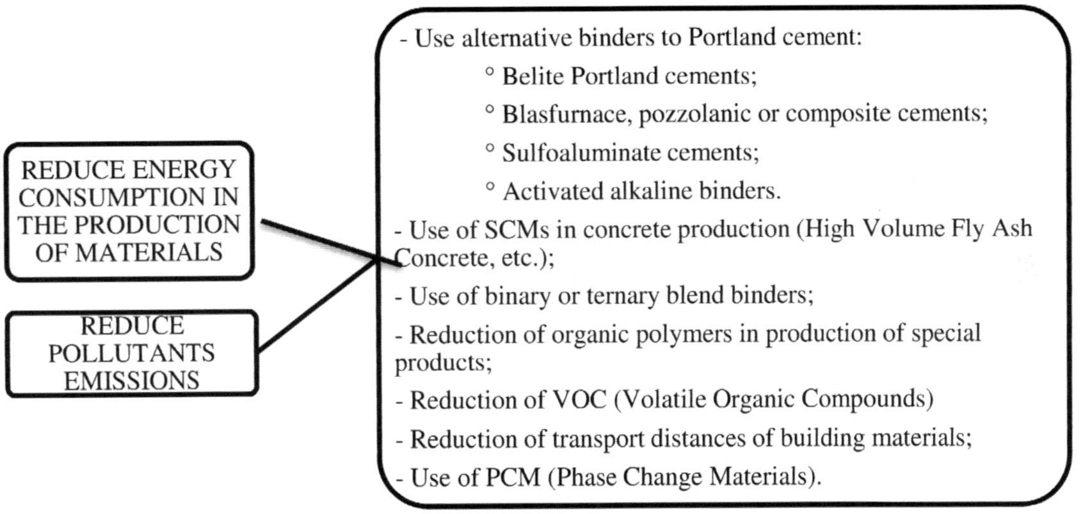

Figure 2. Strategies to reduce both energy consumption and pollutants emissions in production of construction materials

3 Reduction in consumption of non-renewable natural resources

The problem of environmental sustainability cannot be addressed solely on the basis of primary energy consumption and the amount of CO_2 emitted into the atmosphere. For example, the production of aggregates for concrete requires a very low consumption of primary energy (Table 2), about three orders of magnitude lower than that for the production of cement. Moreover, CO_2 emissions are even almost five orders of magnitude lower. In fact, the production of natural aggregates requires only excavation from the rivers and transportation to the concrete mixing plant or, at the most, extraction from the rock quarry bench and subsequent grinding before being used in concrete. On the basis of the GER and the GWP, therefore, it should be concluded that the use of aggregates for the production of concrete is an eco-friendly activity. In reality, the production of aggregates must be considered an activity that does not respect the environment as it determines a consistent consumption of non-renewable resources. Therefore, we can state that among the principles of sustainability in the construction sector, reducing the consumption of natural sands and gravel is one of the basic fundamentals from which one cannot ignore.

Table 2: Global Warming Potential (GWP) and Gross Energy Requirements (GER) of different concrete ingredients

Ingredient	GER (MJ/kg)	GWP (kgCO_2/kg)
Portland cement	3,35	0,98
High-range water reducer	16	1,8
Natural aggregates	$3{,}7 \cdot 10^{-3}$	$7 \cdot 10^{-5}$

The reduction in the consumption of natural aggregates can be pursued through different approaches, all, however, aimed at recovering wastes from various sources, including those produced by the concrete and construction industry itself. This approach also has the advantage of minimizing the number of landfills contributing to a general improvement of the landscape and the livability of our cities. Therefore, the third step of "3-R Strategy" is represented by reduction of natural resources consumption by increasing waste utilization [22–26] in concrete industry. Waste management, in fact, is one of the most important topic of the Green Economy and has emerged as one of the main research issue because in the world, every year, only a quarter of the total waste produced is recycled. In this view, in the next future a consistent use is expected of (Figure 3):

- aggregates arising from demolition of existing concrete structures;
- sand and limestone filler from waste of marble processing;
- fresh concrete in excess returned with truck mixers and washing water in ready-mix concrete plants [27];
- plastic bottles, glass, tires, automotive shredders, crushed asphalt, foundry sands and biomass ashes.

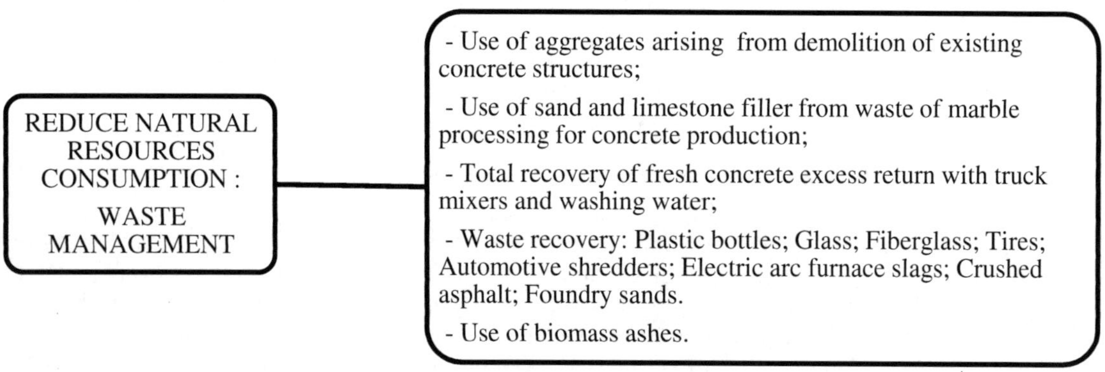

Figure 3. Waste management as key strategy to reduce consumption of natural not renewable resources

The sands and calcareous fillers recovered from marble processing waste are already widely used in proprietary prepacked products for renders, plasters, screeds and repair mortars. There are also numerous studies that testify to the possibility of manufacturing cementitious mixtures with excellent performance by using aggregates from the demolition of reinforced concrete structures, the ruins of bricks, the recycling of glass and PET bottles, of worn tires, the recovery of car bodies, crushed asphalt, electric arc blast (EAF) furnace slag and foundry sands [28]. It is not, however, a goal to list the performance improvements that can be achieved by using waste in concrete production. For example, the use of glass in the form of very fine powder, already widespread in Canada, can solve the complicated problem arising from the use of alkali-reactive aggregates. Furthermore, recycling of EAF slag in the road field represent a valid alternative to natural aggregates to manufacture asphalt-based mixture. In this paragraph, however, we want to focus attention on which virtuous mechanisms should be adopted to encourage the use of waste in place of traditional natural aggregates for cementitious (but also bituminous) concrete production.

However, a consistent and substantial increase in the recycling of wastes can only be achieved when we will be able to move from the culture of "at most - no more than" to "at least - no less than". To better clarify this concept, we take as a reference the Italian legislation on the use of recycled aggregates coming from the demolition of concrete structures or buildings (rubbles). Table 11.2.III D.M. 17.01.2018 [29] allows the use of recycled aggregates from rubbles only for "concrete ??" having a strength class C8/10 (cylinder/cubic characteristic compressive strength equal to 8/10 MPa). Recycling aggregates coming from the demolition of concrete only, on the other hand, can be used to a maximum of 60% in strength class C20/25 cementitious mixtures and "at most" 30% in those with a strength class lower or equal to C30/37. In the ways in which the use of recycled concrete is proposed, it is impossible to conceal how the legislation, in a clear way, "transmits the idea" that it is using a material of poor quality ("garbage!!"). In essence, this approach - "at most - no more than" - in place of that "at least - no less than" seems to go in the opposite direction to that of enhancing waste as a resource. Adopting the approach of "at least – no less than", if someone wants to use an eco-friendly material he has to introduce a minimum percentage of waste because the concrete can be embellished of the (can get) "eco-friendly" title.

A recent survey, within the EU, has highlighted how the reduced consumption of recycled aggregates in the production of ready-mix concrete is due to the lack of demand from designers and construction managers, the lack of sensitivity of the contracting stations to environmental issues and absence of specific rules on concrete manufactured with recycled aggregates. In order to modify this "stall" situation, regarding the actual use of "wastes" in the construction sector, it is necessary (Fig. 4) to implement an overall strategy that includes:

- "introduction of incentive systems and regulations" that allow to increase the percentage of wastes in concrete production. Rules that indicate a minimum percentage of recycled materials to be used to identify concrete as "environmentally friendly". Notwithstanding, obviously, the rheological, elasto-mechanical and durability performances for the mixture in relation to the intended use and to the environmental exposure class in which the concrete structure falls. With regard to this aspect, an approach "at least - no less than" is found in the technical document "SIA 2030: 2010 - Recycling Beton" [30], where the legislation requires that in order to get the "green" title concrete must be contain at least 25% of recycled aggregates. The same rule - at least for what concerns the aggregates coming from the demolition of the concrete - does not limit the use of recycled aggregates concrete in any environmental exposure class. Furthermore, this technical regulatory tool provides all the information necessary for the structural designer to be able to dimension elements in reinforced recycled concrete. These regulatory instruments - exquisitely of a technical nature - need to be accompanied by administrative tools such as the introduction of reward mechanisms for those projects which, in the context of public works tenders, envisage substantial quantities of reused waste. Obviously, the assessment of how much a concrete is eco-friendly must take place on the basis of certain criteria such as, for example, the "carbon foot print" of the mixtures calculated on the basis of reliable and certified databases (GER and GWP) of the ingredients used. Finally, the issue of construction licenses - where demolition of existing buildings and/or infrastructures is planned at the construction site - should be subject to the complete recovery of the rubble and of the demolished concrete. With this in mind, the designer of the work must include in the project a plan to recycle the rubble coming from the demolition to which the contractor will have to strictly follow;
- "strong disincentive to the landfill of waste": increasing the tax for disposal and preventing transportation to landfill of everything that is fuel that, instead, must be conveyed in the incinerators. It should be noted that in most European countries the construction sector contributes more than two thirds of the overall volume of wastes (urban waste accounts for only about 10% of the total). Therefore, providing both technical and administrative

regulatory tools to increase waste consumption in the production of construction materials and making taxing for waste materials in landfills more burdensome could provide the keystone - together with separate collection and recycling urban wastes - to almost completely solve the problem of waste valorization;

- "strong penalties for non-compliance": to complete the actions planned to increase the use of wastes it will be necessary to provide fundamentally administrative sanctions for those who do not comply with the rules and which could involve public offices, designers, contractors (for example, providing a period suspension from the register) and the landfill owners themselves.

Reduction in the consumption of natural resources can also be achieved by a general increase in durability of both masonry and reinforced concrete structures in order to reduce resources for maintenance and refurbishments since repair materials – containing high percentage of both cement and organic polymers - have a strong impact from the environmental point of view (Figure 4). The roads that can be undertaken to achieve this goal are many (Figure 5), but all aimed at preventing the phenomena of degradation and premature deterioration of both reinforcements and concrete, such as improving the models for predicting the onset and propagation of corrosion, optimizing the design of the structures to attain higher robustness, carefully choosing ingredients [31,32] and mixture composition [33–35], using fiber reinforced plastic (FRP) bars in place of traditional steel, or using non-metallic fibers and nanotubes [36,37] to improve both the fragile behavior and the ability of reinforced concrete elements to self-evaluate the stress level.

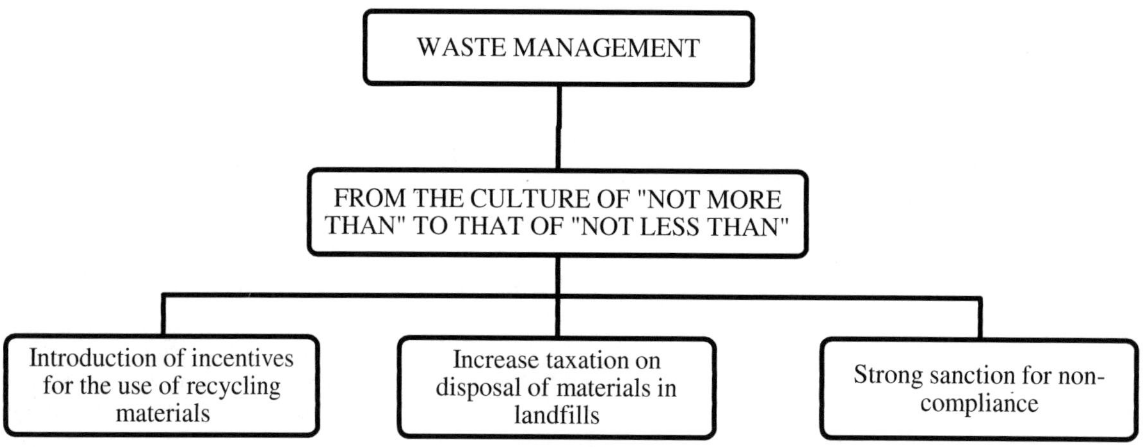

Figure 4. Boost use of wastes moving from the culture of "at most - not more than" to that of "at least - not less than"

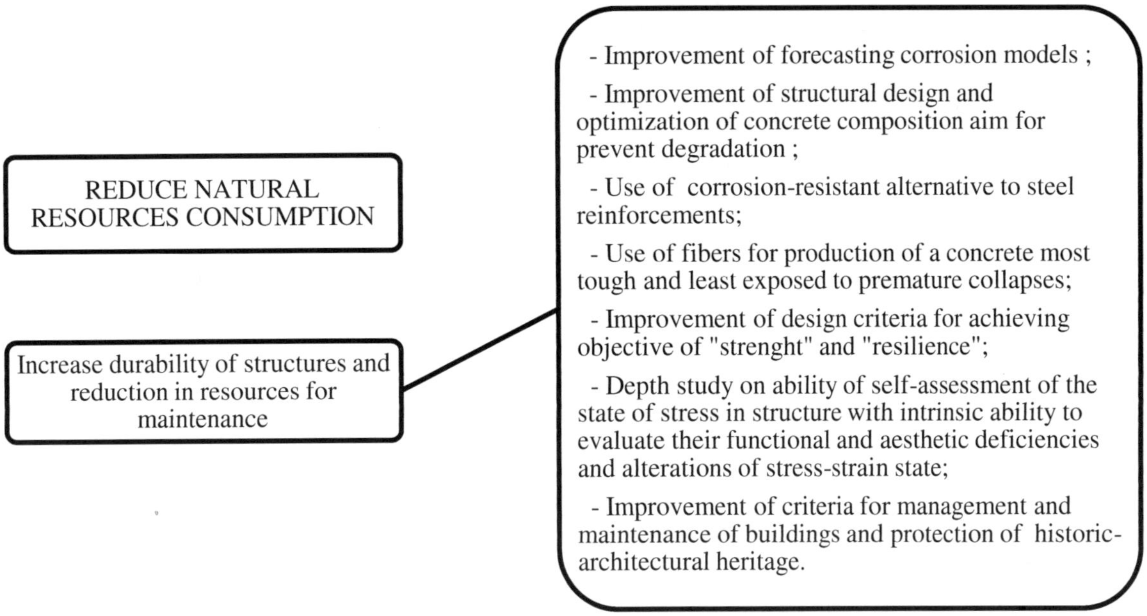

Figure 5. Further strategies to reduce consumption of natural not renewable resources

4 Conclusions

The scientific community is aware the problem of how to make sustainable the construction materials sector appears rather difficult and complicated in the light of the forecast models of growth of the world population over the next 35 years. The challenge for the concrete industry to sustain growth by reducing the strong environmental impact of "cement"-based materials must necessarily be addressed with a multidisciplinary approach – the "3-R Strategy" - that takes into consideration the possibility of reducing energy consumption, pollutant emissions together with an intensive and a generalized use of wastes aimed at drastically reducing the consumption of non-renewable natural resources.

References

[1] L. Barcelo, J. Kline, G. Walenta, E. Gartner, Cement and carbon emissions, Materials and Structures. 47 (2014) 1055–1065. doi:10.1617/s11527-013-0114-5.
[2] G. Habert, N. Roussel, Study of two concrete mix-design strategies to reach carbon mitigation objectives, Cement and Concrete Composites. 31 (2009) 397–402. doi:10.1016/j.cemconcomp.2009.04.001.
[3] M. Schneider, M. Romer, M. Tschudin, H. Bolio, Sustainable cement production-present and future, Cement and Concrete Research. 41 (2011) 642–650. doi:10.1016/j.cemconres.2011.03.019.
[4] C. Chen, G. Habert, Y. Bouzidi, A. Jullien, Environmental impact of cement production: detail of the different processes and cement plant variability evaluation, (2010). doi:10.1016/j.jclepro.2009.12.014.
[5] G.S. U.S., D. of I. U.S., Mineral commodity summaries 2017, 2017. doi:http://dx.doi.org/10.3133/70140094.
[6] J.S. Damtoft, J. Lukasik, D. Herfort, D. Sorrentino, E.M. Gartner, Sustainable development and climate change initiatives, Cement and Concrete Research. 38 (2008) 115–127. doi:10.1016/J.CEMCONRES.2007.09.008.
[7] L. Coppola, T. Bellezze, A. Belli, M.C. Bignozzi, F. Bolzoni, A. Brenna, M. Cabrini, S. Candamano, M.

Cappai, D. Caputo et al., Binders alternative to Portland cement and waste management for sustainable construction—part 1, Journal of Applied Biomaterials & Functional Materials. 16 (2018) 186–202.

[8] M. Ahmaruzzaman, A review on the utilization of fly ash, Progress in Energy and Combustion Science. 36 (2010) 327–363. doi:10.1016/j.pecs.2009.11.003.

[9] E. Özbay, M. Erdemir, H.I. Durmuş, Utilization and efficiency of ground granulated blast furnace slag on concrete properties - A review, Construction and Building Materials. 105 (2016) 423–434.

[10] J.T. Ding, Z. Li, Effects of metakaolin and silica fume on properties of concrete, ACI Materials Journal. 99 (2002) 393–398. doi:10.14359/12222.

[11] L.T. Burak Uzal, P. Kumar Mehta, High-Volume Natural Pozzolan Concrete for Structural Applications, Materials Journal. 104 (n.d.). doi:10.14359/18910.

[12] L. Coppola, D. Coffetti, E. Crotti, Plain and ultrafine fly ashes mortars for environmentally friendly construction materials, Sustainability (Switzerland). 10 (2018). doi:10.3390/su10030874.

[13] C. Shi, A.F. Jiménez, A. Palomo, New cements for the 21st century: The pursuit of an alternative to Portland cement, Cement and Concrete Research. 41 (2011) 750–763. doi:10.1016/j.cemconres.2011.03.016.

[14] L. Coppola, D. Coffetti, E. Crotti, Pre-packed alkali activated cement-free mortars for repair of existing masonry buildings and concrete structures, Construction and Building Materials. (2018). doi:10.1016/j.conbuildmat.2018.04.034.

[15] J.L. Provis, Alkali-activated materials, Cement and Concrete Research. (2016). doi:10.1016/j.cemconres.2017.02.009.

[16] M. Marroccoli, F. Montagnaro, A. Telesca, G.L. Valenti, Environmental implications of the manufacture of calcium sulfoaluminate-based cements, 2nd International Conference on Sustainable Construction Materials and Technologies. 1 (2010) 625–635.

[17] Y. Shen, J. Qian, J. Chai, Y. Fan, Calcium sulphoaluminate cements made with phosphogypsum: Production issues and material properties, Cement and Concrete Composites. 48 (2014) 67–74.

[18] L. Coppola, D. Coffetti, E. Crotti, Use of tartaric acid for the production of sustainable Portland-free CSA-based mortars, Construction and Building Materials. 171 (2018). doi:10.1016/j.conbuildmat.2018.03.137.

[19] L. Coppola, D. Coffetti, E. Crotti, T. Pastore, CSA-based Portland-free binders to manufacture sustainable concretes for jointless slabs on ground, Construction and Building Materials. 187 (2018) 691–698. doi:10.1016/j.conbuildmat.2018.07.221.

[20] L. Coppola, Repair and conservation of reinforced concrete tent-church by Pino Pizzigoni at Longuelo - Bergamo (Italy), International Journal of Architectural Heritage. (2018). doi:10.1080/15583058.2018.1453887.

[21] L. Coppola, D. Coffetti, S. Lorenzi, Cement-Based Renders Manufactured with Phase-Change Materials: Applications and Feasibility, Advances in Materials Science and Engineering. 2016 (2016). doi:10.1155/2016/7254823.

[22] L. Coppola, A. Buoso, D. Coffetti, P. Kara, S. Lorenzi, Electric arc furnace granulated slag for sustainable concrete, Construction and Building Materials. 123 (2016). doi:10.1016/j.conbuildmat.2016.06.142.

[23] L. Coppola, S. Lorenzi, A. Buoso, Electric arc furnace granulated slag as a partial replacement of natural aggregates for concrete production, in: 2nd International Conference on Sustainable Construction Materials and Technologies, 2010.

[24] S.M. Levy, P. Helene, Durability of recycled aggregates concrete: a safe way to sustainable development, (n.d.). doi:10.1016/j.cemconres.2004.02.009.

[25] M. Behera, S.K. Bhattacharyya, A.K. Minocha, R. Deoliya, S. Maiti, Recycled aggregate from C&D waste & its use in concrete - A breakthrough towards sustainability in construction sector: A review, Construction and Building Materials. 68 (2014) 501–516. doi:10.1016/j.conbuildmat.2014.07.003.

[26] L. Coppola, P. Kara, S. Lorenzi, Concrete manufactured with crushed asphalt as partial replacement of natural aggregates, Materiales de Construccion. 66 (2016). doi:10.3989/mc.2016.06515.

[27] L. Coppola, S. Lorenzi, S. Pellegrini, Rheological and mechanical performances of concrete manufactured by using washing water of concrete mixing transport trucks, in: American Concrete Institute, ACI Special Publication, 2015.

[28] L. Coppola, S. Lorenzi, P. Marcassoli, G. Marchese, Concrete production by using cast iron industry by-products | Impiego di sottoprodotti dell'industria siderurgica nel confezionamento di calcestruzzo per opere in c.a. e c.a.p, Industria Italiana Del Cemento. 77 (2007).

[29] Italian DM 17.01.2018 "Norme tecniche per le costruzioni," 2018.
[30] SIA 2030:2010 "Recycling Beton," 2010.
[31] L. Coppola, S. Lorenzi, P. Kara, S. Garlati, Performance and compatibility of phosphonate-based superplasticizers for concrete, Buildings. 7 (2017). doi:10.3390/buildings7030062.
[32] L. Coppola, S. Lorenzi, S. Garlati, P. Kara, The rheological and mechanical performances of concrete manufactured with blended admixtures based on phosphonates, 2016. doi:10.4028/www.scientific.net/KEM.674.159.
[33] L. Coppola, T. Cerulli, D. Salvioni, Sustainable development and durability of self-compacting concretes, in: 11th International Conference on Fracture 2005, ICF11, 2005.
[34] T. Ponikiewski, J. Gołaszewski, The Rheological and Mechanical Properties of High-performance Self-compacting Concrete with High-calcium Fly Ash, Procedia Engineering. 65 (2013) 33–38. doi:10.1016/J.PROENG.2013.09.007.
[35] L. Coppola, D. Coffetti, E. Crotti, Innovative carboxylic acid waterproofing admixture for self-sealing watertight concretes, Construction and Building Materials. 171 (2018) 817–824. doi:10.1016/j.conbuildmat.2018.03.201.
[36] L. Coppola, E. Cadoni, D. Forni, A. Buoso, Mechanical characterization of cement composites reinforced with fiberglass, carbon nanotubes or glass reinforced plastic (GRP) at high strain rates, 2011. doi:10.4028/www.scientific.net/AMM.82.190.
[37] L. Coppola, A. Buoso, F. Corazza, Electrical properties of carbon nanotubes cement composites for monitoring stress conditions in concrete structures, 2011. doi:10.4028/www.scientific.net/AMM.82.118.

AUTHOR INDEX